Fiber Optics and Optoelectronics

Second Edition

PETER K. CHEO

United Technologies Research Center
East Hartford, Connecticut

and

The Hartford Graduate Center
Hartford, Connecticut

Prentice-Hall International, Inc.

ISBN 0-13-312646-3

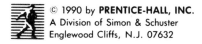 © 1990 by **PRENTICE-HALL, INC.**
A Division of Simon & Schuster
Englewood Cliffs, N.J. 07632

Printed in the United States of America

10 9 8 7 6 5 4 3 2 1

ISBN   0-13-312646-3

Prentice-Hall International (UK) Limited, *London*
Prentice-Hall of Australia Pty. Limited, *Sydney*
Prentice-Hall Canada Inc., *Toronto*
Prentice-Hall Hispanoamericana, S.A., *Mexico*
Prentice-Hall of India Private Limited, *New Delhi*
Prentice-Hall of Japan, Inc., *Tokyo*
Simon & Schuster Asia Pte. Ltd., *Singapore*
Editora Prentice-Hall do Brasil, Ltda., *Rio de Janeiro*
Prentice-Hall, Inc., *Englewood Cliffs, New Jersey*

To

Dorothy
and
my children

Contents

PREFACE   **xiii**

1 INTRODUCTION   **1**

     References   10

2 FIELD RELATIONS FOR DIELECTRIC WAVEGUIDES   **11**

     2.1   A Review of the Basic Laws of Electromagnetics   11
     2.2   Maxwell's Equations   13
     2.3   Solutions of the General Form   14
     2.4   Relations for Planar Waveguides   16
     2.5   Relations for Cylindrical Waveguides   17
     Problems   18
     References   18

3 PLANAR DIELECTRIC WAVEGUIDES   **19**

     3.1   Introduction   19
     3.2   Total Internal Reflection   19
     3.3   Guided-Wave Modes   23
     3.4   Field Expressions for Planar Waveguides   27
     3.5   Power Distribution and the Confinement Factor   29
     3.6   Scaling Rules for One-Dimensional Waveguides   32

3.7   Effective Index Method   37
      Problems   39
      References   40

4   CYLINDRICAL DIELECTRIC WAVEGUIDES   41

4.1   Introduction   41
4.2   Scalar Field Solutions for Step-Index Fibers   42
4.3   Approximation for Weakly Guided Step-Index Fibers   50
4.4   Power Distribution   56
4.5   Exact Solutions for Step-Index Fibers   58
4.6   Ray Analysis for Graded-Index Fibers   60
4.7   Calculations of Guided-Wave Modes
      in the WKB Approximation   66
      Problems   71
      References   72

5   DISPERSION, MODE COUPLING,
       AND LOSS MECHANISMS   73

5.1   Introduction   73
5.2   Group Velocity and Delay   73
5.3   Pulse Broadening   76
5.4   Material Dispersion   79
5.5   Intermodal Dispersion   80
5.6   Mode Coupling in a Multimode Fiber   83
5.7   Pulse Distortion   89
5.8   Scattering and Absorption Losses   92
5.9   Bending Losses   95
      Problems   99
      References   100

6   SINGLE MODE FIBERS   101

6.1   Introduction   101
6.2   Description of the HE_{11} Mode   102
6.3   Doubly-Cladded Step-Index Fibers   105
6.4   Elliptical Core Polarization-Preserving Fibers   114
6.5   Splicing and Coupling Losses   121
      Problems   124
      References   124

7 **GLASS MATERIALS, FIBER FABRICATION, AND CHARACTERIZATION TECHNIQUES**   **125**

   7.1 Introduction   125
   7.2 Glass Materials   125
   7.3 Preform Production   128
   7.4 Fiber Fabrication   132
   7.5 Optical Fiber Coupling   136
   7.6 Index Profile Measurements   141
   7.7 Dispersion Measurements   147
   7.8 Fiber Connection and Splicing   154
       Problems   158
       References   158

8 **LIGHT-EMISSION PROCESSES IN SEMICONDUCTORS**   **159**

   8.1 Introduction   159
   8.2 Quantum Mechanical Description of Semiconductors   160
   8.3 Carrier Distribution and Concentration   169
   8.4 Effects of Doping   173
   8.5 Radiative Transitions and Recombination Rates   176
   8.6 Population Inversion   179
   8.7 Carrier Lifetime   185
   8.8 Gain and Current Relations   188
   8.9 Laser Oscillation   195
   8.10 Optical Modes   197
       Problems   202
       References   203

9 **PROPERTIES AND GROWTH OF SEMICONDUCTOR HETEROJUNCTIONS**   **204**

   9.1 Introduction   204
   9.2 The *pn* Junction   205
   9.3 Forward-Biased Junctions   209
   9.4 Single Heterojunctions   213
   9.5 Double Heterojunctions   224
   9.6 Material Properties and Growth of Semiconductors   230
       Problems   235
       References   236

10 SEMICONDUCTOR LASERS   237

10.1   Introduction   237
10.2   Stripe-Geometry Lasers   238
10.3   Gain-Guided Stripe-Geometry Lasers   243
10.4   Power Spectrum of DH Lasers   250
10.5   Index-Guided Stripe-Geometry Lasers   254
10.6   High-Power and Single Mode Semiconductor Lasers   258
10.7   Long-Wavelength Sources   265
10.8   Distributed Feedback Lasers   268
10.9   Cleaved Coupled-Cavity Semiconductor Lasers   275
10.10 Semiconductor Laser Arrays   278
10.11 Quantum-Well Lasers   286
       Problems   294
       References   295

11 OPTICAL TRANSMITTERS   297

11.1   Introduction   297
11.2   Frequency Response   298
11.3   High-Speed Direct Modulation of Semiconductor Lasers   302
11.4   Bias and Control Circuits   308
11.5   Digital and Analog Codes   311
11.6   External Electro-Optic Modulation   316
11.7   Noise Characteristics   326
11.8   Aspects of Communication Theory   334
       Problems   340
       References   341

12 PHOTODETECTORS   343

12.1   Introduction   343
12.2   pn and pin Photodiodes   344
12.3   Avalanche Photodiodes   348
12.4   Noise in Photodiodes   355
12.5   Effects of Signal Waveforms   361
12.6   Frequency Response of Silicon Photodiodes   367
12.7   High-Speed and Long-Wavelength Photodetectors   369
       Problems   378
       References   379

13 OPTICAL RECEIVERS 380

13.1 Introduction 380
13.2 Receiver Circuits 380
13.3 FET and Bipolar Amplifier Noise 383
13.4 Receiver Design 387
13.5 Signal to Noise Ratio and Error Probability 389
13.6 Receiver Sensitivity 395
Problems 398
References 399

14 OPTICAL FIBER SYSTEMS 400

14.1 Introduction 400
14.2 Preliminary Design Guide 401
14.3 Design Analysis 403
14.4 Telecommunication Systems 409
14.5 In-Service Optical Communication Systems 411
14.6 Optical Local Area Networks 416
14.7 Long-Haul Systems 426
14.8 Coherent Light-Wave Communication Systems 428
14.9 Multiterminal Control and Data Distribution 438
Problems 440
References 440

INDEX 443

Preface

This book is an extensive revision of the first edition of *Fiber Optics: Devices and Systems*, published in 1985. Since then, numerous developments in the field of fiber optic technology have emerged. This new edition presents a large amount of new material, including dispersion-free single-mode fibers, long wavelength, low current threshold and high-power semiconductor lasers, high data rate laser transmitters, broadband and long-wavelength photodetectors, low noise optical receivers, local area network and coherent optical fiber communication systems. As a result, the title has been changed to *Fiber Optics and Optoelectronics* to reflect the increased coverage of this book.

The goal of this book is to present the topic of fiber optic technology in a consistent manner so that students can follow the development of this text with a clear understanding of the subject matter. The presentation has been kept to an introductory level, with emphases on the physical concepts underlying the interpretation of the properties of optical fiber waveguides, semiconductor light emission, and detection devices. I have made an effort to remain as rigorous and up-to-date as possible within the constraint of the presentation level.

This book is intended as a text for first-year graduate level students enrolled in electrical engineering, physics, and applied sciences. Fundamental principles and theories are introduced and developed to the extent that a student can gain a better insight into each topic without losing track of the basic concepts. When mathematics becomes cumbersome, approximate methods or intuitive approaches are introduced to provide students with a semiquantitative picture and a physical interpretation of the process. It is assumed that students have already taken introductory courses in electromagnetic theory, solid-state physics, and quantum mechanics. However, a certain amount of background material has been included so that students who have not taken some of these courses will still be able to follow the text with some extra efforts. Therefore, advanced undergraduate students in engineering and physical sciences, who have already fulfilled sufficient prerequisites, should be able to take this course without much difficulty. It is also my intention to provide general readers with an easy-to-read text, which can help them become acquainted with this subject through self-study.

The first part of this book deals with the principles and applications of optical fibers as data transmission media. Both ray and wave approaches are used to explain the mode structures of optical fibers. Special emphasis is given to their interactions and propagation characteristics. The book's second part reviews and treats in detail the properties and operating characteristics of optical sources and photoreceivers. Special emphases are placed on the emission and regeneration processes in semiconductors and their noise characteristics, which have great impact on system performance. The concluding chapters are devoted to several applications of these components for data transmission and telecommunication purposes. Efforts are made to present materials as explicitly as possible. Occasionally, some details are intentionally omitted as exercises for students, solely for the purpose of stimulating the learning process.

Fabrication techniques for fibers and other optical components are introduced with their specific subject matter. A list of problems is included at the end of each chapter. Optical fiber system designs are discussed only briefly, because it is difficult to elaborate on highly specialized topics, and in many cases, are still in development stages.

Many details and derivations omitted from the first edition have also been added. The major additions to this new edition are:

1. A new chapter (6) on single mode fibers has been added. This chapter includes the theory and design of single mode fibers and deals with modal propagation of dispersion-free single mode fibers and polarization preserving fibers having an elliptical core resulting from birefringence properties. This chapter also deals with splicing and coupling losses, which are major problems associated with single mode fiber communication systems.

2. Chapter 10 now includes extensive material on long-wave, high-power, single frequency semiconductor lasers. Also included are the most recent developments on extremely low threshold current devices utilizing quantum-well materials.

3. Topics on both direct and external high-speed modulation of semiconductor lasers with extremely wide frequency response have been added to Chapter 11—Optical Transmitters. Also added to this chapter are materials on AM, FM and phase noise of semiconductor lasers.

4. The most recent developments on long wavelength and extremely broadband photodetectors have been added to Chapter 12—Photodetectors. The section on detector noise analysis has been expanded.

5. A new chapter (13) on optical receivers has been added. This chapter includes both the theory and design of optical receivers utilizing state-of-the-art GaAs MESFET and bipolar transistor front-ends of preamplifiers. Also included are analyses of receiver sensitivity and receiver system optimization.

6. Two major optical fiber systems have been introduced in Chapter 14. One is the optical local area network and the other is the coherent lightwave communication system. Both system design details and representative examples are given.

No attempt has been made to provide chronology references or to acknowledge the original work by various contributors. References listed in this text are only those directly related to the subject matter and which can provide students with more detailed information. Many review articles are available, most of which provide comprehensive lists of original research papers.

ACKNOWLEDGMENTS

I would like to express my appreciation to Professor W. R. Kolk for his continuing interest and encouragement in this endeavor and to Dr. E. Snitzer, Dr. T. Li, Professor W. S. C. Chang, and my students, who have made contributions by reducing the number of errors and ambiguities occurring in this text. I wish to express my deepest gratitude to my wife, Dorothy, for her unfailing patience in proofreading this text in its entirety. I am also grateful to Mary Villemaire and Marie Johnson for their infinite patience in typing many different versions of this book. Lastly, I wish to express my appreciation to United Technologies Research Center for providing me with excellent word-processing and illustration services.

Peter K. Cheo
Waterford, CT

1

Introduction

Fiber optics has gained prominence in telecommunications, instrumentation, cable television networks, and data transmission and distribution. The major application, however, is in the area of telecommunications. Within this decade, a significant changeover will occur from wires and coaxial cables to optical fibers for telecommunication systems and information services. This anticipated change is dictated almost entirely by economics. The increasing cost and demand for high-data-rate or large-bandwidth-per-transmission channels and the lack of available space in already congested conduits in every metropolitan area are the reasons for this changeover. Furthermore, fiber optical devices interface well with digital data-processing equipment, and their technology is compatible with modern microelectronic technology. For these reasons, it is anticipated that in the future most telephones, television receivers, bank machines, computers, and to a lesser extent, medical and industrial instruments will be linked by optical fibers.

Since 1960 the availability of laser sources has stimulated research into optical communication. However, optical communication was not considered to be practical until 1970, when optical fiber technology had advanced to the point where relatively low-loss (<20 dB/km) fiber could be drawn routinely. Today, fibers with an absorption coefficient $\alpha(\lambda)$ as low as 0.5 dB/km can be manufactured for optical transmission at wavelengths of $\lambda \geq 1.2$ μm. For a complete and up-to-date reference list on fiber optics development, consult the review article by Li (Ref. 1.1). General information can be found in two other articles (Refs. 1.2 and 1.3).

A typical optical fiber system linkage is shown in Figure 1.1. The input data are usually coded using a current pulse network that can directly modulate the light source. The output of light pulses is coupled into a fiber with a lens or simply a butt joint. The optical power received by the photodiode through a certain length of fiber is always substantially reduced from its initial value due to losses through coupling, absorption, scattering, leakage, dispersion, and mode conversion. To maintain reliable, high-fidelity system operation, the power must be sufficient to overcome system losses. The fidelity of the signals transmitted depends on the detectable level of the signal-to-noise (S/N) ratio, which can be estimated from the detection probability function of a given distribution.

For a telecommunication system, the minimum Bit Error Rate (BER) is 10^{-9}, which corresponds to an optical S/N ratio of about 12 dB. The noise equivalent power (NEP) of an optical receiver used in a typical optical fiber circuit depends on the data rate. Therefore, the minimum required power must be determined by also accounting for the data rate. For an avalanche photodiode, the minimum detectable power, which is equivalent to a S/N ratio of unity, can be as low as -45 dBm at a data rate of 400 megabits per second (Mb/s). System analysis of this type requires knowledge of various parameters that govern the performance of the light source and detector, system noise, and propagation characteristics of light in a fiber transmission channel. Various signal processing techniques are required to deal with problems of signal distortion and interference, and statistical methods are often employed to determine system error-detection probability.

This book has four main topics: (1) the theory of optical fiber waveguides and pulse propagation phenomena in fibers; (2) the emission process, structure, and performance characteristics of semiconductor light sources; (3) optical receivers and noise characteristics of semiconductor photodiodes; and, (4) telecommunication and data transmission systems via optical fibers. It provides a self-contained treatment of these topics so that readers can reach a reasonable level of understanding of the fundamentals without relying too much on other sources, including a classroom instructor, for information.

One of the most important components in an optical fiber system is the optical fiber, which is discussed in the next seven chapters. In most cases, it is made of glass material (SiO_2) mixed with various dopants primarily to control the refractive index and reduce the softening point. Most fibers have a cylindrical core with an index n_0 of slightly higher value than that of the cladding material n_c. However, some fibers, primarily those made for optical imaging applications, have a square cross-section. As shown in Figure 1.2, the radii of the core and cladding are denoted by a and b, respectively. For a step-index fiber the refractive index is expressed by

$$n(r) = \begin{cases} n_0 & (r < a) \\ n_c & (a \le r \le b) \end{cases} \tag{1.1}$$

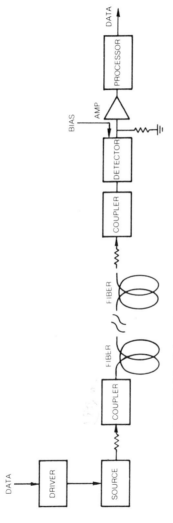

FIGURE 1.1 Schematic of a typical optical fiber data transmission link.

and for a graded-index fiber with a nearly parabolic profile for its core, the refractive index is expressed by

$$n(r) = \begin{cases} n_0 \left[1 - 2\Delta \left(\dfrac{r}{a} \right)^x \right]^{1/2} & (r < a) \\ n_c & (a \leq r \leq b) \end{cases} \tag{1.2}$$

where

$$\Delta = \frac{(n_0 - n_c)}{n_c} \tag{1.3}$$

For a variety of glass fibers with different dopants, $x \simeq 2$.

A single mode fiber has a core radius typically on the order of one optical wavelength λ. A multimode fiber is one whose core radius is sub-

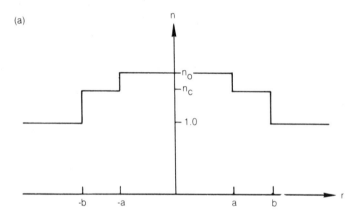

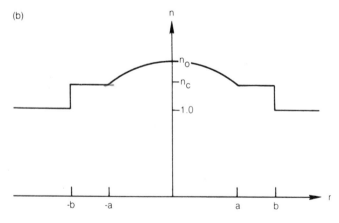

FIGURE 1.2 Index profile for (a) a step-index fiber having a core diameter $2a$ and clad diameter of $2b$ and (b) a graded-index fiber.

stantially larger than λ and is about 50 to 100 μm. In this case, hundreds or even thousands of allowable modes propagate in the guide. As a rule, the number of modes in a step-index fiber is about twice that in a graded-index fiber of the same dimension.

Because each mode possesses a unique group velocity, a short light pulse of energy distributed among these modes will be broadened as it travels through a certain length of fiber. This phenomenon, known as modal dispersion, is treated in greater detail in Chapter 5. To reduce modal dispersion, a fiber with a graded index has been introduced to compensate for the differences in group velocities. Now all modes travel in a graded-index fiber at nearly the same speed. Another alternative for reducing modal dispersion, and thereby increasing the information-carrying capability, is to use a single mode step-index fiber. The penalty paid for using a single mode fiber can be very high because more transmitter power is needed to compensate for the large power loss due to the coupling of light into single mode fibers. It is also more difficult to splice single mode fibers than to splice multimode fibers which have a much larger core diameter.

Propagation in a fiber can be viewed from a simplified ray picture such as that shown in Figure 1.3. If θ_M is the maximum angle beyond which rays that enter the fiber are no longer confined within the fiber, then θ'_M is the critical angle beyond which rays will not be bounded. Snell's law for rays at the air-fiber interface and at the core-cladding interface, gives

$$\sin \theta_M = n_0 \sin \theta'_M \tag{1.4}$$

and

$$n_c = n_0 \cos \theta'_M \tag{1.5}$$

Combining Equations (1.4 and (1.5), yields

$$\sin \theta_M = \sqrt{n_0^2 - n_c^2} \equiv NA \tag{1.6}$$

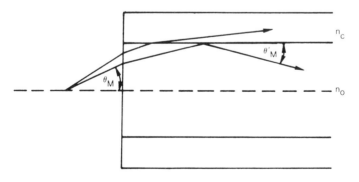

FIGURE 1.3 Ray picture showing total reflected and radiated rays in a glass fiber.

where NA denotes the numerical aperture of the fiber. Equation (1.6) indicates that the collected light power is directly proportional to the NA value. Because the difference between the refractive indices of the core and cladding is usually very small, Equation (1.6) can be expressed as

$$NA \simeq n_0 \sqrt{2\Delta} \tag{1.7}$$

where Δ is given by Equation (1.3). If $\sin \theta_M \simeq \theta_M$ and, from Equation (1.6), NA represents the maximum acceptance angle within which all rays will be guided by the fiber, for NA = 0.2, the maximum angle to capture rays is about 11°. Rays that make larger angles with the fiber axis within the acceptance angle correspond to high-order modes and travel in longer paths than do those propagating along the axis of the fiber. The delay difference between the axial rays and the rays traveling at the maximum angle is

$$\delta\tau = \frac{n_0}{c} \frac{L}{\cos \theta_M'} - \frac{n_0 L}{c} \tag{1.8}$$

Substituting Equation (1.5) into (1.8) yields

$$\delta\tau = \frac{n_0 L(n_0/n_c - 1)}{c} \tag{1.9}$$

Substituting Equation (1.3) into (1.9) for $n_0/n_c - 1$, and noting that $n \simeq n_0 \simeq n_c$ gives

$$\delta\tau = \frac{nL\Delta}{c} \tag{1.10}$$

For $\Delta = 0.01$, Equation (1.10) yields a delay time of 50 ns/km. In a graded-index fiber, the rays traveling away from the axis can gain speed and arrive with a very small delay difference. With careful control of the index profile, one can obtain roughly two orders of magnitude of reduction in the delay difference among all modes. These effects and the effects of material dispersion and mode coupling are discussed in Chapters 5 and 6. Fiber fabrication and measurement techniques are discussed in Chapter 7.

Two types of sources commonly used as transmitters in optical fiber systems are semiconductor light-emitting diodes (LEDs), which are incoherent sources, and laser diodes (LDs). The group of semiconducting materials used in making these sources are GaAs, InAs, InP, AlGaAs, InGaAsP, and so on, with direct bandgap energies extending from 0.7 to 1.6 μm in the spectral region. For most glass materials the spectral region of negligible material dispersion is in the range 1.2 to 1.3 μm. Therefore, the use of longer wavelengths leads to a reduction in material dispersion. Because the effects of material dispersion are at least one order of magnitude smaller than those of modal dispersion, modal dispersion must first be reduced to reach a certain level of data rate. This goal can be reached by employing either a single mode or graded-index fiber. A further increase in data rate can be achieved by using a longer-wavelength source. Other advantages also exist in selecting longer wavelengths. As λ increases, scattering losses decrease as $1/\lambda^4$, as shown in Figure 1.4. However, as λ increases beyond 1.3 μm, the absorption losses arising from the first overtone of the fundamental infrared band structures of the OH radical in the fiber begin to dominate. For the particular fiber shown in Figure 1.4, the OH content was estimated to be in the range of 50 parts in 10^9, which is an exceptionally low level for all practical purposes. Because of low losses and low material dispersion at longer wavelengths, most recent research activity on sources has centered around the InGaAsP and AlGaAsSb quaternary systems, which emit in the wavelength range of 1.3 to 1.8 μm. As a result, very low threshold double-heterostructure InGaAsP lasers with improved surface morphology and reproducibility have been produced. The relatively low threshold current of 1 kA/cm^2 has been achieved for these lasers. This value is only about a factor of 2 to 5 greater than that achieved for the more advanced AlGaAs ternary system. Even though the quaternary system is more complex than the ternary system, present results indicate that quaternary semiconductor sources can be made as efficient and reliable as those made from ternary compounds. In addition to wavelength considerations, other factors, such as component cost, reliability, output power, and coupling efficiency, are also important in formulating the criteria for selecting a source. More details on sources and transmitters are given in Chapters 8, 9, 10 and 11.

The photodetector is another important component in an optical fiber system. A type of detector commonly used in optical fiber systems is the semiconductor photodiode, which is a reverse-biased *pn*-junction device. The two most commonly used photodetectors are the *pin* and avalanche photodiode (APD). They are actually modified *pn*-junction devices with additional layers at slightly different doping levels to provide either more efficient quantum conversion or avalanche gain through ionization. Basically,

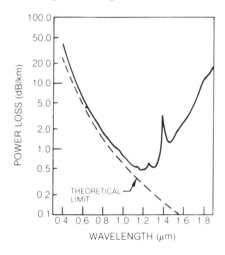

FIGURE 1.4 Loss spectrum of a phosphosilicate glass fiber with a borosilicate cladding having a NA value of 0.18. [After M. Horiguchi and H. Osanai, *Electron. Lett.*, 12, 310 (1976).]

a photon is absorbed in a relatively high E-field region, where an electron–hole pair is created. This process produces current in the detector circuit. To obtain higher quantum efficiency, a device such as a *pin* can provide adequate absorption in the relatively high-resistivity central i region. Another approach for obtaining higher detector currents is to create an avalanche gain effect, as in the case of the APD. In this case, an electron–hole pair may generate tens or hundreds more secondary electron–hole pairs. Because these events occur at random and are of a statistical nature, the noise generated in these devices can be a limiting factor on detectivity. A trade-off between the gain or quantum efficiency and noise exists for these devices. Techniques are available for processing and regeneration of digital postdetection signals through amplification, pulse shaping, equalization, timing extraction, decision, and error detection. More details on detectors and receivers are given in Chapters 12 and 13.

In addition to the discussion of various optical fiber components, standard techniques for signal coding, modulation, and some basic circuits for transmitters and receivers are presented in Chapters 11 and 13, and some examples of commonly used digital and analog fiber systems are presented in Chapter 14. A system designer must choose various components that are best suited for a specific application. These choices should be made based on a trade-off analysis among various system parameters involving optical power, fiber loss, receiver noise, signal type, data rate or bandwidth, bit error rate or signal-to-noise ratio, and the length between terminals or repeater spacing. Once an optimized system configuration is established, the designer must then consider other factors including environmental conditions, cost, reliability, flexibility, size, weight, installation, and maintenance. The procedure must be iterative because most of these factors are interrelated. An added complication is that the cost of optical fiber components is changing rapidly with time. All indications are that the cost of optical fiber components will decrease significantly in time and that the rate of decrease will be determined primarily by the growth or production rate.

One of the attractive features of an optical fiber system is the potentially large repeater spacing for large-capacity data transmission, which can bring about a substantial system cost reduction. Figure 1.5 shows existing and projected repeater spacing as a function of data rate. The solid curve represents the results obtained for existing systems for which AlGaAs lasers emitting at 0.9 μm were used. The average system loss at this wavelength is assumed to be about 4.5 dB/km. If future systems can be built in the region of 1.2 to 1.6 μm, a marked increase in repeater spacing can be expected, as shown in Figure 1.5 by the dashed curves. These curves are obtained by assuming a total loss of 0.7 dB/km, which includes both fiber and splicing losses. Other components used in this estimate are state-of-the-art InGaAsP double-heterostructure lasers and germanium APD detectors with a low ionization factor, $k \simeq 0.1$. The results, as shown in Figure 1.5, indicate that it

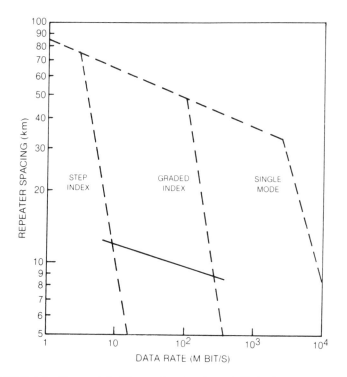

FIGURE 1.5 Measured (solid curve) and estimated (dashed curves) repeater spacing for various types of fibers as a function of data rate. The solid curve represents results obtained from optical fiber systems operating at wavelengths near 0.9 μm.

is possible to achieve a repeater spacing greater than 30 km at 1 gigabit per second by using a single mode fiber with a total dispersion of 1 ps/km and a source that has a spectral width of 2 Å.

Recent advances in single mode fibers and single spectral mode lasers emitting in the region of 1.2 to 1.7 μm has brought on a resurgent interest in coherent lightwave communication systems. A zero-dispersion, single mode fiber is accomplished by compensating material dispersion with waveguide dispersion at a desired wavelength (see Chapter 6). High spectral purity of semiconductor lasers can be obtained using a monolithically extended external cavity (see Chapter 10). Ideally, coherent lightwave systems (see Chapter 14) enable the vast available passband in the low-loss silica window to be efficiently exploited by a combination of high-bit-rate transmission and Optical Wavelength Division Multiplexing (OWDM). System limitations are determined by physical constraints, the transmission medium, and the practical design of the transmitter and receiver. Figure 1.6 is a plot of the length of coherent transmission systems versus the bit-rate for single and multiple channels. The ultimate restriction imposed by the fiber is the nonlinear effect

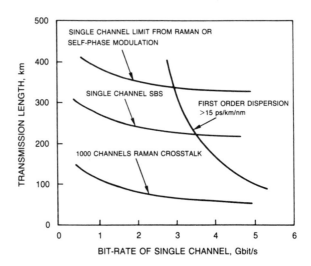

FIGURE 1.6 Potential channel capacity of coherent optical fiber communication systems, indicating the operating range of both single and WDM channels in the 1.5 μm window. (Ref. 1.4. Reprinted with permission of IEEE.)

due to self-phase modulation. With the possibility of a substantial increase in repeater spacing at high data transmission rates, not only the cost of system components, but also installation, operation, and maintenance costs could be significantly reduced. If a system were not limited by its transmission-line length, it would be possible to install repeaters in substations instead of manholes and many advantages could be obtained. For example, the power supply would be readily available at each substation, thus eliminating the need for transmitting electrical power along fiber cable. Also, the convenience of maintaining and repairing optical repeaters at substations, instead of in a manhole environment, could lead to considerable savings in the cost of operations as well as a large improvement in system reliability.

REFERENCES

1.1. T. Li, *IEEE J. Select. Areas Commun.*, *SAC-1*, 356 (1983).

1.2. S. E. Miller, E. A. J. Marcatili, and T. Li, *Proc. IEEE*, *61*, 1703 (1973).

1.3. D. Botez and G. J. Herskowitz, *Proc. IEEE*, *68*, 689 (1980).

1.4. D. W. Smith, *IEEE J. Lightwave Tech.*, *LT-5*, 1466 (1987).

2

Field Relations for Dielectric Waveguides

2.1 A REVIEW OF THE BASIC LAWS OF ELECTROMAGNETICS

Maxwell's equations are the embodiment of four phenomenological laws of electromagnetics. They are:

1. Gauss's Law. Gauss's law is a direct consequence of Coulomb's law of electrostatics and can be expressed in a variety of ways. One way to express Gauss's law is to integrate the electric lines of force emerging from a closed surface S, which is proportional to the entire charges enclosed within the volume V. Explicitly,

$$\oint_S \mathbf{E} \cdot \mathbf{n} \, dS = \frac{1}{\epsilon} \int_V \rho \, dV$$

where ϵ is the permittivity of the medium and ρ is the charge density. Using the divergence theorem, Gauss's law can be rewritten in a differential form as:

$$\nabla \cdot \mathbf{E} = \frac{\rho}{\epsilon} \qquad (2.1)$$

where $\mathbf{E}$ is the electric field vector, ρ the charge density, and ϵ the permittivity, which is related to the dielectric constant of the medium. Equation (2.1) indicates that the electric field diverging from an arbitrarily chosen surface is equal to the charge density enclosed by that surface divided by the permittivity. It indicates further that the electric field must originate from electrical charges with a radially outward line of force. Because the displacement field vector **D** is related to **E** as

$$\mathbf{D} = \epsilon\mathbf{E} \tag{2.2}$$

Equation (2.1) can also be written as

$$\nabla \cdot \mathbf{D} = \rho \tag{2.3}$$

2. Magnetostatic Law. In general the magnetic flux line generated by a circuit of any geometry follows a closed path. In other words, the number of magnetic lines of force emerging from any arbitrary closed surface must be equal to the number that converges into the same surface. Therefore, the net number of flux lines cutting outward from the surface is always zero. Explicitly

$$\oint_S \mathbf{B} \cdot \mathbf{n} \, dS = 0$$

or in a differential form as:

$$\nabla \cdot \mathbf{B} = 0 \tag{2.4}$$

where **B** is the magnetic field vector that defines the magnetic flux diverging from a closed surface and is related to the magnetic field intensity vector **H** through the permeability or susceptivity μ of the medium by the relation

$$\mathbf{B} = \mu\mathbf{H} \tag{2.5}$$

3. Ampère's Law. Ampère's law describes the relationship between **H** and electric current I in a circuit, a phenomenon first observed by Hans C. Oersted. An integral formulation of this law can be given in terms of a line integral of the magnetic field **H** along any closed path encircling a circuit having a value equal to the total current I carried by the circuit. Specifically

$$\oint \mathbf{H} \cdot d\mathbf{r} = I$$

By Stokes's theorem, the equation above can be rewritten in differential form as

$$\nabla \times \mathbf{H} = \mathbf{J} \tag{2.6}$$

where **J** is the current density vector pointing in a direction normal to the surface enclosed by the integration path. For a medium that contains displacement currents, Equation (2.6) must also include the induced current term

$$\nabla \times \mathbf{H} = \mathbf{J} + \frac{\partial \mathbf{D}}{\partial t} \tag{2.7}$$

The second term on the right-hand side of Equation (2.7) is obtained by using the law of conservation of charge with the help of a very useful continuity equation:

$$\frac{\partial \rho}{\partial t} + \nabla \cdot \mathbf{J} = 0 \qquad (2.8)$$

4. Faraday's Law of Induction. The most important law in electrodynamics is Faraday's law, which relates electric field to time-dependent magnetic field. Specifically, it describes the behavior of an induced current in a circuit that has been subjected to a time-varying magnetic field. The integral form of this law can be expressed by

$$\oint \mathbf{E} \cdot d\mathbf{r} = -\frac{\partial}{\partial t} \iint \mathbf{B} \cdot \mathbf{n} \, dS$$

Again using Stokes's theorem, the equation above provides the following differential form of Faraday's law:

$$\nabla \times \mathbf{E} + \frac{\partial \mathbf{B}}{\partial t} = 0 \qquad (2.9)$$

2.2 MAXWELL'S EQUATIONS

Optical power propagating in either a planar or a cylindrical dielectric waveguide is described by the field vectors **E**, **D** and **H**, **B**, thus satisfying Maxwell's equations for a linear, homogeneous and nonconducting medium with no sources. They comprise a set of four relations:

$$\nabla \times \mathbf{E} = -\frac{\partial \mathbf{B}}{\partial t} \qquad (2.10)$$

$$\nabla \times \mathbf{H} = \frac{\partial \mathbf{D}}{\partial t} \qquad (2.11)$$

$$\nabla \cdot \mathbf{D} = 0 \qquad (2.12)$$

$$\nabla \cdot \mathbf{B} = 0 \qquad (2.13)$$

If Equations (2.5) and (2.2) are substituted into (2.10) and (2.11) the following results:

$$\nabla \times \mathbf{E} = -\mu \frac{\partial \mathbf{H}}{\partial t} \qquad (2.14)$$

$$\nabla \times \mathbf{H} = \epsilon \frac{\partial \mathbf{E}}{\partial t} \qquad (2.15)$$

Taking the curl of Equation (2.14) and using Equation (2.15) to eliminate the time derivative of **H** yields

$$\nabla \times (\nabla \times \mathbf{E}) = \nabla(\nabla \cdot \mathbf{E}) - \nabla^2 \mathbf{E} = -\mu\epsilon \frac{\partial^2 \mathbf{E}}{\partial t^2}$$

Because $\nabla \cdot \mathbf{E} = 0$, the equation above reduces to

$$\nabla^2 \mathbf{E} = \mu\epsilon \frac{\partial^2 \mathbf{E}}{\partial t^2} \tag{2.16}$$

and similarly, for **H**, the other wave equation is

$$\nabla^2 \mathbf{H} = \mu\epsilon \frac{\partial^2 \mathbf{H}}{\partial t^2} \tag{2.17}$$

Equations (2.16) and (2.17) are the equations of motion of electromagnetic waves in dielectric waveguides. These waves are represented by the coupled **E** and **H** vectors and propagate with a phase velocity v_p, which is determined by parameters of the medium μ and ϵ as given by the expression $1/\sqrt{\mu\epsilon}$.

2.3 SOLUTIONS OF THE GENERAL FORM

In the following section, the field relations are first developed for a planar dielectric waveguide and then established for a cylindrical dielectric waveguide. This approach is taken because the field components for a planar waveguide involve only elementary functions, which are simpler to manipulate than those describing the fields in cylindrical fibers. However, the mathematical procedures are similar and very instructive when dealing with both types of waveguides.

Figures 2.1 and 2.2 define the coordinate systems used for the two waveguides. In both cases, the z axis is the direction for wave propagation. Therefore, in the case of a planar waveguide, solutions are of the form

$$\mathbf{E} = \mathbf{E}(x, y) \exp[i(\omega t - \beta z)] \tag{2.18}$$
$$\mathbf{H} = \mathbf{H}(x, y) \exp[i(\omega t - \beta z)]$$

and in the case of a cylindrical waveguide,

$$\mathbf{E} = \mathbf{E}(r, \theta) \exp[i(\omega t - \beta z)] \tag{2.19}$$
$$\mathbf{H} = \mathbf{H}(r, \theta) \exp[i(\omega t - \beta z)]$$

where β is the propagation constant of the field and its values are subject to the boundary conditions imposed on $\mathbf{E}(x, y)$ or $\mathbf{E}(r, \theta)$, and others, and ω

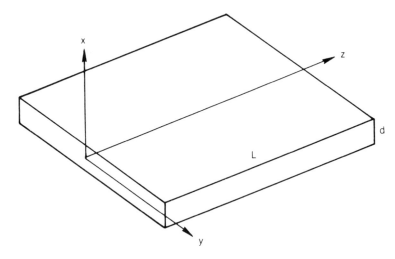

FIGURE 2.1   Planar waveguide in the yz plane having thickness d and length L.

is the angular frequency of the wave, which is related to the phase velocity v_p:

$$v_p = \frac{\omega}{\beta} \tag{2.20}$$

In free space, $\epsilon = \epsilon_0$, $\mu = \mu_0$, and $v_p = c$, and the ratio of the velocity in free space to that in a dielectric medium, where μ is also equal to μ_0, defines the refractive index, $n = \sqrt{\epsilon/\epsilon_0}$.

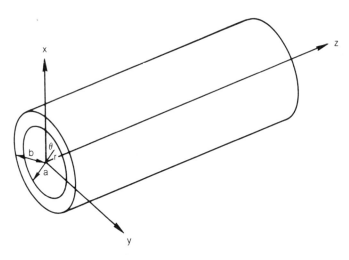

FIGURE 2.2   Cylindrical waveguide with core radius a and clad radius b.

2.4 RELATIONS FOR PLANAR WAVEGUIDES

For waves propagating along the z-direction in an infinite planar waveguide lying in the yz plane, the field component containing $\partial/\partial y$ vanishes. Substituting Equation (2.18) into (2.14) and (2.15) and resolving into component forms gives

TE waves
(E_y, H_x, H_z)

$$\beta E_y = -\mu\omega H_x \tag{2.21}$$

$$-i\beta H_x - \frac{\partial H_z}{\partial x} = i\epsilon\omega E_y \tag{2.22}$$

$$\frac{\partial E_y}{\partial x} = -i\mu\omega H_z \tag{2.23}$$

TM waves
(E_x, E_z, H_y)

$$\beta H_y = \epsilon\omega E_x \tag{2.24}$$

$$i\beta E_x + \frac{\partial E_z}{\partial x} = i\omega\mu H_y \tag{2.25}$$

$$\frac{\partial H_y}{\partial x} = i\epsilon\omega E_z \tag{2.26}$$

From Equations (2.21) to (2.26), note that these field components form two independent groups: one is composed of three coupled components E_y, H_x, and H_z; the other group is composed of E_x, E_z, and H_y coupled waves. Field components associated with the first group correspond to transverse electric (TE) modes and the components associated with the second group correspond to transverse magnetic (TM) modes of an infinite planar waveguide described in a coordinate system as shown in Figure 2.1. With the help of relations (2.21) to (2.26), only two field components (e.g., E_y, the transverse electric field, and H_y, the transverse magnetic field) are required to specify TE and TM waves completely.

By combining Equations (2.21), (2.22), and (2.23) and eliminating H_x and H_z, the following wave equation is obtained for E_y:

$$\frac{\partial^2 E_y}{\partial x^2} + (\omega^2\epsilon\mu - \beta^2)E_y = 0 \tag{2.27}$$

Similarly, by combining Equations (2.24), (2.25), and (2.26) and eliminating E_x and E_z, an identical wave equation is obtained for H_y:

$$\frac{\partial^2 H_y}{\partial x^2} + (\omega^2\epsilon\mu - \beta^2)H_y = 0 \tag{2.28}$$

Equations (2.27) and (2.28) indicate, as expected, that E_y and H_y are plane waves of the form

$$E_y = \exp(ik_x x)$$

$$H_y = \exp(ik_x x) \tag{2.29}$$

where k_x is the propagation constant in the x direction with an amplitude defined by the equation

$$k_x = \sqrt{\omega^2 \epsilon \mu - \beta^2} \tag{2.30}$$

The amplitude of the resultant wave vector $\mathbf{k}$ is

$$|\mathbf{k}| = \sqrt{k_x^2 + \beta^2} = \omega\sqrt{\epsilon\mu} \tag{2.31}$$

Often $\epsilon\mu$ is expressed in terms of $K_e K_m \epsilon_0 \mu_0$, where K_e, the dielectric constant, is approximately equal to n^2, and K_m is approximately equal to unity; and, $\epsilon_0\mu_0 = 1/c^2$, where c is the velocity of light in vacuum. Therefore, Equation (2.31) can be rewritten as

$$\omega\sqrt{\epsilon\mu} = nk_0 \tag{2.32}$$

where $k_0 = 2\pi/\lambda_0$, which is the amplitude of the propagation vector in free space.

2.5 RELATIONS FOR CYLINDRICAL WAVEGUIDES

If Equation (2.19) is substituted into (2.14) and (2.15) and using cylindrical coordinates for $\nabla \times \mathbf{E}$ and $\nabla \times \mathbf{H}$, the following expressions result:

$$\frac{1}{r}\left(\frac{\partial E_z}{\partial \theta} + i\beta r E_\theta\right) = -i\omega\mu H_r \tag{2.33}$$

$$i\beta E_r + \frac{\partial E_z}{\partial r} = i\omega\mu H_\theta \tag{2.34}$$

$$\frac{1}{r}\left(\frac{\partial r E_\theta}{\partial r} - \frac{\partial E_r}{\partial \theta}\right) = -i\omega\mu H_z \tag{2.35}$$

$$\frac{1}{r}\left(\frac{\partial H_z}{\partial \theta} + i\beta r H_\theta\right) = i\omega\epsilon E_r \tag{2.36}$$

$$i\beta H_r + \frac{\partial H_z}{\partial r} = -i\omega\epsilon E_\theta \tag{2.37}$$

$$\frac{1}{r}\left(\frac{\partial r H_\theta}{\partial r} - \frac{\partial H_r}{\partial \theta}\right) = i\omega\epsilon E_z \tag{2.38}$$

From Equations (2.33) to (2.38) r and θ components can be solved in terms of z components to obtain the following relationships:

$$E_r = -\frac{i}{k_r^2}\left(\beta\frac{\partial E_z}{\partial r} + \omega\mu\frac{1}{r}\frac{\partial H_r}{\partial \theta}\right) \tag{2.39}$$

$$E_\theta = -\frac{i}{k_r^2}\left(\frac{\beta}{r}\frac{\partial E_z}{\partial \theta} - \omega\mu\frac{\partial H_z}{\partial r}\right) \tag{2.40}$$

$$H_r = -\frac{i}{k_r^2}\left(\beta\frac{\partial H_z}{\partial r} - \omega\epsilon\frac{1}{r}\frac{\partial E_z}{\partial\theta}\right) \tag{2.41}$$

$$H_\theta = -\frac{i}{k_r^2}\left(\frac{\beta}{r}\frac{\partial H_z}{\partial\theta} + \omega\epsilon\frac{\partial E_z}{\partial r}\right) \tag{2.42}$$

where k_r is the radial component of the **k** vector in the guide and has the amplitude

$$k_r = \sqrt{\omega^2\epsilon\mu - \beta^2} \tag{2.43}$$

From Equations (2.39) to (2.42) note that as in the case of an optical fiber waveguide, field components in cylindrical coordinates are generally not separable. They are considerably more complex and represented not only by the linearly polarized TE and TM modes previously introduced, but also by hybrid HE and EH modes to be discussed in Chapter 4. However, when the difference in refractive index between core and cladding is very small, a simplified approach can be used in which these coupled relations are ignored. In this way a set of solutions are obtained that describe the modes in terms of a set of linearly polarized fields—the LP modes. This problem is discussed further in Chapter 4. A thorough review of the electromagnetic theory can be found in the first eight chapters of the book by Jackson (Ref. 2.1). A similar treatment for planar and cylindrical waveguides can be found in the book by Midwinter (Ref. 2.2).

PROBLEMS

2.1. Derive Equations (2.21) to (2.26).

2.2. Construct a vector diagram of **k** in both rectangular and cylindrical coordinate systems.

2.3. Derive Equations (2.33) to (2.38).

2.4. Derive Equations (2.39) and (2.41).

REFERENCES

2.1. J. D. Jackson, *Classical Electrodynamics,* John Wiley & Sons, Inc., New York, 1962.

2.2. J. E. Midwinter, *Optical Fibers for Transmission,* John Wiley & Sons, Inc., New York, 1979.

3

Planar Dielectric Waveguides

3.1 INTRODUCTION

The conditions necessary for establishing guided waves in a planar wave-guide are that (1) the waves must be totally reflected at upper and lower boundaries, and (2) the total phase shift after two consecutive reflections must be an integer multiple of 2π. To illustrate the effect of these conditions, this chapter shall first treat TE and TM propagating waves at a planar boundary and then extend the analysis to a planar waveguide, which consists of two boundaries. One of the pioneering papers on the theory and techniques for the excitation and propagation of guided-wave modes in a planar wave-guide was the work by Tien and Ulrich (Ref. 3.1).

3.2 TOTAL INTERNAL REFLECTION

Two independent sets of TE and TM waves exist for an infinite planar wave-guide. In this section, optical confinement for both the TE and TM waves in a planar waveguide is established by subjecting these waves to the bound-aries at which total internal reflection occurs. The two TE components that parallel the interface are E_y and H_z, as shown in Figure 3.1. From Equation (2.27), the incident wave in the medium with a reflective index n_1 can be expressed by

$$E_y = A_0 \exp(-ik_1 x)$$

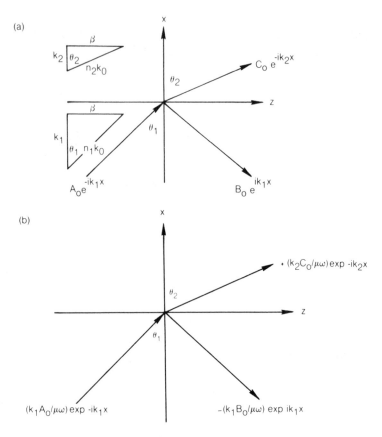

FIGURE 3.1   TE wave at a planar interface: (a) E_y component; (b) H_z component.

The reflected wave in the same medium can be written as

$$E_y' = B_0 \exp(ik_1 x)$$

The transmitted waves into another medium with a refractive index n_2 can be written as

$$E_y'' = C_0 \exp(-ik_2 x)$$

Similarly, using Equation (2.33), H_z for the incident wave in medium 1 is

$$H_z = \frac{i}{\mu\omega} \frac{\partial E_y}{\partial x} = \frac{k_1 A_0}{\mu\omega} \exp(-ik_1 x)$$

The reflected waves in the same medium is

$$H_z' = -\frac{k_1 B_0}{\mu\omega} \exp(ik_1 x)$$

and the transmitted wave into medium 2 is

$$H_z'' = \frac{k_0 C_0}{\mu\omega} \exp(-ik_2 x)$$

The principle of continuity requires that at the boundary $x = 0$, the total transverse field and its derivatives in one medium must be equal to that in the adjacent medium, e.g. $E_y(0) + E_y'(0) = E_y''(0)$. These requirements provide two equations containing three constants A_0, B_0 and C_0 as follows:

$$A_0 + B_0 = C_0 \tag{3.1}$$

$$k_1 A_0 - k_1 B_0 = k_2 C_0 \tag{3.2}$$

Expressing B_0 and C_0 in terms of A_0 provides

$$B_0 = A_0 \frac{k_1 - k_2}{k_1 + k_2} = A_0 r_E \tag{3.3}$$

and

$$C_0 = A_0 \frac{2k_1}{k_1 + k_2} = A_0 t_E \tag{3.4}$$

where r_E and t_E are reflection and transmission coefficients associated with this component. The following relationships between k_1, k_2 and n_1, n_2, θ_1, θ_2 can be obtained from Figure 3.1(a):

$$k_1 = n_1 k_0 \cos\theta_1 \tag{3.5}$$

$$k_2 = n_2 k_0 \cos\theta_2 \tag{3.6}$$

and

$$k_1^2 - k_2^2 = (n_1^2 - n_2^2)k_0^2 \tag{3.7}$$

Total internal reflection, at which $k_2 = 0$, occurs as the angle of incidence θ_1 is increased to a critical value θ_c. As θ_1 increases further, Equation (3.7) indicates that k_2 must take on imaginary values; therefore, for the case of a guided wave

$$k_2 = -i\gamma_2 \tag{3.8}$$

Substituting Equation (3.8) into (3.3) and letting $Z = k_1 + i\gamma_2$ yields

$$B_0 = A_0 \frac{Z}{Z^*} = A_0 \frac{Z^2}{ZZ^*} = A_0 \frac{Z^2}{|Z|^2} = A_0 e^{i2\phi_E} \tag{3.9}$$

where

$$\phi_E = \tan^{-1}\frac{\gamma_2}{k_1} \tag{3.10}$$

Equation (3.9) indicates that when a wave is totally internally reflected, the phase angle of the reflected wave differs from that of the incident wave by $2\phi_E$. This phenomenon, known as the Goos-Haenchen shift, is caused by the penetration of the wave into the less dense medium before it is totally reflected. To incur a phase shift of $2\phi_E$, it can be shown that the wave must extend into Medium 2 with a penetration depth d equal to $1/\gamma_2$.

Now, extend the treatment to TM waves by examining the behavior of the H_y and E_z components at the planar interface of two dielectric media with $n_1 k_0 = \omega\sqrt{\epsilon_1\mu}$ and $n_2 k_0 = \omega\sqrt{\epsilon_2\mu}$, as shown in Figure 3.2. By equating the incident, reflected, and refracted H_y and E_z waves at $x = 0$, the following is obtained:

$$A_0 + B_0 = C_0 \tag{3.11}$$

$$\frac{k_1 A_0}{n_1^2} - \frac{k_1 B_0}{n_1^2} = \frac{k_1 C_0}{n_2^2} \tag{3.12}$$

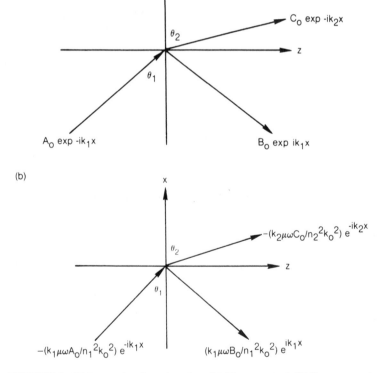

FIGURE 3.2   TM wave at a planar interface: (a) H_y component; (b) E_z component.

From Equations (3.11) and (3.12), B_0 and C_0 can be expressed in terms of A_0 as

$$B_0 = A_0 \frac{k_1 n_2^2 - k_2 n_1^2}{k_1 n_2^2 + k_2 n_1^2} = A_0 r_M \qquad (3.13)$$

and

$$C_0 = A_0 \frac{2 n_2^2 k_1}{k_1 n_2^2 + k_2 n_1^2} = A_0 t_M \qquad (3.14)$$

In the case of total internal reflection, which occurs if $k_2 = i\gamma_2 n_1^2$ is substituted into Equation (3.13), the coefficient B_0 for TM waves can be expressed as

$$B_0 = A_0 e^{i2\phi_M} \qquad (3.15)$$

where

$$\phi_M = \tan^{-1} \frac{\gamma_2/n_2^2}{k_1/n_1^2} \qquad (3.16)$$

Equations (3.15) and (3.16) indicate that a phase shift of $2\phi_M$ for the TM waves will again occur when the angle of incidence is increased beyond the critical angle.

3.3 GUIDED-WAVE MODES

The treatment described above can be extended to include both the upper and lower boundaries of a planar waveguide. In the case of TE waves, the phase shift for E_y and H_z components will be determined after reaching the condition satisfying total internal reflection at both upper and lower boundaries. Total internal reflection implies that the amplitude A of the incident E_y wave must be equal to the amplitude E of the reflected wave [see Figure 3.3(a)], provided that no loss in the media occurs. As shown in Figure 3.3, expressions for the incident, reflected, and evanescent waves are obtained with the help of Equations (2.23) and (2.27). The term "evanescent" rather than "refracted" is used here, because under the condition that $\theta > \theta_c$, k_2 and k_3 become imaginary and must be replaced by $-i\gamma_2$ and $-i\gamma_3$, respectively. Therefore, the waves that extend beyond the boundaries switch from oscillatory form to exponential decaying form. For simplicity a coordinate system is chosen such that the lower boundary of the waveguide lies in the $x = 0$ plane and the upper boundary lies in the $x = d$ plane, as shown in Figure 3.3.

By equating the fields in Medium 1 to those in Medium 3 at $x = 0$, and the fields in Medium 1 to those in Medium 2 at $x = d$, a set of four

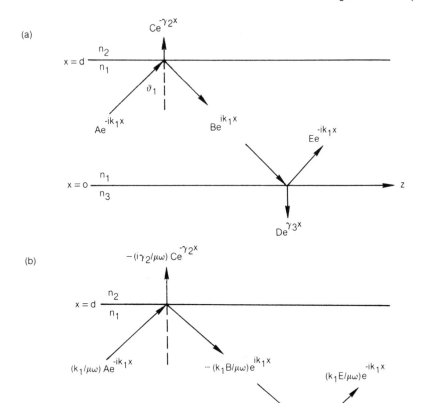

FIGURE 3.3 Guided TE mode in an asymmetric waveguide with thickness d:
(a) E_y component; (b) the H_z component.

equations results

$$B + E = D \tag{3.17}$$

$$-k_1 B + k_1 E = i\gamma_3 D \tag{3.18}$$

$$Ae^{-ik_1d} + Be^{ik_1d} = Ce^{-\gamma_2d} \tag{3.19}$$

$$k_1 Ae^{-ik_1d} - k_1 Be^{ik_1d} = -i\gamma_2 Ce^{-\gamma_2d} \tag{3.20}$$

By eliminating D from Equation (3.17) and (3.18), the following is obtained:

$$E = B\frac{k_1 + i\gamma_3}{k_1 - i\gamma_3} = Be^{i2\phi_{13}} \tag{3.21}$$

where

$$\phi_{13} = \tan^{-1} \frac{\gamma_3}{k_1} \tag{3.22}$$

is the phase shift of the guided wave after total internal reflection at the boundary between 1 and 3.

By eliminating C from Equations (3.19) and (3.20), the following is yielded:

$$B = A e^{-i2k_1 d} \frac{k_1 + i\gamma_2}{k_1 - i\gamma_2} = A e^{i(2\phi_{12} - 2k_1 d)} \tag{3.23}$$

where

$$\phi_{12} = \tan^{-1} \frac{\gamma_2}{k_1} \tag{3.24}$$

The phase shift of the wave after total internal reflection at the boundary between 1 and 2 is equal to $2\phi_{12} - 2k_1 d$. Applying the condition that the total phase shift after two consecutive reflections must be equal to an integer multiple of 2π, an equation, which is commonly called the eigen value equation, is obtained. This equation is characteristic of the waveguide by providing a discrete set of values for k_1 of the guided-wave TE modes as follows:

$$k_1 = \frac{1}{d} (\phi_{12} + \phi_{13} - m\pi) \tag{3.25}$$

where $m = 0, 1, 2, \ldots$. Substituting Equation (3.25) into (2.30), where $k_x = k_1$, β values for various propagating TE modes can thus be determined.

By the same analogy, an expression for the k_1 values of TM modes can be derived. In this case, the analysis of H_y and E_z components must be extended to include both boundaries, as shown in Figure 3.2. The major difference between the results of TE and TM modes is due to the susceptivity associated with these fields. In the TE case, E_y and H_z are related by the coupling coefficient $k_1/\mu\omega$, where μ is a constant; in the TM case, H_y and E_z are related by $k_1/\epsilon\omega$, where ϵ takes on a different value for the different medium. Therefore, the eigenvalue equation for TM modes has a slightly modified form:

$$k_1 = \frac{1}{d} \left(\tan^{-1} \frac{\gamma_2 n_1^2}{k_1 n_2^2} + \tan^{-1} \frac{\gamma_3 n_1^2}{k_1 n_3^2} - m\pi \right) \tag{3.26}$$

where $m = 0, 1, 2, \ldots$.

The number of modes as given by Equation (3.25) for TE waves and Equation (3.26) for TM waves is limited and depends on the thickness of the guide. For a very thin guide, only a finite number of modes are allowed to propagate, while others fall beyond cutoff. This phenomenon can be explained by the fact that only a few angles of incidence $\theta > \theta_c$ satisfy the phase-shift condition. As d approaches zero, the lowest-order mode is always allowed because its phase-shift angle also approaches zero as both θ and θ_c approach $90°$.

Modal dispersion relationships and their cutoff conditions as functions of waveguide thickness can best be illustrated using a graphic method. Only the graphical solution for TE modes of a symmetric waveguide ($n_2 = n_3$; $\gamma_2 = \gamma_3$) with a thickness d is presented. Chapter 10 explains that such a waveguide structure is commonly used for semiconductor lasers. For convenience, the origin of the coordinate system is set at the center of the planar waveguide. The eigenvalue equation (3.25) for a symmetric waveguide can be reduced to the form

$$\frac{k_1 d}{2} = \tan^{-1}\left(\frac{\gamma_2}{k_1}\right) - \frac{m\pi}{2}$$

If m is an even integer

$$\frac{k_1 d}{2} \tan \frac{k_1 d}{2} = \frac{\gamma_2 d}{2} \qquad \text{(even } m) \tag{3.27}$$

If m is an odd integer, $\tan(k_1 d/2 - m\pi/2) = \cot(k_1 d/2)$, therefore

$$\frac{k_1 d}{2} \cot \frac{k_1 d}{2} = -\frac{\gamma_2 d}{2} \qquad \text{(odd } m) \tag{3.28}$$

Equations (3.7) and (3.8) give an equation of a circle in the $k_1 d/2$ and $\gamma_2 d/2$ plane as expressed by

$$\left(\frac{k_1 d}{2}\right)^2 + \left(\frac{\gamma_2 d}{2}\right)^2 = (n_1^2 - n_2^2)\left(\frac{k_0 d}{2}\right)^2 \tag{3.29}$$

where k_1 and γ_2, in accordance with the vector diagram shown in Figure 3.1(a), are related to β in the following ways:

$$k_1^2 = n_1^2 k_0^2 - \beta^2 \tag{3.30}$$

$$\gamma_2^2 = \beta^2 - n_2^2 k_0^2 \tag{3.31}$$

Equations (3.30) and (3.31) show that the requirement for guided modes is

$$n_2^2 k_0^2 < \beta^2 < n_1^2 k_0^2 \tag{3.32}$$

From Equation (3.29) the family of circles shown in Figure 3.4 is obtained for three different waveguide thicknesses: 0.2, 1.0, and 1.5 μm. These values are representative of the active-layer thickness of single- and double-heterostructure AlGaAs lasers, emitting at $\lambda = 0.9$ μm. Typical values for n_1 and n_2 of these devices are 3.590 and 3.385, respectively. In Figure 3.4, the dashed curves are plots of Equations (3.27) and (3.28). The points of intersection uniquely define the values of k_1 and γ_2 and also determine the value

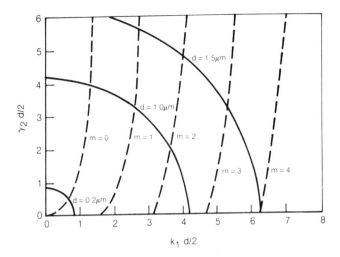

FIGURE 3.4 Graphical solution of the eigenvalue equation for TE modes in a symmetric $Al_{0.3}Ga_{0.7}As$ double-heterostructure laser. λ is assumed to be 0.9 μm.

of β for various modes of the waveguide. For $d = 0.2$ μm, only one point of intersection exists between the circle and the $m = 0$ curve, corresponding to only one eigenvalue for the lowest-order TE_0 mode, while all other modes are beyond cutoff. The TE_0 mode has no cutoff no matter how small d becomes. For $d = 1$ μm, TE_0, TE_1, and TE_2 modes exist with corresponding $k_1 d/2$ values 1.3, 2.5, and 3.6, respectively. For $d = 1.5$ μm four modes exist in the guide and the fifth TE_4 mode is just at cutoff, as indicated by the intersection point occurring at $\gamma_2 d/2 = 0$. From the intersections of the $(k_1 d/2) \tan(k_1 d/2)$ curves with the $k_1 d/2$ axis that occur at $m\pi/2$, the cutoff condition for the waveguide thickness is

$$d_c = \frac{1}{2} \frac{m\lambda}{(n_1^2 - n_2^2)^{1/2}} \tag{3.33}$$

where $m = 0, 1, 2, \ldots$.

3.4 FIELD EXPRESSIONS FOR PLANAR WAVEGUIDES

A planar waveguide has two independent modes, TE and TM, as described by the wave equations (2.27) and (2.28), respectively. The TE mode has its electric polarization transverse to the direction of propagation and is specified by the E_y component. The TM mode has its magnetic polarization transverse to the propagation direction and is specified by the H_y component. Two other components, H_z and H_x for the TE mode, and E_x and E_z for the

TM mode, can be generated from E_y and H_y. Because the same exponential factor, $\exp[i(\omega t - \beta z)]$, appears in all components, it will be omitted from all expressions in the following treatment. Considered here is only a symmetric waveguide. Again, for this case it is more convenient to set the origin of the coordinate system at the center of the waveguide and deal with even and odd modes separately.

Within the thickness of the guide $|x| < d/2$, the standing wave of the TE mode has a form that satisfies the wave equation (2.27), as given by

$$E_y(x) = \begin{cases} A_e \cos k_1 x & m = 0, 2, \ldots \\ A_0 \sin k_1 x & m = 1, 3, \ldots \end{cases} \tag{3.34}$$

Equation (2.23) provides

$$H_z(x) = \begin{cases} -\dfrac{ik_1 A_e}{\mu\omega} \sin k_1 x & m = 0, 2, \ldots \\[2mm] \dfrac{ik_1 A_0}{\mu\omega} \cos k_1 x & m = 1, 3, \ldots \end{cases} \tag{3.35}$$

For the field outside the guide, $|x| > d/2$,

$$E_y(x) = \begin{cases} A_e \cos \dfrac{k_1 d}{2} \exp\left[-\gamma_2 \left(|x| - \dfrac{d}{2} \right) \right] & m = 0, 2, \ldots \\[4mm] \dfrac{x}{|x|} A_0 \sin \dfrac{k_1 d}{2} \exp\left[-\gamma_2 \left(|x| - \dfrac{d}{2} \right) \right] & m = 1, 3, \ldots \end{cases} \tag{3.36}$$

and

$$H_z(x) = \begin{cases} -i \dfrac{x}{|x|} \dfrac{\gamma_2 A_e}{\mu\omega} \cos \dfrac{k_1 d}{2} \exp\left[-\gamma_2 \left(|x| - \dfrac{d}{2} \right) \right] & m = 0, 2, \ldots \\[4mm] -i \dfrac{\gamma_2 A_0}{\mu\omega} \sin \dfrac{k_1 d}{2} \exp\left[-\gamma_2 \left(|x| - \dfrac{d}{2} \right) \right] & m = 1, 3, \ldots \end{cases} \tag{3.37}$$

The variation of the electric field E_y as a function x is shown in Figure 3.5(a) for a symmetric waveguide of thickness 1 μm. For even modes, the maximum field occurs at the center. For odd modes, the field amplitude always has a zero at the center. All field amplitudes are modulated by the exponential term $\exp[i(\omega t - \beta z)]$. Figure 3.5(b) shows the intensity distribution for these modes. These curves, shown in Figure 3.5(b), correspond to the variation of the square of the electric field as a function of x.

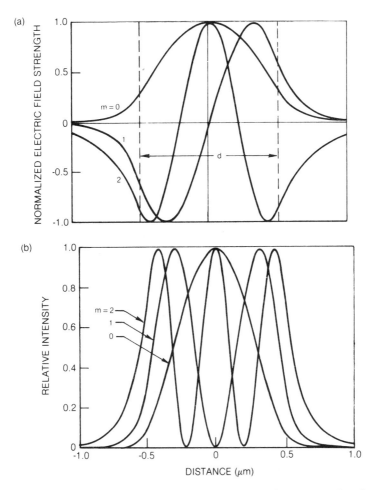

FIGURE 3.5 (a) Plot of normalized electric field for the first three modes of a symmetric waveguide with parameters $n_1 = 3.590$, $n_2 = 3.385$, and $\lambda = 0.9$ μm; (b) corresponding intensity distribution, which is the square of electric field strength.

3.5 POWER DISTRIBUTION AND THE CONFINEMENT FACTOR

Figure 3.5 shows that the electric field and power of a guided-wave mode can extend beyond the geometric boundary of the guide and the amount is dependent on the thickness and mode number. Chapter 10 discusses the importance of determining the fraction of an optical mode within the guiding layer of a semiconductor laser in which strong interaction exists between the field and gain medium. This section evaluates the spatial extent of the light intensity within the guide. It is measured by the confinement factor Γ, which is defined by the ratio of light intensity within the layer to the total light intensity.

The light intensity I is proportional to the magnitude of the Poynting vector **P**, defined by

$$\mathbf{P} = \mathbf{E} \times \mathbf{H} \tag{3.38}$$

where **E** and **H** are usually expressed in the form of complex functions as given by Equation (2.18). For an infinite planar waveguide, the power propagating along the z-axis can be obtained by integrating the z-component of the Poynting vector P_z over the entire thickness of the guide. However, actual fields are the real part of the complex quantity with sinusoidally varying functions oscillating in time. The energy density associated with these fields can be obtained by taking the time average of the product of the real parts of **E** and **H** vectors. This value is equivalent to one-half the real part of the product of one vector and the complex conjugate of the other. Therefore, the time average of the Poynting vector along the z-axis is

$$P_z = \frac{1}{2} \int (E_x H_y - H_x E_y) \, dx \tag{3.39}$$

For TE modes, E_x and H_y vanish and substituting Equation (2.21) into (3.39) gives

$$P_z = \frac{1}{2} \frac{\beta}{\mu \omega} \int E_y^2 \, dx \tag{3.40}$$

Now, if the real part of E_y is taken for even modes of TE waves as given by Equations (3.34) and (3.36), the fraction of light intensity inside the guide as denoted by the integral I_{in} is obtained and can be expressed by

$$I_{in} = \frac{\beta A_e^2}{\mu \omega} \int_0^{d/2} \cos^2 k_1 x \, dx \tag{3.41}$$

The fraction of light intensity outside the guide is denoted by the integral I_{out}:

$$I_{out} = \frac{\beta A_e^2}{\mu \omega} \int_{d/2}^{\infty} \cos^2 k_1 \frac{d}{2} \exp\left[-2\gamma_2 \left(x - \frac{d}{2}\right)\right] dx \tag{3.42}$$

A parameter Γ, which is commonly called the confinement factor is defined as the ratio of light intensity within the waveguide to total intensity. Explicitly

$$\Gamma \equiv I_{in}/(I_{in} + I_{out})$$

For even TE modes,

$$\Gamma = \left\{ 1 + \frac{\cos^2(k_1 d/2)}{\gamma_2 \left[d/2 + \left(\dfrac{1}{k_1} \right) \sin(k_1 d/2) \cos(k_1 d/2) \right]} \right\}^{-1} \tag{3.43}$$

From Equation (3.43) it is interesting to note that when $\gamma_2 = 0$, Γ approaches zero; therefore, the entire power resides in the cladding. This result is consistent with the cutoff condition for modes. Because of the difference in k_1 and γ_2 values among various modes, Γ values vary significantly with m values; however, the differences in Γ values for TE and TM waves of the same m value remain negligibly small. As the guiding-layer thickness increases from zero, the Γ value for the lowest-order mode increases rapidly and reaches a saturation level. A further increase in thickness allows the growth of the next-higher-order mode. This behavior is shown in Figure 3.6. The confinement factor is plotted for various m values as a function of the guiding-layer thickness of a symmetric planar waveguide, with $n_1 = 3.590$, $n_2 = 3.385$, and $\lambda = 0.9 \ \mu m$.

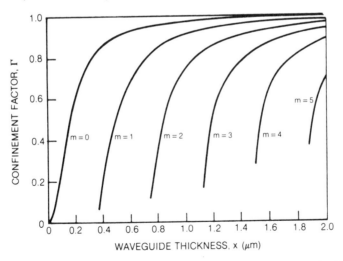

FIGURE 3.6 Variation of confinement factor Γ as a function of waveguide thickness for various mode orders in a symmetric planar waveguide, having $n_1 = 3.590$ and $n_2 = 3.385$. λ is assumed to be 0.9 μm.

3.6 SCALING RULES FOR ONE-DIMENSIONAL WAVEGUIDES

Generally speaking, five independent parameters are involved in determining the propagation constant β of infinitely extending (one-dimensional) planar waveguides. They are: the refractive indices of the substrate n_s, the guiding layer n_g, the covering layer n_c, the guiding layer thickness d, and the free space propagation constant k_0. It is desirable to reduce these parameters by introducing (Ref. 3.4) a normalized frequency parameter V, an asymmetric parameter a, and a normalized guiding index b:

$$V = k_0 d (n_g^2 - n_s^2)^{1/2} \tag{3.44}$$

$$a = (n_s^2 - n_c^2)/(n_g^2 - n_s^2) \tag{3.45}$$

$$b = (N_{\text{eff}}^2 - n_s^2)/(n_g^2 - n_s^2) \tag{3.46}$$

where

$$N_{\text{eff}} = \beta/k_0 \tag{3.47}$$

The values of the propagation constant β for various guided-wave modes can be determined from Equations (3.25) and (3.30). These relations characterize the modal dispersion of the waveguide and are generally referred to as dispersion relations. To quantify these modal dispersion relations, the phase shifts must be determined at the upper and lower boundaries. To do so, the decay constants γ_s and γ_c of the evanescent waves as given by Equation (3.31) must be known. The effective guiding layer thickness, which extends beyond the geometric guiding layer thickness d, is given by

$$d_{\text{eff}} = d + \frac{1}{\gamma_s} + \frac{1}{\gamma_c} \tag{3.48}$$

Using these definitions, the following discussion will deal separately with the TE and the TM modes.

With TE modes, phase shifts at the upper and lower boundaries are given by

$$\tan \phi_s = \gamma_s/k \tag{3.49}$$

$$\tan \phi_c = \gamma_c/k \tag{3.50}$$

These normalized parameters V, a, and b can be related by the following expression:

$$V(1 - b)^{1/2} = m\pi + \tan^{-1}[b/(1 - b)]^{1/2}$$
$$+ \tan^{-1}[(a + b)/(1 - b)]^{1/2} \tag{3.51}$$

Equation (3.51) indicates that the normalized guide index b depends on only two independent parameters, a and V. The a value can vary from zero for symmetric guides ($n_s = n_c$), to infinity for strong asymmetry if $n_g \simeq n_s$.

Table 3.1 lists the refractive indices of three commonly used waveguides. In all cases $n_s \geq n_c$. At cutoff, $b = 0$ and $N_{eff} = n_s$. For the case far above the cutoff, $\gamma \to \infty$, $N_{eff} = n_g$ and $b = 1$. To illustrate, examine the case near cutoff ($b = 0$) for the fundamental TE$_0$ mode. Equation (3.51) yields

$$V_0 = \tan^{-1}(a)^{1/2} \qquad (3.52)$$

which indicates that for a symmetric guide ($a = 0$) no cutoff exists for the lowest-order TE mode. For highly asymmetrical guides ($a \to \infty$), the cutoff value for V_0 is $\pi/2$. For higher-order modes ($m > 0$) the cutoff occurs at

$$V_m = V_0 + m\pi \qquad (3.53)$$

When the mode number m is large ($V_0 \ll m\pi$), Equation (3.53) reduces to the well-known formula Equation (3.33), for the number of allowed guided modes. Figure 3.7 is a plot of Equation (3.51) showing the dependence of the normalized guide index b on the normalized frequency V of the first three modes for various values of a. Figure 3.8 shows the same curves on an expanded scale for the TE$_0$ mode with various values of a. In this figure, the limiting V value for single mode operation is marked by V_1.

To investigate modal dispersion relations in more detail, the slope from Equation (3.51) is calculated and is:

$$\frac{\partial b}{\partial V} = \frac{2(1 - b)}{W} \qquad (3.54)$$

where W representing the normalized guide thickness is

$$W = V + \frac{1}{b^{1/2}} + \frac{1}{(a + b)^{1/2}} \qquad (3.55)$$

It is related to the effective guide thickness d_{eff} by the expression:

$$W = k_0 d_{eff}(n_g^2 - n_s^2)^{1/2} \qquad (3.56)$$

Equation (3.56) is analogous to the definition of V as given by Equation (3.44). At cutoff ($b = 0$), the slope $(\partial b/\partial V)_{b=0} = 0$. Equation (3.54) indicates that W approaches infinity, which is consistent with Equation (3.55). Far

TABLE 3.1   Asymmetry Measures for TE (a_{TE}) and TM (a_{TM}) Modes

Waveguide	n_s	n_g	n_c	a_{TE}	a_{TM}
GaAs/GaAlAs DH lasers	3.55	3.6	3.55	0	0
Outdiffused LiNbO$_3$	2.214	2.215	1	881	21,206
Sputtered glass	1.515	1.62	1	3.9	27.1

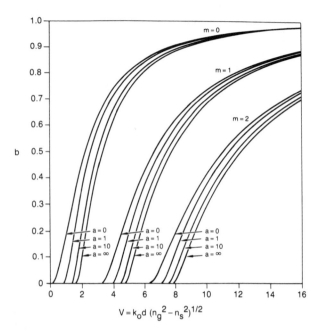

FIGURE 3.7 Normalized guide index b as a function of V for TE modes. (From Ref. 3.4. Reprinted with permission of OSA, 1974.)

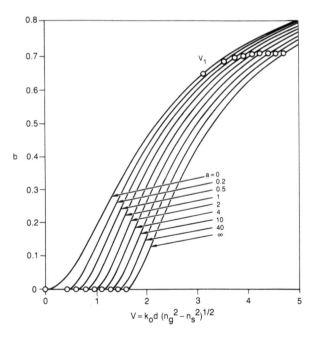

FIGURE 3.8 Normalized guide index b for the TE_0 mode with various degree of waveguide asymmetry. (From Ref. 3.4. Reprinted with permission of OSA, 1974.)

from cutoff ($b \simeq 1$), Equation (3.55) indicates that

$$W = V + 1 + (1 + a)^{1/2} \qquad (3.57)$$

where the asymptote for $a = 0$ is $W = V + 2$ and for $a = \infty$ is $W = V + 1$. Figure 3.9 is a plot of W as a function of V for various values of a for the fundamental mode. The envelope of the $W(V)$ curves shows a quite broad minimum, with the smallest value of $W_{min} = 4.4$ occurring at $V = 2.55$ for $a = \infty$. W_{min} does not change appreciably until $a = 1$. For a symmetric guide ($a = 0$), $W_{min} = 4.93$ at $V = 1.73$.

The situation for TM modes is analogous but is somewhat more complex than that for TE modes. The normalized guide index b for TM modes is defined as

$$b = [(N_{eff}^2 - n_s^2)/(n_g^2 - n_s^2)][n_g^2/n_s^2 q_s] \qquad (3.58)$$

which is equivalent to

$$N_{eff}^2 = [n_g^2(1 - b) + n_s^2 b]q_s \qquad (3.59)$$

where the reduction factor q_s is

$$q_s = \frac{N_{eff}^2}{n_g^2} + \frac{N_{eff}^2}{n_s^2} - 1 = \frac{n_s^2/n_g^2}{(1 - b) + bn_s^4/n_g^4} \qquad (3.60)$$

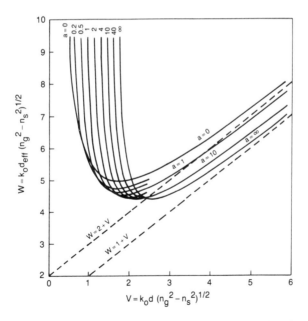

FIGURE 3.9   Normalized effective guide thickness W as a function of V for the TE$_0$ mode. (From Ref. 3.4. Reprinted with permission of OSA, 1974.)

Even with these complicated definitions, the dispersion curves maintain the same range for $b = 0$ at cutoff and $b = 1$ at far from cutoff. Also, the simple linear relation of N_{eff} with b obtained from Equation (3.46) as given by

$$N_{eff} \simeq n_s + b(n_g - n_s) \tag{3.61}$$

remains valid for both the TE and TM modes in the case of $n_g \simeq n_s$. The normalized forms of the dispersion relation for TM modes becomes:

$$V[q_s^{1/2} n_g/n_s](1 - b)^{1/2}$$

$$= m\pi + \tan^{-1}\left(\frac{b}{1 - b}\right)^{1/2} + \tan^{-1}\left[\frac{b + a(1 - bc)}{1 - b}\right]^{1/2} \tag{3.62}$$

where the asymmetry measure for TM modes has been defined as:

$$a = (n_g^4/n_c^4)[(n_s^2 - n_c^2)/(n_g^2 - n_s^2)] \tag{3.63}$$

and

$$c = (1 - n_s^2/n_g^2)(1 - n_c^2/n_g^2) \tag{3.64}$$

Note that two parameters are needed to specify the asymmetric feature of the waveguide structure for TM modes. Furthermore, each parameter, e.g., a, c, and q depends only on the index ratio n_s/n_g and n_c/n_g, which can be used as two independent parameters. When the index difference between the guide and substrate is small ($n_s/n_g \simeq 1$), Equation (3.60) yields $(n_g/n_s^2)q_s^2 = 1$ and $c = 0$. With these approximations, Equation (3.62) assumes a form that is exactly the same as that of the corresponding Equation (3.51) for TE modes. Therefore, the universal dispersion curves given in Figures 3.7 and 3.8 apply to TM modes also, with the only difference between these two cases being the different definitions and values of asymmetry. From Table 3.1 note the considerable differences in a values between TE and TM modes for asymmetric waveguides.

To investigate modal dispersion relations even further, again consider the cases near cutoff and far from cutoff. At cutoff, $b = 0$, and the terms $q_s^{1/2} n_g/n_s$ and c drop out for any value of $n_g - n_s$. As a result, exactly the same relations, Equations (3.52) and (3.53) are obtained for the cutoff frequencies V_0 and V_m as in the case of TE modes. Detailed numerical calculations of the dispersion curves from Equation (3.62) indicate that b values between the frequency range $0 \le V \le 10$ of the fundamental TM_0 mode for a series of values of a between 0 to infinity differ very slightly ($<1\%$) from those calculated by neglecting the term bc in Equation (3.62). If $bc = 0$ Equations (3.51) and (3.62) can be equated to obtain the relation

$$V_{TE} = V_{TM}(q_s^{1/2} n_g/n_s) \tag{3.65}$$

This relation allows the use of the calculated results shown in Figures 3.7

and 3.8 for TM modes even when $n_g - n_s$ is large. For a given b value, the value of V_{TE} can be read off and V_{TM} can be calculated from Equation (3.65). The effective guide thickness for TM modes is defined as

$$d_{\text{eff}} = d + \frac{1}{q_s \gamma_s} + \frac{1}{q_c \gamma_c} \tag{3.66}$$

where the reduction factor q_c is

$$q_c = q_s \left[(1 - b) \left(1 + \frac{n_g^2}{n_c^2} - \frac{n_g^2}{n_s^2} \right) + b \frac{n_s^2}{n_c^2} \right] \tag{3.67}$$

In terms of the normalized thickness W and frequency V, Equation (3.67) can be rewritten as

$$W q_s^{1/2} \left(\frac{n_g}{n_s} \right) = V q_s^{1/2} \left(\frac{n_g}{n_s} \right) + \left[q_s \left(\frac{n_s^2}{n_g} \right) b^{1/2} \right]^{-1}$$

$$+ \left[q_c \left(\frac{n_c^2}{n_g} \right) (b + a(1 - bc)) \right]^{-1} \tag{3.68}$$

In the case when $n_g - n_s$ is small, Equation (3.68) can be approximated by the expression:

$$q_c \frac{n_c^2}{n_s^2} = 1 - (1 - b) \left(1 - \frac{n_s^2}{n_g^2} \right) \left(1 + \frac{n_c^2}{n_g^2} \right) \tag{3.69}$$

In the approximation that $q_c(n_c^2/n_s^2) = 1$, Equations (3.55) and (3.68) become identical in form. Consequently the effective thickness curves in Figure 3.9 can be used for TM modes when n_g and n_s is small. For large index differences between the substrate and guide, W values can be obtained directly from Equation (3.68). Again, the term bc can be neglected without creating errors in W greater than 1%.

3.7 EFFECTIVE INDEX METHOD

Many optical waveguides have configurations considerably more complex than the one-dimensional structure discussed in previous sections. For example, channel waveguides formed by the diffusion of Ti transition metals into LiNb0$_3$ or LiTa0$_3$ are commonly used for making integrated optical circuits in which many passive and active components such as directional couplers, modulators, and switches can be fabricated. Another frequently used waveguide is the buried channel double heterostructure produced by the epitaxial growth of GaAlAs or InGaAsP of which most index-guided semiconductor lasers are constructed. Exact calculations of modal dispersion in these two-dimensional waveguides usually involve rather extensive computer time. In this section, an approximation known as the effective index method (Ref. 3.5) is introduced. This method makes use of the scaling rules of the one-dimensional waveguide dispersion relations already established in Section 3.6 to estimate modal dispersion relations in 2-D waveguides without carrying out extensive computer calculations. For a buried channel waveguide as shown in Figure 3.10(a), the 2-D waveguide shown can be dissected into two 1-D waveguides, as shown in Figures 3.10(b) and (c). This process can be accomplished by first letting the long dimension (along the y-axis) of the rectangular structure approach infinity, and subsequently, by constructing the second planar waveguide along the x-axis. For the first guide, the effective index n_{eff} can be readily obtained for the TE or TM modes from one of the universal modal dispersion curves (Figures 3.7 and 3.8). The additional confinement imposed by the finite width W can be accounted for by forming an equivalent planar waveguide, as shown in Figure 3.10(c), with an effective index n_{eff} assigned to the guiding layer. The cladding layer indices must be kept at their original values, n_3 and n_4. The effective index n'_{eff} of the TE and TM modes of this equivalent guide can then be obtained from the universal curves. This procedure is particularly simple if the index difference between n_0 and n_2 is large but small between n_0 and all others. In this case, the normalized dispersion curves $b(a, V)$ for TE modes can be used in both x and y directions with an appropriate change in the asymmetric measure a. However, if the index differences ($n_0 - n_3$ and $n_0 - n_4$) are large, the polarization characteristics of the mode must be retained. In this case, the normalized modal dispersion curves for both TE and TM modes must be employed.

To illustrate, consider the buried channel waveguide shown in Figure 3.10(a), with the channel width W and the depth D. For simplicity, assume

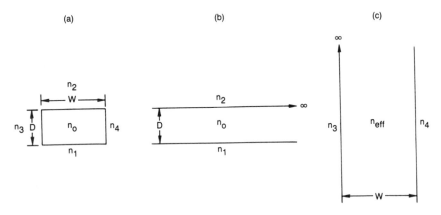

FIGURE 3.10   (a) Configuration of a 2-D buried channel waveguide. (b) A planar waveguide obtained by extending the width of a channel to infinity as an approximation. (c) An equivalent 1-D waveguide by choosing n_{eff} to be the guiding layer index. n_{eff} is the effective index of the mode in Guide (b).

that $n_1 = n_2 = n_3 = n_4 = n$. This assumption implies that $a_{TE} = a_{TM} = 0$. For TE modes, the normalized frequency V and normalized index b for the 1-D guide shown in Figure 3.10(b) are

$$V = k_0 D(n_0^2 - n^2)^{1/2} \qquad (3.70)$$

and

$$b = (n_{\text{eff}}^2 - n^2)/(n_0^2 - n^2) \qquad (3.71)$$

For given values of n_0, n, k_0, and D, the value of b can be obtained from the dispersion curve of Figure 3.7. The normalized frequency V' and the normalized guide index b' of the equivalent waveguide, as shown in Figure 3.10(c) are

$$V' = k_0 W(n_{\text{eff}}^2 - n^2)^{1/2} = Vb^{1/2}W/D \qquad (3.72)$$

$$b' = (n_{\text{eff}}'^2 - n^2)/(n_{\text{eff}}^2 - n^2) \qquad (3.73)$$

With given values of b and W, n_{eff} can be obtained from Equation (3.71) and V' from Equation (3.72). Again, using the universal dispersion curve of Figure 3.7, b' and, subsequently, $n_{\text{eff}}' = \beta/k_0$ can be determined. Combining Equations (3.71) and (3.73) n'_{eff} can be expressed in terms of b and b' as

$$n_{\text{eff}}' = [n^2 + bb'(n_0^2 - n^2)]^{1/2} \qquad (3.74)$$

For small index difference $\Delta n = (n_0 - n)$:

$$n_{\text{eff}}' = n + bb'\Delta n \qquad (3.75)$$

Equation (3.75) provides a reasonably accurate estimate of the propagation constant β of the 2-D waveguide mode. For TM modes the corresponding V_{TM} value can be obtained from Equation (3.65) and the process is repeated using the universal curves. In summary, the effective index method is very useful for a reasonably accurate analysis of 2-D waveguides having a variety of index profiles and can be used to obtain the modal effective index, propagation constant, and/or transverse momentum components k and γ with the help of a pocket-sized calculator. This method has been used extensively in designs of integrated optical circuits and semiconductor lasers, which are discussed in Chapter 10.

PROBLEMS

3.1. Show that if $\theta_1 + \theta_2 = \pi/2$, r_M defined in Equation (3.13) is zero or t_M is unity. This result is known as Brewster's condition for the TM wave.

3.2. Derive Equations (3.21) and (3.23).

3.3. Establish the eigenvalue for TM modes of a symmetric planar waveguide (e.g., $n_2 = n_3$).

3.4. Obtain the eigenvalue equation for TM modes of a asymmetric planar waveguide.

3.5. Construct the vector diagram of **k** for a guided wave both inside and outside the guide.

3.6. From the curves of Figure 3.4, calculate the β values for the 1-μm guide.

3.7. Calculate the fraction of light intensity inside the guiding layer for even TE modes.

3.8. Calculate the fraction of light intensity outside the guiding layer for even TE modes.

3.9. Calculate the Γ factor for odd TE modes.

3.10. Using the effective index method, analyze the modal dispersion relations for a buried rectangular ($W = 2D$) GaAs/GaAlAs double heterostructure laser ($n_g = 3.6$ and all the surrounding indices are equal with $n = 3.55$). Plot b' as a function of V' for TE_0 and TE_1 modes.

REFERENCES

3.1. P. K. Tien, and R. Ulrich, *J. Opt. Soc. Am.*, *60*, 1325 (1970).

3.2. J. E. Midwinter, *Optical Fibers for Transmission*, John Wiley & Sons, Inc., New York, 1979.

3.3. N. C. Casey, Jr. and M. B. Panish, *Heterostructure Lasers*, Part A: *Fundamental Principles*, Academic Press, New York, 1978.

3.4. H. Kogelnik and V. Ramaswamy, *Appl. Opt. 13*, 1857 (1974).

3.5. G. B. Hocker and W. K. Burns, *Appl. Opt. 16*, 113 (1977).

4

Cylindrical
Dielectric
Waveguides

4.1 INTRODUCTION

Exact solutions of Maxwell's equations can be obtained for a cylindrical dielectric waveguide with a step-index profile as shown in Figure 1.2(a), by using appropriate boundary conditions applied to the transverse field components. However, the results are considerably more complex than those obtained for planar waveguides, because in this case, all field components are coupled, and for each axial propagation constant β, two sets of nearly degenerate modes exist. To illustrate this degeneracy, visualize the situation from a ray picture by following two opposite helical paths in fiber. Because of circular symmetry, these modes with opposite helicity have almost the same β value. These helical waves are called hybrid EH or HE modes. The notation EH implies that the z component of the E field is larger than that of the H field. The converse is true for HE modes. Note that both E_z and H_z are usually very small compared with other components. The polarization associated with these modes is circular and rotating either clockwise or counterclockwise in accordance with the helicity. Other groups of modes associated with meridional rays are TE and TM linearly polarized waves. In these cases the field components have no angular or azimuthal dependence. In addition to guided-wave modes, radiation and leaky-wave modes exist, which are also solutions of Maxwell's equations, but are not treated.

This chapter begins with a simplified analysis to establish a simple eigenvalue equation for a weakly guiding fiber ($n_0 \simeq n_c$). With these results approximate solutions are obtained for two cases: one is close to cutoff and the other is far from cutoff. Then the exact eigenvalue equation by which a complete characterization of hybrid modes can be achieved with considerable complexity is established. In all cases the results indicate that an enormous number of modes exist in a multimode fiber, each of which propagates at a distinct group velocity. If a short pulse of light is launched at one end of a multimode fiber, the width of this pulse is broadened primarily as a result of modal dispersion. However, a graded-index fiber, if made with a correct index profile, can greatly suppress modal dispersion and consequently reduce group delay. An alternative approach is, of course, to use a single mode, step-index fiber. But the disadvantage of using such a fiber is the difficulty encountered in efficiently splicing and optically coupling a fiber that has a core diameter of only a few micrometers. Dispersion, coupling, and splicing will be discussed in Chapter 5.

This chapter also introduces some important properties related to the propagation characteristics of light in both step-index and graded-index fibers. Because of the complexity of wave equations, which contain a term functionally dependent on the core radius, it is impossible to obtain exact solutions for the case of graded-index fibers. Instead, this problem is treated using simple ray analysis. With this approach light paths can be traced precisely for either meridional or skew rays inside the fiber with a given graded-index profile. Conceptually, visualize a bundle of rays that strike a cylindrical boundary at different angles, each of which represents a mode in the wave description. Lower-order modes strike the boundary at angles near grazing and therefore travel very close to the core of the fiber. If the core index is constant, lower-order modes will travel at faster velocities than higher-order modes. To slow down lower-order modes or speed up higher-order modes, a parabolic index profile such as the one shown in Figure 1.2(b) can be used. However, the exact shape of this profile must be determined for the least dispersion. This chapter will discuss a procedure for selecting a proper index profile in ray analysis by requiring that all rays propagate with nearly the same period. Following the ray analysis, a method known as the Wentzel–Kramers–Brillouin (WKB) approximation is described, by which β values can be determined for graded-index fibers. An explicit expression for β is needed to evaluate the effects of modal dispersion on the information-carrying capacity of a fiber.

4.2 SCALAR FIELD SOLUTIONS FOR STEP-INDEX FIBERS

For simplicity this section will first solve the wave equation for a cylindrical waveguide with a step-index profile in the approximation of a scalar field by neglecting the complications of coupled fields. In this approximation, each transverse component ψ of the electric field obeys the scalar Helmholtz equation, which can be derived directly from Equation (2.16) or (2.17). By eliminating the time dependence, the scalar field equation becomes

$$[\nabla^2 + k_0^2 n^2(r)]\psi = 0 \tag{4.1}$$

where ∇^2 is the Laplacian operator. For cylindrical coordinates

$$\nabla^2 = \frac{d^2}{dr^2} + \frac{1}{r}\frac{d}{dr} + \frac{1}{r^2}\frac{d^2}{d\theta^2} + \frac{d^2}{dz^2} \tag{4.2}$$

Because of the axial and circular symmetry of the fiber, a solution of Equation (4.1) can be assumed in the form

$$\psi = \psi(r) \exp[i(l\theta + \beta z)] \tag{4.3}$$

where l is the azimuthal eigenvalue and β is the propagation wave number along the z axis of a fiber with a core radius a. Substituting Equation (4.3) into (4.1), yields for the case of a step-index profile:

$$\frac{d^2\psi}{dr^2} + \frac{1}{r}\frac{d\psi}{dr} + \left(k_0^2 n_0^2 - \beta^2 - \frac{l^2}{r^2}\right)\psi = 0 \qquad r \leq a \tag{4.4}$$

$$\frac{d^2\psi}{dr^2} + \frac{1}{r}\frac{d\psi}{dr} + \left(k_0^2 n_c^2 - \beta^2 - \frac{l^2}{r^2}\right)\psi = 0 \qquad r > a \tag{4.5}$$

The equations above can be simplified by defining

$$u^2 \equiv (k_0^2 n_0^2 - \beta^2)a^2 \tag{4.6}$$

$$\gamma^2 \equiv (\beta^2 - k_0^2 n_c^2)a^2 \tag{4.7}$$

An important parameter V for the fiber is

$$V = (u^2 + \gamma^2)^{1/2} = k_0 a(n_0^2 - n_c^2)^{1/2} \tag{4.8}$$

$$= \frac{2\pi a}{\lambda}(NA)$$

where NA is the numerical aperture of a step-index fiber.

The solutions of Equations (4.4) and (4.5) are well-known. For specific solutions, the first kind of Bessel function of order l for Equation (4.4) and the second kind of modified Bessel function of order l for Equation (4.5) are chosen:

$$\psi(r) = AJ_l\left(\frac{ur}{a}\right) \qquad r < a \tag{4.9}$$

$$\psi(r) = BK_l\left(\frac{\gamma r}{a}\right) \qquad r > a \tag{4.10}$$

where J_l is given in Problem 4.2 and K_l is given by the following equation

$$K_l(x) = \frac{\pi}{2} i^{n+1}[J_l(ix) + iY_l(ix)]$$

where

$$Y_l(x) = \frac{2}{\pi}\left[\left(\log\frac{x}{2} + 0.5772\right)J_l(x) - \frac{1}{2}\sum_{k=0}^{n-1}\frac{(n-k-1)!}{k!}\left(\frac{x}{2}\right)^{2k-n}\right.$$
$$\left. + \frac{1}{2}\sum_{k=0}^{\infty}(-1)^{k+1}\frac{[\varphi(k) + \varphi(k+n)]}{k!(n+k)!}\left(\frac{x}{2}\right)^{2k+n}\right]$$

and $\varphi(k)$ is given by the equation

$$\varphi(k) = \sum_{m=1}^{k}\frac{1}{m} = 1 + \frac{1}{2} + \cdots + \frac{1}{k}$$

The choice of these functions is clear from the following asymptotic forms. For $x \ll 1$

$$J_l(x) = \frac{1}{l!}\left(\frac{x}{2}\right)^l \qquad l = 0, 1, \ldots \qquad (4.11)$$

$$K_l(x) = (l - 1)!2^{l-1}x^{-l} \qquad l \geq 1 \qquad (4.12)$$

and for $x \gg 1$

$$J_l(x) = \sqrt{\frac{2}{\pi x}}\cos\left[x - \frac{\pi(2l + 1)}{4}\right] \qquad (4.13)$$

$$K_l(x) = \sqrt{\frac{\pi}{2x}}e^{-x}\left(1 + \frac{4l^2 - 1}{8x}\right) \qquad (4.14)$$

These results indicate that J_l and K_l are well-behaved functions inside and outside the boundary of a cylindrical waveguide, respectively. Other Bessel functions are not suitable for representing this waveguide. For further analysis, the following recurrence relations for these functions (with argument x) and some asymptotic forms that are useful are

$$J_{-l} = (-1)^l J_l \qquad (4.15)$$

$$J_l' = \frac{1}{2}(J_{l-1} - J_{l+1}) = \pm J_{l\mp 1} \mp \frac{lJ_l}{x} \qquad (4.16)$$

$$J_{l\mp 1} = \frac{2lJ_l}{x} - J_{l\pm 1} \qquad (4.17)$$

$$J_{l\mp 2} = \frac{2(l \mp 1)J_{l\mp 1}}{x} - J_l \qquad (4.18)$$

$$K_l = K_{-l} \tag{4.19}$$

$$K_l' = -\frac{1}{2}(K_{l-1} + K_{l+1}) = \mp\frac{lK_l}{x} - K_{l\mp1} \tag{4.20}$$

$$K_{l\mp1} = \mp\frac{2lK_l}{x} + K_{l\pm1} \tag{4.21}$$

$$K_{l\mp2} = \mp\frac{2(l\mp1)K_{l\mp1}}{x} + K_l \tag{4.22}$$

For $x \ll 1$

$$\frac{K_0}{K_1} = x \ln\frac{2}{1.782x} \tag{4.23}$$

$$\frac{K_{l-1}}{K_l} = \frac{x}{2(l-1)} \qquad l \geq 2 \tag{4.24}$$

$$\frac{K_{l+1}}{K_l} = \frac{2l}{x} \qquad l \geq 1 \tag{4.25}$$

For $x \gg 1$

$$\frac{K_{l\mp1}}{K_l} = 1 + \frac{1\mp2l}{2x} \tag{4.26}$$

The continuity of ψ and its derivative at the boundary of the core and cladding (e.g., $r = a$) leads to a simple eigenvalue equation of the form

$$\frac{uJ_l'(u)}{J_l(u)} = \frac{\gamma K_l'(\gamma)}{K_l(\gamma)} \tag{4.27}$$

Substituting Equations (4.16) and (4.20) into (4.27) two equivalent eigenvalue equations of the form are obtained:

$$\frac{uJ_{l\pm1}(u)}{J_l(u)} = \pm\frac{\gamma K_{l\pm1}(\gamma)}{K_l(\gamma)} \tag{4.28}$$

For various values of l, Equation (4.28) provides the corresponding values of u and γ, either of which leads to the corresponding value for the propagation wave number β, lying within the range

$$n_c < \frac{\beta}{k_0} < n_0 \tag{4.29}$$

As shown in the next section, this is a special case of a cylindrical dielectric waveguide for which $n_0 \simeq n_c$. Due to the oscillatory nature of Bessel functions, for every value of lm, allowed solutions exist for β. Therefore, each allowed β value is characterized by two integers l and m. The first integer, l, is associated with two circular functions, cos $l\theta$ and sin $l\theta$, and the second integer, m, corresponds to the mth root of the eigenvalue equation. By convention, $HE_{l,m}$ modes are those whose longitudinal electric fields dominate and $EH_{l,m}$ modes are those whose longitudinal magnetic fields dominate. If the longitudinal fields are zero, as in the case of very weakly guiding fibers, the LP modes are linearly polarized. Note that the eigenvalue equation (4.28) contains $\pm$ signs. The plus sign usually corresponds to $HE_{l+1,m}$ modes and the minus sign corresponds to $EH_{l-1,m}$ modes.

Equation (4.28) implies that $HE_{l+1,m}$ modes are degenerate with $EH_{l-1,m}$ modes, because the same equation applies for both modes. However, for $l = 0$, $HE_{1,m}$ modes have special significance because $EH_{-1,m}$ modes do not exist. Therefore, $HE_{1,m}$ modes are nondegenerate. In the following, the β value for the HE_{11} mode are calculated for only two limiting cases. For the first case, a simple relationship is established between β and V near cutoff. Near cutoff, $\gamma = 0$. For the second case, $\gamma = \infty$. This situation describes the mode far above cutoff.

If $\gamma = 0$, Equation (4.28) indicates that $J_{l\pm1}(u) = 0$. If $l = 0$, the lowest root of $J_{l\mp1}(u)$ is $u = 0$. This value means that no cutoff exists for the lowest-order HE_{11} mode in a fiber. The cutoff for the next-order mode in this group ($l = 0$) is at $u_m = 3.832$. However, before this mode, the first cutoff of another group ($l = 1$) will occur at $u_m = 2.405$, which is the first root of $J_0(u)$. Table 4.1 is a list of cutoff frequencies for a few lower-order modes. Figure 4.1 shows the oscillatory behavior of J_0 and J_1 functions and the roots of these two functions.

Equation (4.28) for $l = 0$ can be written as

$$\frac{uJ_1(u)}{J_0(u)} = \frac{\gamma K_1(\gamma)}{K_0(\gamma)}$$

Using Equation (4.23) for the limiting value near cutoff gives

$$\frac{J_0(V)}{VJ_1(V)} = -\ln\frac{1.782\gamma}{2}$$

where the approximation $u \simeq V$ is made. Therefore

$$\gamma = 1.122 \exp\left[-\frac{J_0(V)}{VJ_1(V)}\right] \tag{4.30}$$

TABLE 4.1   First Four Zeroes of J_0 and J_1 Functions

		m		
l	1	2	3	4
0	0	3.832	7.016	10.173
1	2.405	5.520	8.654	11.790

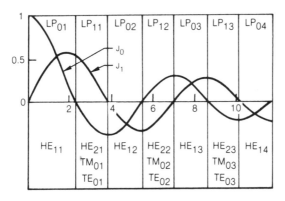

FIGURE 4.1   Plot of the Bessel J_0 and J_1 functions.

This approximation is as good as the exact solution (Ref. 4.1) for values of V ranging from zero to 1, and is off only by about 10% as V approaches 2.5.

For the case in which the mode is far from cutoff, let $\gamma \rightarrow \infty$, Equation (4.26) yields

$$\lim_{\gamma \to \infty} \frac{K_0(\gamma)}{K_1(\gamma)} \simeq 1$$

Using this limiting value and replacing γ by V, Equation (4.28) becomes

$$VJ_0(u) = uJ_1(u) \tag{4.31}$$

Further simplification can be made by eliminating J_0 and J_1 from Equation (4.31). Differentiating Equation (4.31) with respect to V and using the following identities

$$J_1'(x) = J_0 - \frac{1}{x} J_1(x)$$

$$J_0'(x) = -J_1(x)$$

yields

$$\frac{du}{dV} \left[u + V \frac{J_1(u)}{J_0(u)} \right] = 1$$

Now eliminating $J_1(u)/J_0(u)$ from the equation above and again using Equation (4.31) yields

$$\frac{du}{u} (u^2 + V^2) = dV$$

In the limit as $\gamma \rightarrow \infty$, $u \ll V$; therefore, the u^2 term can be neglected in the equation above and a simple expression for u can be obtained after integrating the equation above from (u_{0m}, V) to $(u_{0m}^{\infty}, \infty)$, giving

$$u_{0m} = u_{0m}^{\infty} \exp\left(-\frac{1}{V}\right) \tag{4.32}$$

where u_{0m}^{∞} is the root of $J_0(u)$ for V values far from cutoff. Similar relationships can be obtained for higher-order modes by extending the foregoing analysis for $l > 1$. In the limit as $\gamma \to \infty$, the following eigenvalue equation results:

$$\gamma J_1(u) = u J_{l+1}(u)$$

Again taking the derivative of the equation above with respect to V and eliminating the Bessel functions from the equation by using appropriate identities yields

$$\frac{du}{dV} = \frac{u}{V[V - 2(l - 1)]}$$

Upon integration the approximate solution in the limit as $\gamma \to \infty$ for HE modes $(l > 1)$ is

$$u_{lm} = u_{lm}^{\infty}\left[1 - \frac{2(l - 1)}{V}\right]^{1/2(l-1)} \tag{4.33}$$

where u_{lm}^{∞} is the mth root of $J_l(u)$. For large values of m, u_{lm}^{∞} can be approximated by the expression

$$u_{lm}^{\infty} \simeq (l + 2m)\frac{\pi}{2} \tag{4.34}$$

Figure 4.2 is a plot of u and γ as a function of V for the HE_{11} mode. The approximate solutions are shown by dashed curves. Figure 4.2 shows that the results given by the approximation near cutoff coincide with the exact solution for $V \leq 1$, and the approximation far from cutoff is very good for $V \geq 1.5$. The next two higher-order modes are the transverse electric (TE_{01}) and transverse magnetic (TM_{01}) modes. They are the two lowest cylindrically symmetric modes. The electric vectors are oriented as shown in Figure 4.3(a).

In general, TE_{0m} and TM_{0m} are two linearly polarized sets of modes with polarization vectors perpendicular to each other. Therefore, they can easily be distinguished from each other by using a polarizer to analyze the orientation of the electric vectors. The next higher-order mode is HE_{21}. It is a hybrid mode with a twofold degeneracy, which is a consequence of circular symmetry as shown in Figure 4.3(b). The HE_{21} mode has two possible field configurations as shown in Figure 4.3(b) and they can be obtained simply by rotating one 90° with respect to the other. Because the TE_{01}, TM_{01}, and HE_{21} modes have nearly the same β-values and cutoff characteristics

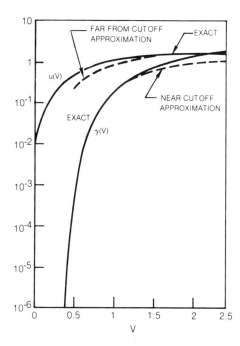

FIGURE 4.2 Plot of u and γ as a function of V for the HE_{11} mode.

(see Figure 4.5), they usually occur simultaneously. Figure 4.3(c) shows four independent linear combinations of these three modes. The first two combinations have their resultant electric vector parallel to the null line, whereas the second two combinations have their resultant electric vector perpendicular to the null line. Again it is possible to distinguish TE_{0m} and TM_{0m} modes from HE_{2m} modes by rotating an analyzer placed at the entrance to

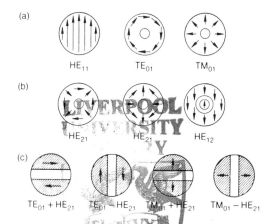

FIGURE 4.3 Electric vectors of several lowest-order modes of a step-index fiber.

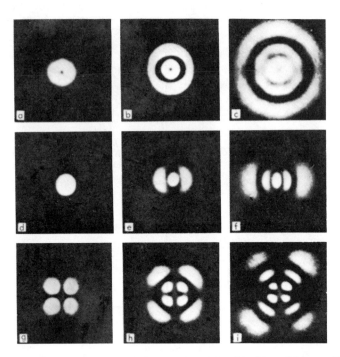

FIGURE 4.4 Images of several lower-order modes: (a), (b), and (c) are either TE$_{0m}$, TM$_{0m}$, or HE$_{2m}$ (m = 1, 2, 3); (d) is the HE$_{11}$ mode; (e) is a linear combination of HE$_{12}$ with either HE$_{31}$ or EH$_{11}$; (f) is a linear combination of HE$_{13}$ with either HE$_{32}$ or EH$_{12}$; (g), (h), and (i) are either EH$_{1m}$ or HE$_{3m}$ (m = 1, 2, 3). (Courtesy of Dr. Elias Snitzer, United Technologies Research Center.)

the fiber. As the analyzer rotates, the null line rotates in the same direction for l = 0 modes but in the opposite direction for HE$_{2m}$ modes.

The third group of higher-order modes are the HE$_{12}$, EH$_{11}$, and HE$_{31}$ modes with the same cutoff value at u_c = 3.832. As the diameter of the core increases further (V > 3.832), more and more modes are introduced with values of β much closer to each other. As a result, spatial resolution of these modes becomes extremely difficult to achieve. Figure 4.4 shows a few typical lower-order mode patterns, some of which are observed with the help of polarizers.

4.3 APPROXIMATION FOR WEAKLY GUIDED STEP-INDEX FIBERS

This section contains a more rigorous analysis of the modes in a cylindrical dielectric waveguide with a very small refractive index difference between the cladding and core by following the treatment of Gloge (Ref. 4.2). He recognized that the E and H fields in a weakly guided fiber ($\Delta \ll 1$) are very weakly coupled. In other words, transverse field components are essentially linear-polarized and longitudinal field components are very small compared to transverse field components. Therefore, hybrid EH_{lm} and HE_{lm} modes break up into two degenerate groups, each of which consists of two linearly polarized LP modes, which are orthogonal to each other for a given $l(l > 0)$. For $l = 0$, the degeneracy is removed and only one set of orthogonal modes exists, e.g., TE_{0m} and TM_{0m}. For this reason, it is convenient to express the field expressions in Cartesian coordinates rather than in cylindrical coordinates. If E_y is chosen to represent the state of polarization for the TE modes, then $E_x \ll E_y$. The state of polarization for TM modes must be represented by H_x because in this approximation $H_y \ll H_x$. This approximation can be verified by examining the transverse magnetic field components using Maxwell's Equation (2.14). Note that the z-dependence is given by $\exp(i\beta z)$, therefore

$$H_x = \frac{i}{\mu\omega} \left(\frac{\partial E_z}{\partial y} - \frac{\partial E_y}{\partial z} \right) = \frac{i}{\mu\omega} \frac{\partial E_z}{\partial y} + \frac{n}{Z_0} E_y \qquad (4.35)$$

$$H_y = \frac{i}{\mu\omega} \left(\frac{\partial E_x}{\partial z} - \frac{\partial E_z}{\partial x} \right) \simeq -\frac{i}{\mu\omega} \frac{\partial E_z}{\partial x} \qquad (4.36)$$

where

$$Z_0 = \frac{\mu\omega}{k_0} \qquad (4.37)$$

which is the impedance in free-space. The following shows that E_z as well as H_z are very small compared to E_y and H_x in the approximation $\Delta \ll 1$. Consequently, $H_y \ll H_x$. Using Maxwell's equations (2.14) and (2.15), the longitudinal field components can be expressed as

$$E_z = \frac{1}{i\omega\epsilon} \left(\frac{\partial H_y}{\partial x} - \frac{\partial H_x}{\partial y} \right) \simeq \frac{iZ_0}{k_0 n^2} \frac{\partial H_x}{\partial y} \qquad (H_y \ll H_x) \qquad (4.38)$$

$$H_z = \frac{i}{\mu\omega} \left(\frac{\partial E_y}{\partial x} - \frac{\partial E_x}{\partial y} \right) \simeq \frac{i}{k_0 Z_0} \frac{\partial E_y}{\partial x} \qquad (E_x \ll E_y) \qquad (4.39)$$

Now E_y is assumed to be of the form

$$E_y = \begin{cases} \dfrac{E_l}{J_l(u)} J_l \left(\dfrac{u}{a} r \right) \cos l\theta & (r < a) \\[4ex] \dfrac{E_l}{K_l(\gamma)} K_l \left(\dfrac{\gamma}{a} r \right) \cos l\theta & (r > a) \end{cases} \qquad (4.40)$$

where E_l is the electric field strength of the lth mode at the core-clad interface. In Equation (4.40), cos $l\theta$ is arbitrarily assigned to this set of modes. In general, there are two independent sets of modes; one goes with cos $l\theta$ and the other goes with sin $l\theta$. For $l = 0$, there are two independent sets of linearly polarized modes which are orthogonal with respect to each other. Substituting Equations (4.40) and (4.35) into (4.38) and (4.39), and making use of the following relations derived from Jacobian determinants:

$$\frac{\partial}{\partial x} = \cos\theta \frac{\partial}{\partial r} - \frac{\sin\theta}{r}\frac{\partial}{\partial\theta}$$

$$\frac{\partial}{\partial y} = \sin\theta \frac{\partial}{\partial r} + \frac{\cos\theta}{r}\frac{\partial}{\partial\theta}$$

(4.41)

the following is obtained:

$$H_z \simeq -\frac{iE_l}{2k_0 a Z_0}\begin{cases} u\dfrac{J_{l+1}(ur/a)}{J_l(u)}\cos(l+1)\theta - u\dfrac{J_{l-1}(ur/a)}{J_l(u)}\cos(l-1)\theta \\ \\ \hspace{4cm}(r < a) \\ \\ \gamma\dfrac{K_{l+1}(\gamma r/a)}{K_l(\gamma)}\cos(l+1)\theta + \gamma\dfrac{K_{l-1}(\gamma r/a)}{K_l(\gamma)}\cos(l-1)\theta \\ \\ \hspace{4cm}(r > a) \end{cases}$$

(4.42)

Observe that Equation (4.42) contains factors u/k_0a and γ/k_0a, which are both of the order $\sqrt{\Delta}$. The same factors are also found in E_z field expressions. This result has led to the assumption of linearly polarized modes. If Equation (4.35) is simplified to the following form

$$H_x \simeq \frac{n}{Z_0}E_y$$

then Equation (4.38) gives

$$E_z \simeq -\frac{iE_l}{2k_0 a}\begin{cases} \dfrac{u}{n_0}\dfrac{J_{l+1}(ur/a)}{J_l(u)}\sin(l+1)\theta + \dfrac{u}{n_0}\dfrac{J_{l-1}(ur/a)}{J_l(u)}\sin(l-1)\theta \\ \\ \hspace{4cm}(r < a) \\ \\ \dfrac{\gamma}{n_c}\dfrac{K_{l+1}(\gamma r/a)}{K_l(\gamma)}\sin(l+1)\theta - \dfrac{\gamma}{n_c}\dfrac{K_{l-1}(\gamma r/a)}{K_l(\gamma)}\sin(l-1)\theta \\ \\ \hspace{4cm}(r > a) \end{cases}$$

(4.43)

From these longitudinal components, E_z and H_z, the transverse components in the cylindrical coordinate system can be generated using Equations (2.40) and (2.42), as given by

$$
E_\theta = \begin{cases}
\dfrac{1}{2} E_l \dfrac{J_l(ur/a)}{J_l(u)} [\cos(l+1)\theta + \cos(l-1)\theta] & \text{for } r < a \quad (4.44) \\[4ex]
\dfrac{1}{2} E_l \dfrac{K_l(\gamma r/a)}{K_l(\gamma)} [\cos(l+1)\theta + \cos(l-1)\theta] & \text{for } r > a \quad (4.45)
\end{cases}
$$

$$
H_\theta = \begin{cases}
-\dfrac{1}{2} \dfrac{E_l n_0}{Z_0} \dfrac{J_l(ur/a)}{J_l(u)} [\sin(l+1)\theta - \sin(l-1)\theta] & \text{for } r < a \quad (4.46) \\[4ex]
-\dfrac{1}{2} \dfrac{E_l n_0}{Z_0} \dfrac{K_l(\gamma r/a)}{K_l(\gamma)} [\sin(l+1)\theta - \sin(l-1)\theta] & \text{for } r > a \quad (4.47)
\end{cases}
$$

Equating the tangential components H_z at the boundary $r = a$ and letting $n_0 = n_c$ yields

$$
u \frac{J_{l+1}(u)}{J_l(u)} \cos(l+1)\theta + u \frac{J_{l-1}(u)}{J_l(u)} \cos(l-1)\theta
$$

$$
= \gamma \frac{K_{l+1}(\gamma)}{K_l(\gamma)} \cos(l+1)\theta - \gamma \frac{K_{l-1}(\gamma)}{K_l(\gamma)} \cos(l-1)\theta
$$

From the equation above

$$
u \frac{J_{l\pm1}(u)}{J_l(u)} = \pm\gamma \frac{K_{l\pm1}(\gamma)}{K_l(\gamma)} \tag{4.48}
$$

which is identical to the two eigenvalue equations given by Equation (4.28). Other transverse components lead to the same eigenvalue equations in this approximation.

If $n_0 \neq n_c$, this degeneracy ceases to exist. Each LP$_{lm}$ mode breaks up into modes with the terms $(l + 1)\theta$, which can be identified as HE$_{l+1,m}$, and modes with terms $(l - 1)\theta$, which are labeled EH$_{l-1,m}$ or TE$_{0m}$ and TM$_{0m}$. For all practical purposes Equation (4.28) is sufficiently accurate for calculating the propagation constant β of these LP modes. Again, remember that the subscripts on HE$_{lm}$ and EH$_{lm}$ refer to the lth order and mth rank, where l is an integer and is associated with the circular functions $\sin l\theta$ or $\cos l\theta$; m is also an integer, which identifies the successive roots of $J_l = 0$. Physically, the values l and m represent the number of angular and radial antinodes in the field pattern, respectively.

It is useful to establish an explicit functional relationship between u and V for higher-order modes ($l > 0$), following the analysis of Snyder (Ref. 4.3). Taking the derivative of Equation (4.48) with respect to V and using the lower-case ($l - 1$) yields

$$\frac{du}{dV} = \frac{-1}{J_{l-1}K_l}\left[\frac{d}{dV}(\gamma K_{l-1}J_l) + u\frac{d}{dV}K_lJ_{l-1}\right] \tag{4.49}$$

And, using Equation (4.8) gives

$$\frac{d\gamma}{dV} = \left(V - u\frac{du}{dV}\right)\Big/\gamma$$

The derivatives in Equation (4.49) can be expressed as

$$\frac{d}{dV}(\gamma K_{l-1}J_l) = \frac{du}{dV}\left[\gamma K_{l-1}J_l' - \frac{uJ_l(\gamma K_{l-1}' + K_{l-1})}{\gamma}\right]$$
$$+ \frac{V(\gamma K_{l-1}' + K_{l-1})J_l}{\gamma}$$

and

$$\frac{d}{dV}(J_{l-1}K_l) = \left(K_lJ_{l-1}' - \frac{uJ_{l-1}K_l'}{\gamma}\right)\frac{du}{dV} + \frac{VJ_{l-1}K_l'}{\gamma}$$

After considerable algebraic manipulation and using Bessel functional relationships, the following results:

$$\frac{du}{dV} = \frac{X}{Y} \tag{4.50}$$

where

$$X = -\frac{lV^2J_{l-1}K_l}{\gamma^2} - \frac{V^2J_{l-1}K_{l-1}}{\gamma} + \frac{lV^2K_{l-1}J_l}{u\gamma} \tag{4.51}$$

$$Y = \frac{V}{\gamma J_{l-1}}\left[uK_{l-1}\left(\frac{2lJ_l}{uJ_{l-1}} - 1\right) - \frac{\gamma K_lJ_l}{J_{l-1}}\right] \tag{4.52}$$

With the help of the eigenvalue equation (4.48), Equations (4.51) and (4.52) can be simplified to

$$X = -\frac{V^2J_{l-1}K_{l+1}}{\gamma} \tag{4.53}$$

$$Y = -\frac{uV}{\gamma}\left(K_{l+1} - \frac{K_l^2}{K_{l-1}}\right) \tag{4.54}$$

Substituting Equations (4.53) and (4.54) into (4.50) give

$$\frac{du}{dV} = \frac{u}{V}\left[1 - \frac{K_l^2(\gamma)}{K_{l+1}(\gamma)K_{l-1}(\gamma)}\right] \tag{4.55}$$

Again, evaluate the equation above for two cases: near cutoff and far from cutoff. For $\gamma \to \infty$

$$\frac{K_l^2(\gamma)}{K_{l+1}(\gamma)K_{l-1}(\gamma)} \simeq 1 - \frac{1}{V} \tag{4.56}$$

Equation (4.56) is valid for all l values. If $\gamma \to 0$, the values of $K_l^2(\gamma)/K_{l+1}(\gamma)K_{l-1}(\gamma)$ are listed below for a few lower-order LP modes:

$$\lim_{\gamma \to 0} \frac{K_l^2(\gamma)}{K_{l+1}(\gamma)K_{l-1}(\gamma)} = \begin{cases} 0 & \text{for } l = 0, l = 1 \\ \\ \dfrac{l-1}{l} & \text{for } l > 1 \end{cases}$$

It has been shown by Gloge (Ref. 4.2) that the approximation

$$\frac{K_l^2}{K_{l+1}K_{l-1}} \simeq 1 - (\gamma^2 + l^2 + 1)^{-1/2} \tag{4.57}$$

provides a reasonable fit for all values of V. It should be noted that for a given value of l, a set of u_c values exists, which are the roots of $J_l(u)$. Because u values are limited within a narrow range between successive roots of adjacent Bessel functions, γ^2 may be replaced by $V^2 - u_c^2$, where u_c is the cutoff value for the lth mode. With this approximation, Equation (4.55) can be solved for u as a function of V using Equation (4.57), thus giving the following expression:

$$u(V) = \frac{u_c}{s} \exp\left[\arcsin\frac{s}{u_c} - \arcsin\left(\frac{s}{V}\right)^2\right] \tag{4.58}$$

where

$$s = (u_c^2 - l^2 - 1)^{1/2} \tag{4.59}$$

Equation (4.58) is good for all LP modes except the LP_{01} (HE_{11}) mode. In the case of the HE_{11} mode, the functional relationship is given by (Ref. 4.2)

$$u(V) = \frac{(1 + \sqrt{2})V}{1 + (4 + V^4)^{1/4}} \tag{4.60}$$

Using Equations (4.58), (4.60), and (4.6), the propagation constant β can be plotted for a few lower-order modes in a step-index fiber as a function of V in Figure 4.5. A comparison of this approximation with the exact solution

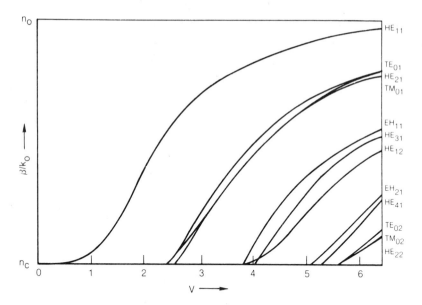

FIGURE 4.5 Plot of axial wave number for several lower-order modes in a step-index fiber as a function of the parameter V. (From Ref. 4.2.)

shows that the difference between these two treatments is too small to be displayed in this figure.

4.4 POWER DISTRIBUTION

Using the scalar approximation, the modal power distribution in the core and the cladding can be computed by carrying out the following integrals:

$$P_{\text{core}} = 2\pi \int_0^a \psi^2 r \, dr \tag{4.61}$$

$$P_{\text{cladding}} = 2\pi \int_a^\infty \psi^2 r \, dr \tag{4.62}$$

where ψ is assumed to be normalized to unity at the boundary and can be expressed as follows:

$$\psi = J_l^{-1}(u)J_l\left(\frac{ur}{a}\right) \qquad r < a \tag{4.63}$$

$$\psi = K_l^{-1}(\gamma)K_l\left(\frac{\gamma r}{a}\right) \qquad r > a \tag{4.64}$$

Using the identity for integrals of Bessel functions gives

$$P_{\text{core}} = \pi a^2[1 - \bar{J}_l(u)] \tag{4.65}$$

$$P_{\text{cladding}} = \pi a^2[\bar{K}_l(\gamma) - 1] \tag{4.66}$$

where

$$\bar{J}_l(u) = \frac{J_{l-1}(u)J_{l+1}(u)}{J_l^2(u)} \tag{4.67}$$

$$\bar{K}_l(\gamma) = \frac{K_{l-1}(\gamma)K_{l+1}(\gamma)}{K_l^2(\gamma)} \tag{4.68}$$

The eigenvalue equation (4.28) can be expressed in terms of $\bar{J}_l$ and $\bar{K}_l$ as

$$u^2\bar{J}_l(u) = -\gamma^2\bar{K}_l(\gamma) \tag{4.69}$$

By adding Equations (4.65) and (4.66) and using Equation (4.69), the expression for the total power P_T is

$$P_T = \pi a^2 \frac{V^2}{u^2} \bar{K}_l(\gamma) \tag{4.70}$$

The fractional power in the cladding can be obtained from Equations (4.66)

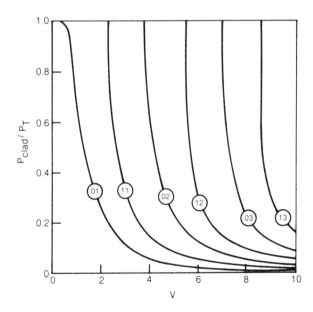

FIGURE 4.6 Plot of normalized optical power in the cladding as a function of V for a few LP modes. (From Ref. 4.2.)

and (4.70) and expressed by

$$\frac{P_{\text{cladding}}}{P_T} = \frac{u^2}{V^2}\left[1 - \frac{1}{\overline{K_l(\gamma)}}\right] \tag{4.71}$$

and the fractional power in the core is

$$\frac{P_{\text{core}}}{P_T} = 1 - \frac{P_{\text{cladding}}}{P_T} \tag{4.72}$$

It is interesting to examine the power distribution for different modes near cutoff using the calculated results for the LP mode given in Equations (4.58) and (4.60). Figure 4.6 is a plot of the fractional power inside the cladding as a function of V for a few lower-order LP modes in a step-index fiber. For the first two lowest-order modes, the power flow is mostly in the cladding near cutoff. For very large values of l however, the power remains in the core even at or just beyond cutoff.

4.5 EXACT SOLUTIONS FOR STEP-INDEX FIBERS

The treatment discussed above turns out to be valid only for the special case of $n_0 \simeq n_c$. To treat this problem exactly, this section starts with appropriate E_z and H_z field expressions and develops appropriate expressions for other field components using the relationships as given by Equations (2.39) to (2.42).

For the components inside the core:

$$E_z = \frac{AJ_l(ur/a)e^{il\theta}}{J_l(u)} \tag{4.73}$$

$$H_z = \frac{BJ_l(ur/a)e^{il\theta}}{J_l(u)} \tag{4.74}$$

and for those components outside the core:

$$E_z = \frac{AK_l(\gamma r/a)e^{il\theta}}{K_l(\gamma)} \tag{4.75}$$

$$H_z = \frac{BK_l(\gamma r/a)e^{il\theta}}{K_l(\gamma)} \tag{4.76}$$

These E_z and H_z field components can be expressed in the forms given by Equations (4.73) to (4.76), because the boundary condition that E_z and H_z are continuous at $r = a$ has been met. In doing so, the number of constants is reduced to A and B. To determine A and B, apply the continuity requirement on the other tangential components E_θ and H_θ at the boundary. From Equations (2.39) to (2.42), appropriate expressions for the field components in the core of a cylindrical fiber can be derived as follows:

$$E_r = -i \frac{a^2}{u^2 J_l(u)} \left[\frac{\beta u A}{a} J_l'(ur/a) + \frac{i\omega\mu B}{r} J_l(ur/a) \right] e^{il\theta} \tag{4.77}$$

$$E_\theta = -i \frac{a^2}{u^2 J_l(u)} \left[-\frac{u\omega\mu B}{a} J_l'(ur/a) + i \frac{\beta l A}{r} J_l(ur/a) \right] e^{il\theta} \tag{4.78}$$

$$H_r = -i \frac{a^2}{u^2 J_l(u)} \left[\frac{u\beta B}{a} J_l'(ur/a) - i \frac{\omega\epsilon_1 l A}{r} J_l(ur/a) \right] e^{il\theta} \tag{4.79}$$

$$H_\theta = -i \frac{a^2}{u^2 J_l(u)} \left[\frac{\epsilon_1 u\omega A}{a} J_l'(ur/a) + i \frac{\beta l B}{r} J_l(ur/a) \right] e^{il\theta} \tag{4.80}$$

For the field components in the cladding of a cylindrical fiber

$$E_r = i \frac{a^2}{\gamma^2 K_l(\gamma)} \left[\frac{\beta\gamma A}{a} K_l'(\gamma r/a) + i \frac{\omega\mu l B}{r} K_l(\gamma r/a) \right] e^{il\theta} \tag{4.81}$$

$$E_\theta = i \frac{a^2}{\gamma^2 K_l(\gamma)} \left[-\frac{\gamma\omega\mu B}{r} K_l'(\gamma r/a) + i \frac{\beta l A}{r} K_l(\gamma r/a) \right] e^{il\theta} \tag{4.82}$$

$$H_r = i \frac{a^2}{\gamma^2 K_l(\gamma)} \left[\frac{\gamma\beta B}{a} K_l'(\gamma r/a) - i \frac{\omega\epsilon_2 l A}{r} K_l(\gamma r/a) \right] e^{il\theta} \tag{4.83}$$

$$H_\theta = i \frac{a^2}{\gamma^2 K_l(\gamma)} \left[\frac{\gamma\epsilon_2 A}{a} K_l'(\gamma r/a) + i \frac{\beta l B}{r} K_l(\gamma r/a) \right] e^{il\theta} \tag{4.84}$$

By equating Equations (4.78) and (4.82) for E_θ and similarly Equations (4.80) and (4.84) for H_θ at $r = a$, two homogeneous equations are obtained, both containing two unknown constants A and B:

$$i \frac{l\beta}{a} \left(\frac{1}{u^2} + \frac{1}{\gamma^2} \right) A - \frac{\omega\mu}{a} \left[\frac{1}{u} \frac{J_l'(u)}{J_l(u)} + \frac{1}{\gamma} \frac{K_l'(\gamma)}{K_l(\gamma)} \right] B = 0 \tag{4.85}$$

$$\frac{\omega\epsilon_1}{a} \left[\frac{n_0^2}{u} \frac{J_l'(u)}{J_l(u)} + \frac{n_c^2}{\gamma} \frac{K_l'(\gamma)}{K_l(\gamma)} \right] A + i \frac{l\beta}{a} \left(\frac{1}{u^2} + \frac{1}{\gamma^2} \right) B = 0 \tag{4.86}$$

A nontrivial solution exists only if the determinant of the coefficient vanishes. From this determinant comes the desired eigenvalue equation:

$$\left[\frac{n_0^2}{n_c^2} \frac{\gamma^2}{u^2} \frac{J_l'(u)}{J_l(u)} + \gamma \frac{K_l'(\gamma)}{K_l(\gamma)} \right] \left[\frac{\gamma^2}{u^2} \frac{J_l'(u)}{J_l(u)} + \gamma \frac{K_l'(\gamma)}{K_l(\gamma)} \right]$$

$$= \left[l \left(\frac{n_0^2}{n_c^2} - 1 \right) \beta n_c k_0 \left(\frac{a}{u} \right)^2 \right]^2 \tag{4.87}$$

For a given waveguide, this equation will give a set of discrete values for β

falling within the range

$$n_c \le \frac{\beta}{k_0} \le n_0$$

Note that when n_0 is very close to n_c, Equation (4.87) reduces to the simple form given by Equation (4.28). More detailed treatment can be found in the book by Marcuse (Ref. 4.1).

4.6 RAY ANALYSIS FOR GRADED-INDEX FIBERS

Chapter 1 pointed out that a graded-index fiber having a nearly parabolic index profile is a good choice for optical data transmission because it reduces significantly the modal dispersion that exists in multimode step-index fibers. To demonstrate this fact, a simple approach of geometrical ray analysis is used. This analysis seeks an appropriate index profile that can simultaneously satisfy the synchronization condition for all rays. The path of a light ray in a medium of varying index can be described by the Eikonal equation, which can be derived directly from Maxwell's equations:

$$|\nabla S| = n(r) \tag{4.88}$$

where S is a surface of constant phase along a ray path s. The vector ∇S specifies the direction of energy flow and is equivalent to the Poynting vector as previously introduced. Let $\mathbf{r}$ be a position vector. Then $d\mathbf{r}/ds$ is a unit vector normal to the surface S. In vector form, Equation (4.88) can be written as

$$\frac{d\mathbf{r}}{ds} = \frac{1}{n} \nabla S \tag{4.89}$$

If Equation (4.89) is differentiated with respect to s and the Eikonal equation is used again, an equivalent form of the ray equation in terms of $\mathbf{r}$ is obtained as follows:

$$\frac{d}{ds}\left(n\frac{d\mathbf{r}}{ds}\right) = \frac{d}{ds}\nabla S$$

Because

$$\frac{d}{ds} = \frac{d\mathbf{r}}{ds} \cdot \nabla$$

It follows that

$$\frac{d}{ds}\nabla S = \frac{d\mathbf{r}}{ds} \cdot \nabla(\nabla S)$$

Substituting Equation (4.89) into the equation above gives

$$\frac{1}{n} \nabla S \cdot \nabla (\nabla S) = \frac{1}{2n} \nabla (\nabla S)^2$$

Substituting Equation (4.88) into the equation above gives

$$\frac{d}{ds} \left(n \frac{d\mathbf{r}}{ds} \right) = \frac{1}{2n} \nabla n^2 = \nabla n \qquad (4.90)$$

It is desirable to express Equation (4.90) in cylindrical coordinates. Let $\mathbf{e}_r$, $\mathbf{e}_\theta$, and $\mathbf{k}$ be three unit vectors in the cylindrical coordinate system. Then, $\mathbf{r} = r\mathbf{e}_r + z\mathbf{k}$, and

$$\frac{d\mathbf{r}}{ds} = \frac{dr}{ds} \mathbf{e}_r + r \frac{d\mathbf{e}_r}{ds} + \frac{dz}{ds} \mathbf{k} \qquad (4.91)$$

and

$$\frac{d\mathbf{e}_r}{ds} = \frac{d\theta}{ds} \mathbf{e}_\theta \qquad (4.92)$$

$$\frac{d\mathbf{e}_\theta}{ds} = - \frac{d\theta}{ds} \mathbf{e}_r \qquad (4.93)$$

Now, Equation (4.90) can be rewritten by using the relationships above as follows:

$$\left[\frac{d}{ds} \left(n \frac{dr}{ds} \right) - nr \left(\frac{d\theta}{ds} \right)^2 \right] \mathbf{e}_r + \left[n \frac{dr}{ds} \frac{d\theta}{ds} + \frac{d}{ds} \left(nr \frac{d\theta}{ds} \right) \right] \mathbf{e}_\theta$$

$$+ \left[\frac{d}{ds} \left(n \frac{dz}{ds} \right) \right] \mathbf{k} = \nabla n \qquad (4.94)$$

Equation (4.94) provides three equations of motion, of which the θ and z components offer two time-invariant relationships as a result of the fact that n is only a function of r:

$$z \text{ component:} \qquad n \frac{dz}{ds} = \text{constant} \equiv E \qquad (4.95)$$

$$\theta \text{ component:} \qquad nr^2 \frac{d\theta}{ds} = \text{constant} \equiv I \qquad (4.96)$$

A combination of these two relationships yields another time-invariant quality:

$$r^2 \frac{d\theta}{dz} = \frac{I}{E} \equiv l \qquad (4.97)$$

The quantities dr/ds, $d\theta/ds$, and dz/ds are directly related to the directional cosines of the incident and refracted rays as defined in Figure 4.7. The r component, on the other hand, gives the equation

$$\frac{d}{ds}\left(n\frac{dr}{ds}\right) - nr\left(\frac{d\theta}{ds}\right)^2 = \frac{dn}{dr} \tag{4.98}$$

Equation (4.98) is one of the equations of ray propagation. However, it is not in a convenient form. To make use of those invariant relationships, Equations (4.95) and (4.96), and also to make the following transformation, $d/ds = (dz/ds)d/dz = (E/n)d/dz$, Equation (4.98) is written into the following form:

$$\frac{d^2r}{dz^2} - \frac{l^2}{r^3} - \frac{1}{2E^2}\frac{d}{dr}n^2 = 0 \tag{4.99}$$

Equation (4.99) can be further reduced to a first-order differential equation:

$$\frac{dr}{dz} = \sqrt{\frac{n^2}{E^2} - \frac{l^2}{r^2} - 1} \tag{4.100}$$

Equations (4.100) and (4.97) provide two integrals that describe the ray path completely for a given $n(r)$ if the initial direction of the ray is specified. They are

$$z = z_0 + \int_{r_0}^{r} \frac{E\,dr}{\sqrt{n^2 - l^2E^2/r^2 - E^2}} \tag{4.101}$$

$$\theta = \theta_0 + \int_{r_0}^{r} \frac{El\,dr}{r^2\sqrt{n^2 - l^2E^2/r^2 - E^2}} \tag{4.102}$$

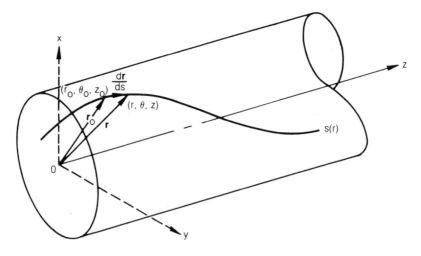

FIGURE 4.7   Path of a skew ray pasing through the point (r_0, θ_0, z_0) and the point (r, θ, z) in a cylindrical graded-index fiber.

Note that Equations (4.101) and (4.102) both involve a quadratic form

$$n^2 - \frac{l^2 E^2}{r^2} - E^2 \tag{4.103}$$

where n is a slowly varying function of r and has a maximum value of n_0 at $r = 0$ and a minimum value of n_c at $r = a$. As shown in Figure 4.8, these integrals exist only within a range defined by $R_{\min}$ and $R_{\max}$, which are the roots of the quadratic form (4.103). Now the two physical quantities are defined as

$$P \equiv 2E \int_{R_{\min}}^{R_{\max}} \frac{dr}{\sqrt{n^2 - l^2 E^2/r^2 - E^2}} \tag{4.104}$$

$$\Theta \equiv 2E \int_{R_{\min}}^{R_{\max}} \frac{l \, dr}{r^2 \sqrt{n^2 - l^2 E^2/r^2 - E^2}} \tag{4.105}$$

where P is the total distance and Θ is the total angular change over one period for a ray specified by the initial values E and l. Therefore, the integrals representing P and Θ are important for the study of time dispersion of the ray path.

For rays to be confined within the core, the E values must be limited to $n_c < E < n_0$. This condition actually defines the acceptance angle of the fiber. Equation (4.105) shows that when $l = 0$, $\Theta = 0$. This result indicates that all rays must intersect the axis of the fiber. They are called meridional rays and resemble TE_{om} and TM_{om} modes in fibers in the wave description. For $l > 0$, the rays skew around the fiber axis and are confined within an

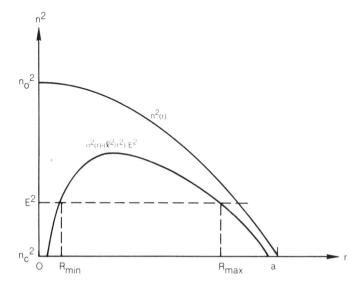

FIGURE 4.8   Region in which skew rays are confined in a graded-index fiber.

annular zone bounded by $R_{\min}$ and $R_{\max}$. These rays closely resemble the EH_{lm} and HE_{lm} modes.

The integrals given by Equations (4.104) and (4.105) specify completely all rays in a fiber and are dependent on the function $n(r)$ representing the refractive index. The interest here is to find a graded index profile such that minimum modal dispersion is achieved. The requirements for this condition to be true are that both P and Θ be independent of the values of E and l, which specify the directional cosines and mode orders. In other words, the period for all rays to travel from $R_{\min}$ to $R_{\max}$ and the angular change over the period must remain constant. This statement implies that

$$\left(\frac{\partial P}{\partial E}\right)_{E_{\mathrm{ave}}} = \left(\frac{\partial \Theta}{\partial E}\right)_{E_{\mathrm{ave}}}$$

and

$$\left(\frac{\partial P}{\partial l}\right)_{l_{\mathrm{ave}}} = \left(\frac{\partial \Theta}{\partial l}\right)_{l_{\mathrm{ave}}}$$

where E_{ave} and l_{ave} are the average values of E and l in the range of interest. These conditions can be realized if an index profile $n(r)$ is found such that both P and Θ are independent of E and l.

Examine the following three types of index profile:

Case 1—Parabolic profile:

$$n(r) = n_0[1 - \epsilon^2 r^2]^{1/2} \tag{4.106}$$

Case 2—α profile:

$$n(r) = n_0 \left[1 - 2\Delta \left(\frac{r}{a}\right)^{\alpha} \right]^{1/2} \tag{4.107}$$

Case 3—Hyperbolic or "selfoc" profile:

$$n(r) = n_0 \operatorname{sech} \frac{\Delta r}{a} \tag{4.108}$$

For case 1, the results of the integrals are

$$P = \frac{2\pi E}{\epsilon n_0} \quad \text{and} \quad \Theta = \pi$$

These results indicate that Θ is independent of both E and l, and P is now independent of l but is still proportional to E. Therefore, the parabolic profile is not the best choice. Because P is proportional to E, the parabolic profile does not yield the least modal dispersion. In fact, the period of a ray propagating in this fiber increases with the decreasing angle defined by that ray and the z axis.

For case 2, consider the values of Δ to be very small so that all rays are symmetrically spread about the axis. This situation implies that $l = 0$ and $R_{\min} = 0$. In this approximation Equation (4.104) is written as

$$P(E, 0) = 2 \frac{E}{n_0} \left(1 - \frac{E^2}{n_0^2}\right)^{-1/2} \int_0^{R_{\max}} \frac{dr}{\sqrt{1 - 2\Delta \left(1 - \frac{E^2}{n_0^2}\right)^{-1} \left(\frac{r}{a}\right)^\alpha}}$$

Now, let

$$v = (2\Delta)^{1/\alpha} \left(1 - \frac{E^2}{n_0^2}\right)^{-1/\alpha} \left(\frac{r}{a}\right)$$

then, for $r = R_{\max}$, $v = 1$ and

$$dr = a(2\Delta)^{-1/\alpha} \left(1 - \frac{E^2}{n_0^2}\right).$$

Substituting the above gives

$$P(E, 0) = 2a(2\Delta)^{-1/\alpha} \left(\frac{E}{n_0}\right) \left(1 - \frac{E^2}{n_0^2}\right)^{1/\alpha - 1/2} C(\alpha) \qquad (4.109)$$

where

$$C(\alpha) = \int_0^1 \frac{dv}{\sqrt{1 - v^\alpha}}$$

Because $C(\alpha)$ is a function of α only, it can be treated as a constant when taking the partial derivative of P with respect to E and evaluate it at E_{ave}. E_{ave} is the average value between n_0 and $n_0\sqrt{1 - 2\Delta}$, and is given by

$$E_{\mathrm{ave}} \simeq n_0 \left(1 - \frac{\Delta}{2}\right).$$

Substituting E_{ave} into $\partial P/\partial E$ gives

$$\left(\frac{\partial P}{\partial E}\right)_{E_{\mathrm{ave}}} = \frac{2a}{n_0} (2)^{-1/\alpha} C(\alpha)(\Delta)^{-1/2} \left[1 - \frac{2}{\Delta} \left(\frac{1}{\alpha} - \frac{1}{2}\right) \left(1 - \frac{\Delta}{2}\right)^2\right]$$

Letting $(\partial P/\partial E)_{E_{\mathrm{ave}}} = 0$ or equating the quantity in the bracket of the above equation to zero yields

$$\Delta = 2 \left(\frac{1}{\alpha} - \frac{1}{2}\right) \left(1 - \frac{\Delta}{2}\right)^2$$

From this expression, the following is obtained:

$$\alpha = 2 - 2\Delta \qquad (4.110)$$

Equation (4.110) indicates that with a slight correction (2Δ) in the index from the parabolic profile, the path of all meridional rays in such a fiber is periodic with the period independent of the angle of incidence. The same result can also be obtained for all skew rays ($l \neq 0$), but the analysis is omitted. The wave description shows that a direct correspondence exists between the mode angle and mode number. The result of this ray analysis reflects that all modes traveling in a fiber with an appropriate α profile as given by $2 - 2\Delta$ should have nearly the same group velocity. Even though a graded-index fiber is multimode, it has a minimum modal dispersion, and a maximum data-carrying capacity.

For case 3 it can be shown that all ray paths are periodic in z with the period

$$P = \frac{\pi}{\Delta} \qquad (4.111)$$

Equation (4.111) indicates that the period of all ray paths in a hyperbolic index fiber is also independent of the angle of incidence.

4.7 CALCULATION OF GUIDED-WAVE MODES IN THE WKB APPROXIMATION

The WKB (Wentzel–Kramers–Brillouin) method, which is well-known in quantum mechanics, can be applied to obtain some useful solutions of otherwise very complex dielectric waveguide problems for graded-index fibers. This approximation takes advantage of the fact that wave functions in the guide change very slowly, because the variation in the refractive index is very small over a distance on the order of one optical wavelength. This method will be used to determine (1) the total number of modes, and (2) the propagation constant β_{lm} associated with the corresponding mode. The establishment of an explicit expression for β_{lm} is necessary for the study of modal dispersion characteristics in Chapter 5.

Recall that the wave equation for a graded-index fiber is of the form

$$\frac{d^2\psi}{dr^2} + \frac{1}{r}\frac{d\psi}{dr} + \left[n^2(r)k_0^2 - \beta^2 - \frac{l^2}{r^2} \right] \psi = 0 \qquad (4.112)$$

where $n(r)$ in general is a slowly varying function of r. Therefore, a general solution is assumed of the form

$$\psi(r) = \exp i\phi(r) \qquad (4.113)$$

where $\phi(r)$ can be expressed in a power series in terms of k_0^{-1} as follows:

$$\phi(r) = \phi_0(r) + k_0^{-1}\phi_1(r) + \frac{1}{2} k_0^{-2}\phi_2(r) + \cdots \qquad (4.114)$$

Equation (4.113) implies that if n is a constant, the wave function reduces to a simple plane wave function, e.g., $\phi_0 = k_0 r$ and $\phi_1, \phi_2, \ldots$ are all zero.

By substituting (4.113) into (4.112) and collecting terms in accordance with the power of $1/k$, two equations are obtained belonging to the zeroth and first order:

$$\left(\frac{d\phi_0}{dr}\right)^2 - \left[k_0^2 n^2(r) - \beta^2 - \frac{l^2}{r^2}\right] = 0 \tag{4.115}$$

and

$$ik_0 \left(\frac{d^2\phi_0}{dr^2} + \frac{1}{r}\frac{d\phi_0}{dr}\right) - 2\frac{d\phi_0}{dr}\frac{d\phi_1}{dr} = 0 \tag{4.116}$$

Equation (4.115) indicates that

$$\phi_0(r) = \int_{r_1}^{r_2} \left[k_{r1}^2 n^2(r) - \beta^2 - \frac{l^2}{r^2}\right]^{1/2} dr \tag{4.117}$$

Substituting Equation (4.117) into (4.116) gives

$$\phi_1(r) = \frac{i}{4}\ln\left[r^2 n^2(r) - \frac{\beta^2 r^2}{k_0^2} - \frac{l^2}{k_0^2}\right] \tag{4.118}$$

Because ϕ_1 is a logarithm of $d\phi_0/dr$, it is not, in general, small compared with ϕ_0, therefore both ϕ_0 and ϕ_1 must be retained. However, ϕ_2, which involves dn/dr is small if dn/dr is small. The same condition applies to higher approximations, e.g., ϕ_3, ϕ_4, ... Substituting ϕ_0 and ϕ_1 into Equation (4.113) shows that $\phi_1(r)$ provides an approximate solution for the amplitude function of $\psi(r)$ and ϕ_0 is the phase function of $\psi(r)$. Furthermore, the real limits of the integral in Equation (4.117) are two points inside the fiber at which the integrand vanishes and they represent the "caustic" or turning points, which separate regions of oscillatory and evanescent field variation, as illustrated in Figure 4.9. In other words, within the region bound by r_1 and r_2, as shown in Figure 4.9, ϕ_0 is real, therefore, guided-wave modes exist and can have the allowed β values. At the caustic point, the function ϕ_1 in Equation (4.118) possesses a pole and hence the first-order WKB approximation fails at these points.

To establish an expression for β, Equation (4.117) is used. As seen before, one of the conditions for establishing guided-wave modes is that on two consecutive reflections, the total phase angle must be an integer multiple of 2π. By integrating Equation (4.117) from r_1 to r_2, only one-half of the cycle for a skew ray ($l \neq 0$) is obtained; therefore

$$m\pi = \int_{r_1}^{r_2} \left[k_0^2 n^2(r) - \beta^2 - \frac{l^2}{r^2}\right]^{1/2} dr \tag{4.119}$$

Where the integer m is associated with the mth mode number. Equation (4.119) is useful for determining the number of modes in a given range.

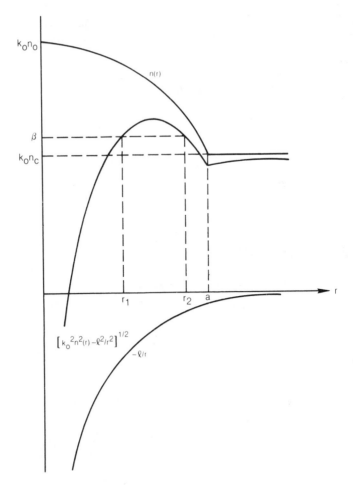

FIGURE 4.9   Region in which hybrid modes are confined in a graded-index fiber.

However, some comments and corrections to this equation must be made. First, the validity of the zeroth WKB approximation is good only for the ray optics picture, in which phase changes at the caustic or turning points r_1 and r_2 are ignored. Second, twofold degeneracies are associated with each lm mode: one goes with the clockwise or counterclockwise rotation and the other one goes with the orientations of the linear polarization. These degeneracies increase the mode number by a factor of four from what has been already accounted for by Equation (4.119). Third, for each m value, a set of l numbers exists with an upper limiting value l_{max} at which the wave is no longer bound. Because the largest l value for a given m occurs for the mode near its cutoff point, β is replaced by $k_0 n_c$. Therefore

$$l_{\max} = k_0 r[n^2(r) - n_c^2]^{1/2} \tag{4.120}$$

By taking these corrections into account and treating l as a continuous variable, the sum over l can be replaced by an integral. Thus, the total number of modes can be obtained from Equation (4.119) as follow:

$$M = \frac{4k_0}{\pi} \int_0^a \int_0^{l_{\max}} \left[n^2(r) - n_c^2 - \frac{l^2}{k_0^2 r^2} \right]^{1/2} dl \, dr$$

After integrating over l and using Equation (4.120) the following results:

$$M = k_0^2 \int_0^a [n^2(r) - n_c^2] r \, dr \tag{4.121}$$

In Equation (4.121), β has been replaced by $k_0 n_c$ and the integral has been extended over the entire core.
If the following is substituted into Equation (4.121):

$$n^2(r) = n_0^2 \left[1 - 2\Delta \left(\frac{r}{a} \right)^\alpha \right]$$

$$n_c^2 = n_0^2(1 - 2\Delta)$$

the total number of modes is obtained after integrating over r

$$M = \frac{\alpha}{\alpha + 2} n_0^2 k_0^2 a^2 \, \Delta \tag{4.122}$$

Equation (4.122) gives an expression for calculating the total number of modes in a graded-index fiber with an α profile. This expression can be extended to a step-index fiber by letting $\alpha = \infty$. Equation (4.122) gives

$$M = n_0^2 k_0^2 a^2 \Delta = \frac{V^2}{2} \tag{4.123}$$

an expression for calculating the total number of modes in a step-index fiber with a given V value. Because $\alpha \approx 2$, the results of Equations (4.122) and (4.123) indicate that the number of modes in a graded-index fiber is only about one-half of those in a step-index fiber of identical core diameter.
The number of modes having their corresponding β values falling within the limit from $n_0 k_0$ to β_m can be obtained by using Equation (4.121) as

$$m = \int_0^{r_m} [n^2(r) k_0^2 - \beta_m^2] r \, dr \tag{4.124}$$

where

$$r_m = a \left[\frac{1 - (\beta_m/n_0 k_0)^2}{2\Delta} \right]^{1/\alpha} \tag{4.125}$$

Upon integration of Equation (4.124)

$$m = a^2 \,\Delta n_0^2 k_0^2 \, \frac{\alpha}{\alpha + 2} \left(\frac{n_0^2 k_0^2 - \beta_m^2}{2\Delta k_0^2 n_0^2} \right)^{(\alpha+2)/\alpha} \tag{4.126}$$

By combining the results of Equations (4.126) and (4.123) an expression for β_m is obtained in terms of total number of the m-fold degenerated group

$$\beta_m = k_0 n_0 \left[1 - 2\Delta \left(\frac{m}{M} \right)^{2a/(a+2)} \right]^{1/2} \tag{4.127}$$

To account for the m-fold degeneracy of β_m, a factor of two is introduced in the exponent for the second term in the bracket of Equation (4.127). This correction is necessary because for a given β-value, m-number of modes share the same β-value (see Figure 4.5). This m-fold degeneracy must be accounted for each β_m, therefore, a total of $\sum m$ modes is possible. For large values of m, $\sum m \simeq m^2$. The same rule also applies to the normalization factor by replacing M with M^2.

The expression for β_m, which will be used in Chapter 5 for the calculation of modal dispersion, is useful in gaining some understanding of physical properties of fibers. First, the modal spacing in a fiber can be computed by differentiating β_m with respect to m which yields

$$\frac{d\beta_m}{dm} = \left(\frac{\alpha}{\alpha + 2} \right)^{1/2} \frac{2\sqrt{\Delta}}{a} \left(\frac{m}{M} \right)^{(\alpha-2)/(\alpha+2)} \tag{4.128}$$

Equation (4.128) indicates that, for a parabolic index profile ($\alpha = 2$) modal spacing is independent of the mode number. In the case of a step-index fiber ($\alpha = \infty$), Equation (4.128) yields

$$\frac{d\beta_m}{dm} = \frac{2\sqrt{\Delta}}{a} \left(\frac{m}{M} \right) \tag{4.129}$$

This result indicates that the modal spacing in a step-index fiber increases linearly with increasing mode order, which is one of the major distinctions between these two types of fiber. Another interesting application of Equation (4.127) is to relate β_m with the z component of $\mathbf{k}$:

$$\cos \theta_m = \left[1 - 2\Delta \left(\frac{m}{M} \right)^{2\alpha/(\alpha+2)} \right]^{1/2} \tag{4.130}$$

where θ_m is the maximum angle that the wave vector $\mathbf{k}$ makes with the z axis at $r = 0$. From Equation (4.130):

$$\sin \theta_m = \sqrt{2\Delta} \left(\frac{m}{M} \right)^{\alpha/(\alpha + 2)} \tag{4.131}$$

The expression in (4.131) indicates that a direct correspondence exists between the mode angle and mode number. In the case of a parabolic index fiber ($\alpha = 2$)

$$\sin \theta_m = \sqrt{\frac{2\Delta m}{M}} \tag{4.132}$$

whereas in the case of a step-index fiber ($\alpha = \infty$)

$$\sin \theta_m = \sqrt{2\Delta} \left(\frac{m}{M} \right) \tag{4.133}$$

A comparison of these results indicates that modes in a graded-index fiber are more confined to the core of the guide than those in a step-index fiber. At cutoff, $m = M$ so that $n_0 \sin \theta_c = n_0 \sqrt{2\Delta}$, which is precisely the definition of numerical aperture of a step-index fiber. Expressions of (4.132) and (4.133) can be used to calculate the field patterns radiated from the ends of these fibers.

PROBLEMS

4.1. Give geometric interpretations of u and γ as defined by Equations (4.6) and (4.7).

4.2. Derive the identity given by Equation (4.16), using the series definition of $J_l(x)$ given by

$$J_l(x) = \sum_{n=0}^{\infty} \frac{(-1)^n}{n! \Gamma(n + l + 1)} \left(\frac{x}{2} \right)^{2n + l}$$

where $\Gamma(x)$ is the gamma function.

4.3. Using the scalar field approximation, derive the eigenvalue equation given by Equation (4.28).

4.4. Using recurrence relations for J_l and K_l, show that the two eigenvalue equations given by Equation (4.28) are equivalent.

4.5. Summarize the cutoff conditions for TM_{om}, HE_{1m}, and HE_{lm} modes.

4.6. For large V, show that

$$u(V) = u(\infty)e^{-1/V}$$

4.7. Derive the expressions for the fractional power residing in the core and cladding as given by Equations (4.65) and (4.66). *Hint:*

$$\int_0^r r J_{l \mp 1}(ar) J_{l \mp 1}(ar) \, dr = \frac{r^2}{2} [J_{l \mp 1}{}^2(ar) - J_l(ar) J_{l \mp 2}(ar)]$$

and

$$\int_r^\infty rK_{l\mp1}(ar)K_{l\mp1}(ar)\,dr = \frac{r^2}{2}[K_{l\mp2}(ar)K_l(ar) - K_{l\mp1}^2(ar)]$$

4.8. Show that the period P of a ray propagating in a parabolic index fiber is equal to $2\pi E/n_0\epsilon$.

4.9. Derive Equations (4.121) and (4.122).

4.10. Derive Equation (4.126).

REFERENCES

4.1. D. Marcuse, *Theory of Dielectric Waveguides*, Academic Press, Inc., New York, 1974.

4.2. D. Gloge, *Appl. Opt.*, *10*, 2252 (1971).

4.3. A. W. Snyder, *IEEE Trans. Microwave Theory Tech.*, *MIT-17*, 1130 (1969).

4.4. M. Eve, *Opt. Quantum Electron.*, *8*, 285 (1976).

5

Dispersion, Mode Coupling, and Loss Mechanisms

5.1 INTRODUCTION

Many limiting factors that originate from the geometric and physical nature of glass fibers have a profound effect on the information-carrying capacity of optical fiber waveguides. This chapter deals with various effects of fiber dispersions and imperfections. It also shows that in addition to modal dispersion, other factors, such as the dispersive properties of the glass and the spectral distribution of the source, can lead to a significant change in α values for graded-index fibers and can cause further pulse broadening. In practice many structural imperfections exist in fibers that introduce losses through scattering and absorption of optical power. Imperfections in a multimode fiber also create a random coupling of modes that in effect can produce pulse narrowing at the cost of power reduction through leakage into unguided radiative modes. To take advantage of this pulse-narrowing effect, the core-cladding interface must be prepared very carefully to minimize radiation loss.

5.2 GROUP VELOCITY AND DELAY

Chapter 4 derived the expressions for β values associated with the modes of a fiber of a given index profile. In a multimode fiber, the power of a short input signal pulse is distributed among a set of modes, each of which propagates at a phase velocity $v_B = \omega/\beta$. The output signal pulse is expected to be distorted as a result of modal dispersion of the fiber. Even in a single mode fiber, chromatic dispersion of a fundamental mode can cause waveform distortion of the signal pulse (see Chapter 6), because in reality there is no monochromatic source. Even for a laser source, there is a finite spectral width or a spread of frequencies associated with the laser output (see Chapter 10). In general, if the transmission medium is dispersive, the phase velocity is not the same for each frequency component of the wave. Therefore, the wave velocity must be represented by a group velocity v_g, which is shown in the following to take on the form $v_g = d\omega/d\beta$. To derive this expression for v_g, consider a special case for which the pulse energy propagates in a planar dispersive waveguide. The E-field of this system is the sum of a large number of plane waves characterized by different β values, each of which travels at a different phase velocity ω/β. Within the spectral range of interest, e.g. $\omega = \omega(\beta)$,

$$E(z, t) = \sum_n A_n \exp i[\beta_n z - \omega(\beta)t]$$

If the values β_n are very close together and separated by $d\beta$, and A_n may assume the values of any given function of β at these points, the above E field can be written in the limit of an integral of the form

$$E(z, t) = \int_{-\infty}^{\infty} A(\beta) \exp i[\beta z - \omega(\beta)t] \qquad (5.1)$$

where the amplitude $A(\beta)$ is determined by a linear superposition of different frequency components. If the distribution of $A(\beta)$ is sharply peaked around a value β_0, $\omega(\beta)$ can be expanded around β_0 in a power series as

$$\omega(\beta) = \omega_0 + \frac{d\omega}{d\beta}(\beta - \beta_0) + \cdots \qquad (5.2)$$

where $d\omega/d\beta$ is evaluated at $\beta = \beta_0$. Substituting Equation (5.2) into (5.1) and neglecting the higher-order terms in the expansion, gives

$$E(z, t) \simeq e^{i[\beta_0(d\omega/d\beta) - \omega_0]t} \int_{-\infty}^{\infty} A(\beta)e^{i[z - (d\omega/d\beta)t]\beta} \, d\beta \qquad (5.3)$$

The inverse transform of Equation (5.3) evaluated at $t = 0$ is

$$A(\beta) = \int_{-\infty}^{\infty} E(z', 0)e^{-i\beta z'} \, dz' \qquad (5.4)$$

where $z' = z - (d\omega/d\beta)t$. Apart from an overall phase factor, the pulse travels along the waveguide undistorted in shape with a group velocity

$$v_g = \frac{d\omega}{d\beta} \qquad (5.5)$$

The propagation characteristic of a pulse in a fiber can be analyzed by examining the group delay time for a given mode β_{lm}. Consider that all modes are excited by a short pulse at the input. Each mode transports an equal amount of energy to the end of the fiber. As they recombine, the short pulse is expected to suffer a certain distortion, depending on the $\beta - \omega$ characteristics of each mode and the dispersion in the fiber. The power profile of the inpulse reponse can be measured by several methods (see Chapter 7). To compute the impulse response, the group delay τ_g is defined in terms of the group velocity v_g as given by Equation (5.5):

$$\tau_g = L \frac{d\beta_{lm}}{d\omega} = \frac{L}{c} \frac{d\beta_{lm}}{dk_0} \tag{5.6}$$

where β_{lm} depends on both waveguide parameters and the wavelength λ of the source. Once τ_g is known, the impulse reponse can be established by a superposition of the energy of all lm modes that arrive between τ and $\tau + d\tau$.

From Equations (4.6), (4.7), and (4.8) an expression is obtained for a multimode step-index fiber in the form

$$\frac{\beta^2/k_0^2 - n_c^2}{n_0^2 - n_c^2} = 1 - \frac{u^2}{V^2} \equiv b \tag{5.7}$$

where a quantity b is defined in terms of V. For small index differences, Equation (5.7) becomes

$$b \simeq \frac{\beta/k_0 - n_c}{n_0 - n_c} \tag{5.8}$$

Because β and b are proportional, the quantity b can be regarded as a normalized propagation constant. From Equation (5.8), β is expressed in terms of b as

$$\beta = n_c k_0 (b\Delta + 1) = n_c k_0 \left[1 + \Delta \left(1 - \frac{u^2}{V^2} \right) \right] \tag{5.9}$$

Substituting Equation (5.9) into (5.6) and letting $V = kan\sqrt{2\Delta}$, an expression for the group delay is obtained:

$$\tau_g = \frac{L}{c} \left(\frac{d}{dk_0} nk_0 + n\Delta \frac{d}{dV} Vb \right) \tag{5.10}$$

Equation (5.10) ignores the difference in the dispersive effect between the core and cladding materials by omitting the subscript. Also the products of Δ with $(k_0/n)dn/dk_0$ and $d\Delta/dk_0$ terms are omitted because they are small compared with all other terms.

The first part of Equation (5.10) characterizes the material dispersion

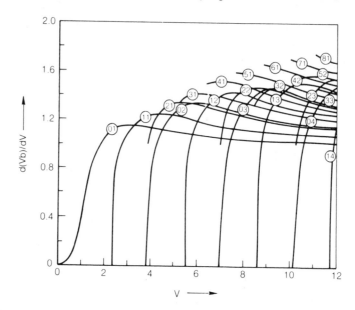

FIGURE 5.1   Relative group delay among various guided-wave modes in a weakly guided step-index fiber is plotted as a function of V. (From Ref. 5.1.)

for all modes. The second term, which represents the group delay caused by waveguide dispersion, is governed by the derivative $d(Vb)/dV$.

Using Equation (5.7):

$$\frac{d}{dV} Vb = 1 + \frac{u^2}{V^2}\left(1 - 2\frac{V}{u}\frac{du}{dV}\right) \tag{5.11}$$

Substituting the result of Equation (4.55) into (5.11) yields

$$\frac{d}{dV} Vb = 1 + \frac{u^2}{V^2}\left[1 - 2\frac{K_l^2(\gamma)}{K_{l+1}(\gamma)K_{l-1}(\gamma)}\right] \tag{5.12}$$

The function $d(Vb)/dV$ is plotted in Figure 5.1. For large values of V, it approaches unity for all modes. At cutoff, $d(Vb)/dV = 0$ for $l = 0$, 1, and equals $2(1 - 1/l)$ for $l \geq 2$. Figure 5.1 also shows that the mode of largest-order l has the largest group delay. The difference between this and the lowest mode is approximately $1 - 2/l$. For large V, the group spread is approximately equal to $(1 - 2/V)(n_0 - n_c)L/c$ for a multimode step-index fiber.

5.3 PULSE BROADENING

This section examines the influence of source distribution on optical transmission properties of short pulses through a fiber. When a light pulse is coupled into a fiber, its power is distributed among all modes. If no coupling among modes occurs, the impulse response $P_{lm}(\lambda, z)$ of an lm mode for a spectral component λ at a position z is time-independent and is determined primarily by the spectral distribution of the source $S(\lambda)$ and the process of excitation. It is therefore reasonable to assume the distribution function to be a product of two independent functions of the form

$$P_{lm}(\lambda, z) = S(\lambda)P_{lm}(z_0) \tag{5.13}$$

The effect of a finite spectral width on the pulse width can be significant after a short light pulse propagates a long distance L in a fiber. The root-mean-square (rms) pulse width $\sigma(z)$ in terms of group delay is defined by the expression

$$\sigma(z) = [\langle \tau_{lm}^2 \rangle - \langle \tau_{lm} \rangle^2]^{1/2} \tag{5.14}$$

where the notation $\langle \cdot \rangle$ represents the average value of the quantity in question. To perform the average, only the time-independent impulse response function is needed. Integrating over the spectral width gives

$$\langle \tau_{lm}^2 \rangle = \int \sum_{lm} P_{lm}(z_0)S(\lambda)\tau_{lm}^2 \, d\lambda \tag{5.15}$$

and

$$\langle \tau_{lm} \rangle = \int \sum_{lm} P_{lm}(z_0)S(\lambda)\tau_{lm} \, d\lambda \tag{5.16}$$

where the summation in Equations (5.15) and (5.16) is extended over all guided modes. For the steady-state situation, τ_{lm} is independent of z and is proportional only to L. Equation (5.14) can be used to determine the information-carrying capacity B in bits per second as defined by

$$B = \frac{1}{4\sigma} \tag{5.17}$$

To simplify the numerical calculation of Equation (5.14), it is instructive to expand the group delay τ_{lm} in a Taylor series about λ_0:

$$\tau_{lm}(\lambda) = \tau_{lm}(\lambda_0) + \tau'_{lm}(\lambda - \lambda_0) + \tfrac{1}{2}\tau''_{lm}(\lambda - \lambda_0)^2 + \cdots \tag{5.18}$$

where primes denote the derivatives with respect to λ and have been evaluated at λ_0.

Substituting Equation (5.18) into (5.14) and neglecting the higher-order terms, the first term on the right-hand side of Equation (5.14) can be written as

$$\langle \tau_{lm}^2 \rangle = \sum_{lm} P_{lm}(z_0)\left\{ \tau_{lm}^2(\lambda_0) + \frac{\sigma_s^2}{2\lambda_0^2}[2\tau_{lm}(\lambda_0)\lambda_0^2\tau''_{lm} + 2\lambda_0^2\tau'^2_{lm}] \right\} \tag{5.19}$$

where σ_s is the rms spectral width of the source and is defined as

$$\sigma_s^2 = \int_0^\infty (\lambda - \lambda_0)^2 S(\lambda) \, d\lambda \qquad (5.20)$$

and the mean value λ_0 is given by

$$\lambda_0 = \int_0^\infty \lambda S(\lambda) \, d\lambda \qquad (5.21)$$

These definitions assume normalized spectral distribution [e.g., $\int S(\lambda) \, d\lambda = 1$].

The second term on the right-hand side of Equation (5.14) can be written as

$$\langle \tau_{lm} \rangle^2 = [\sum P_{lm} \tau_{lm}(\lambda_0)]^2 + \left(\frac{\sigma_s}{\lambda_0}\right)^2 [\sum P_{lm} \tau_{lm}(\lambda_0) \sum P_{lm} \lambda_0^2 \tau_{lm}'']$$

$$+ \left(\frac{\sigma_s}{2\lambda_0}\right)^2 [\sum P_{lm} \lambda_0^2 \tau_{lm}'']^2 \qquad (5.22)$$

By introducing the notation for a weighted-average value, for example,

$$\langle \tau(\lambda_0) \rangle \equiv \sum P_{lm} \tau_{lm}(\lambda_0)$$

into Equations (5.19) and (5.22), the rms pulse width $\sigma(z)$ can be expressed by rearranging the terms into two distinct groups, commonly referred to as the intermodal and intramodel dispersions:

$$\sigma(z) = (\sigma_{\text{intermodal}}^2 + \sigma_{\text{intramodal}}^2)^{1/2} \qquad (5.23)$$

where

$$\sigma_{\text{intermodal}}^2 = \langle \tau^2(\lambda_0) \rangle - \langle \tau(\lambda_0) \rangle^2$$

$$+ \left(\frac{\sigma_s}{\lambda_0}\right)^2 [\langle \lambda_0^2 \tau''(\lambda_0) \tau(\lambda_0) \rangle - \langle \lambda_0^2 \tau''(\lambda_0) \rangle \langle \tau(\lambda_0) \rangle] \qquad (5.24)$$

and

$$\sigma_{\text{intramodal}}^2 = \left(\frac{\sigma_s}{\lambda_0}\right)^2 \langle \lambda_0^2 \tau'(\lambda_0)^2 \rangle \qquad (5.25)$$

To calculate pulse width, both the spectral function of the source and the impulse response function of the modes must be known. Usually, functional forms can be assumed to approximate these distributions. Before carrying this analysis further, some observations must be made concerning the origins of various terms in Equations (5.24) and (5.25) and their order of importance must be considered.

Intermodal broadening is a result of the delay difference among modes. The leading term that is independent of the source spectrum is the dominating term, but its magnitude can be reduced significantly if a graded-index fiber with $\alpha = 2 - 2\Delta$ is chosen. The remaining term in Equation (5.24), which contains the zeroth and second order of derivatives, is only a small correction, whch is proportional to the square of the relative spectral width σ_s/λ_0. Typically, σ_s is about 10^{-3} μm for an injection laser source and about 2×10^{-2} μm for an incoherent LED source, all of which operate at ≈ 0.9 μm. Intramodal broadening represents an average spreading of each mode with a value also proportional to the square of σ_s/λ_0. It contains only the first derivative of the group delay.

5.4 MATERIAL DISPERSION

Intramodal broadening arises from two distinct effects: One is caused by material dispersion and the other is contributed to by the waveguide structure. The separation of these two effects has been made by the expression of the group delay as given by Equation (5.10). A fiber can have a properly graded-index profile such that the second term in Equation (5.10) is zero and waveguide dispersions can be eliminated completely. After again differentiating Equation (5.10) with respect to λ and substituting into Equation (5.25), an expression for the pulse-broadening effect is obtained, which can only be attributed to material dispersion:

$$\sigma_{material} = \frac{L}{c} \sigma_s \lambda \frac{d^2 n}{d\lambda^2} \tag{5.26}$$

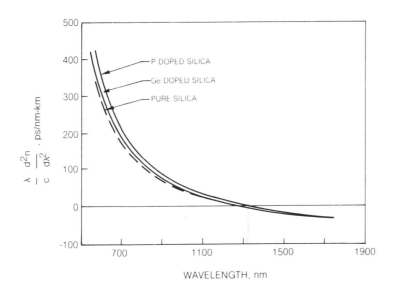

FIGURE 5.2   Material dispersion measurements for pure and doped silica glasses. [After D. N. Payne and A. H. Hartog, *Electron. Lett., 13,* 627 (1977).]

This expression represents the ultimate information-carrying capability for a graded-index fiber without modal and waveguide dispersion.

It is possible to eliminate the broadening effect caused by material dispersion by choosing a proper wavelength such that $d^2n/d\lambda^2 = 0$. Figure 5.2 shows the effect of dispersion of most silicate fibers as a function of wavelength and also shows that the inflection point of the $n(\lambda)$ curve occurs at a λ value lying between 1.2 and 1.3 μm. For this reason and because the minimum attenuation loss also occurs in this wavelength region (see Figure 1.4), much research and development effort has gone into improving the performance of the light source and photodetection in this wavelength region, which lies beyond the emitting range of GaAs. It must be realized that the effect of material dispersion can be very serious only in fibers having very low modal dispersion.

5.5 INTERMODAL DISPERSION

The effect of intermodal dispersion on pulse width can be analyzed by calculating the group delay time among the propagating modes and using Equation (5.24) to calculate the rms value of the pulse width σ. This calculation has been done to some extent in the treatment of a step-index fiber in Section 5.2. Here, a graded-index fiber with an α profile is considered. Equation (4.127) provides

$$\beta_m = k_0 n_0 \left[1 - 2\Delta \left(\frac{m}{M}\right)^{2\alpha/(\alpha+2)} \right]^{1/2}$$

where $M(\alpha) = [\alpha/(\alpha + 2)]a^2 k_0^2 n_0^2 \Delta$. Substituting β_m into Equation (5.6) and neglecting the Δ^2 and higher-order terms gives

$$\tau_m = \frac{L}{c}(n - \lambda n')\left[1 + \Delta \frac{\alpha - 2 - \delta}{\alpha + 2}\left(\frac{m}{M}\right)^{2\alpha/(\alpha+2)} \right] + O(\Delta^2) \quad (5.27)$$

where

$$\delta = -\frac{2n\lambda}{n - \lambda n'}\frac{\Delta'}{\Delta} \quad (5.28)$$

To calculate the rms value of the pulse width σ, τ_m^2 and τ_m must be weighted and averaged using Equations (5.19) and (5.22). It is necessary to acquire an expression for the impulse response function P_{lm}. Olshansky and Keck (Ref. 5.2) have obtained an expression for σ by assuming that the impulse function is a constant for all modes and that the summation in these equations can be replaced by an integral over all m modes. The result is

$$\sigma = \frac{L\Delta}{2c}(n - \lambda n')\frac{\alpha}{\alpha + 1}\left(\frac{\alpha + 2}{3\alpha + 2}\right)^{1/2}\left[\left(\frac{\alpha - 2 - \delta}{\alpha + 2}\right)^2\right.$$

$$+ \frac{2(\alpha - 2 - \delta)(3\alpha - 2 - 2\delta)(\alpha + 1)}{(2\alpha + 1)(\alpha + 2)^2}\Delta \qquad (5.29)$$

$$\left. + \frac{2(3\alpha - 2 - 2\delta)^2(2\alpha + 2)^2\Delta^2}{(5\alpha + 2)(3\alpha + 2)(\alpha + 2)^2}\right]^{1/2}$$

Equation (5.29) gives an optimum α value such that σ is a minimum by letting $d\sigma/d\alpha = 0$. After considerable algebraic manipulation the following results:

$$\alpha_{optimum} = 2 + \delta - \frac{(3 + \delta)(4 + \delta)}{5 + 2\delta}\Delta \qquad (5.30)$$

If the effect of spectral width (e.g., $\delta = 0$) is ignored, Equation (5.30) indicates that $\alpha_{opt} = 2 - (12/5)\Delta$. This result is in good agreement with that calculated by the ray analysis given by Equation (4.110). When $\delta \neq 0$ the result of Equation (5.30) deviates from that predicted by Equation (4.110). This difference is caused by introducing a finite spectral width for the source. Because the value of δ depends on fiber material, the α value can vary over a rather wide range of wavelength, as shown in Figure 5.3.

Equation (5.23) has been plotted in Figure 5.4 as a function of α for

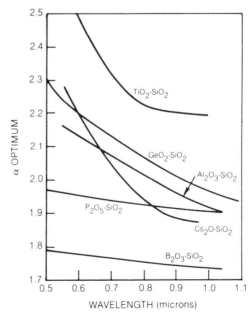

FIGURE 5.3 Optimum value of α as a function of λ for several types of composite glass. [After H. M. Presby and I. P. Kaminow, *Appl. Optics*, *15*, 3029 (1976).]

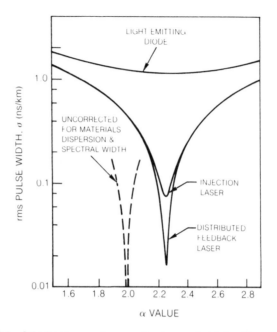

FIGURE 5.4 Calculated rms pulse spreading in a graded-index fiber as a function of the parameter α assuming a spectral width of 15, 1, and 0.2 nm for a LED, an injection laser, and a DFB laser, respectively. The uncorrected curve assumes no material dispersion and negligible spectral width. (From Ref. 5.2.)

three light sources with different spectral widths. The three curves represent an LED, injection laser, and distributed feedback (DFB) laser, having typical σ_s values of 150, 10, and 2 Å, respectively. The fiber parameters used in this calculation are $n_0 = 1.460$, $n_c = 1.452$, $\lambda n' = -0.014$, $\lambda^2 n'' = 0.02$, and $\lambda \Delta' = -0.0008$. For the LED source, a pulse broadening of less than 1.5 ns/km can be achieved if α is within 25% of the optimum value. For the injection laser, an α value within 5% of its optimum value will yield a pulse width of less than 0.2 ns/km. For a DFB laser, a width of less than 0.05 ns/km is predicted for fibers with optimum α values. The dashed curve represents the rms width when both material dispersion and the spectral width of the source are ignored.

For a step-index fiber, the rms width can be obtained directly from Equation (5.29) by letting α = ∞:

$$\sigma_{step} = \frac{L\Delta}{2\sqrt{3}c}(n - \lambda n')\left(1 + 3\Delta + \frac{24}{5}\Delta^2\right) \tag{5.31}$$

The difference in pulse width between step-index and graded-index fibers is fairly large and the ratio of the two σ values is approximately equal to

$$\frac{\sigma_{\text{graded}}}{\sigma_{\text{step}}} \simeq \frac{\Delta}{\sqrt{2}} (1 - 3\Delta) \qquad (5.32)$$

The result of Equation (5.29) is obtained with the assumption that all modes of the fiber are excited equally. This assumption is just a simplification to obtain an estimation of pulse width. In reality, mode distribution and impulse response can vary substantially and depend strongly on source distribution and coupling schemes. For example, it is possible to obtain a narrower pulse width by varying the incident beam waist or the launching position of a focused source than by using an unfocused incoherent source. On the other hand, an incoherent source can provide a shorter pulse width than an unfocused laser source. However, these variations are usually very small and lead to only small corrections in the results obtained from the simple model above.

5.6 MODE COUPLING IN A MULTIMODE FIBER

Random coupling of modes in a multimode fiber can have a profound effect on pulse shape as the pulse propagates in a multimode fiber. For a Gaussian input pulse, the shape of the output pulse will remain approximately Gaussian; however, its width increases proportionally to the square root of the fiber length. This result is in a sharp contrast to the usual result that $\sigma \propto L$ for the case when mode coupling is ignored. In other words, a reduction in pulse dispersion can be achieved by inducing mode coupling in a multimode fiber. In practice, this reduction must be compensated for by a certain loss penalty. Because the couplings between guided modes not only transfer energy among themselves but also transfer a certain amount of energy to the radiation modes, radiation losses are unavoidable.

Figure 5.5 shows the mode spacing and regions of various modes. It is apparent that the spacing between guided-wave modes in β space decreases with decreasing mode number, and coupling is strongest among its nearest neighbors. For the same reason, radiation loss probably occurs through the coupling of higher-order modes. The reduction in pulse width can be viewed as a redistribution of power as a result of random mode

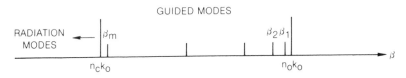

FIGURE 5.5   Regions defining guided-wave modes and radiation modes.

coupling, during which the power contained in lower-order modes is transferred to higher-order modes, and vice versa. In the meantime, the power carried by higher-order modes is also transferred to the radiation modes. This transfer constitutes a loss in the system. As a result of this loss mechanism, a center of gravity of the power distribution is established around an average group velocity v_g, which more closely represents lower-order modes with negligible power transfer losses. This redistribution of power is possible because the power outflow from these modes has been compensated for by the power inflow from lower-order modes.

In a steady-state situation, the coupled power equations assume the form

$$\frac{dP_m}{dz} = -(a_m + b_m)P_m + \sum_n c_{mn}P_n \tag{5.33}$$

where a_m is the power-loss coefficient of the mth mode that transfers directly to the radiation modes, and

$$b_m = \sum_n c_{mn} \tag{5.34}$$

and c_{mn} are the power coupling coefficiency among various modes and depend on the imperfections of the waveguide. In the time-dependent case, the derivative dP_m/dz from a stationary frame to a moving coordinate system traveling at velocity v_m is transformed by the expression:

$$\frac{dP_m}{dz} = \frac{\partial P_m}{\partial z} + \frac{1}{v_m}\frac{\partial P_m}{\partial t} \tag{5.35}$$

Consequently, the time-dependent coupled power equations can be written as

$$\frac{\partial P_m}{\partial z} + \frac{1}{v_m}\frac{\partial P_m}{\partial t} = -(a_m + b_m)P_m + \sum_n c_{mn}P_n \tag{5.36}$$

The first term on the right-hand side of Equation (5.36) represents the power loss of the mth mode. The second term represents the power gain through random coupling. Exact solutions of this set of coupled equations are complex. The model that follows is that given by Gloge (Ref. 5.3), who treats this problem by assuming a modal continuum rather than thousands of discrete modes. In this model the mode coupling problem can be described by a diffusion process.

First consider fibers in which the steady-state mode distribution does not include modes close to cutoff. The output power from a fiber can be expressed as a function of time and output angle, which is related directly to the mode number for the case of a short input pulse. Of most interest is the impulse response that can be obtained by integrating over all angles at the output. Because Δ is very small for all fibers of interest, the maximum angle limited by the condition for critical internal reflection can be approximated by

$$\theta_{max} \simeq \sqrt{2\Delta} \qquad (5.37)$$

In a multimode fiber, the modes are so densely packed that their distribution can be considered as continuous. The state of the fiber at a point z and time t can then be described by a distribution $P(\theta, z, t)$, where θ is a continuous variable.

Using the model above, Equation (5.36) can be modified by replacing the two terms on the right-hand side with the following:

1. The loss term is expressed by $-A\theta^2 P$, where A is the loss coefficient measured in $m^{-1} rad^{-2}$, and the attenuation effect is assumed to be proportional to the square of the characteristic angle. This assumption is reasonable, because loss is expected to be greater for higher-order modes.

2. Mode coupling is found to occur essentially between closely adjacent modes and for this reason, takes the form of a diffusion process. The increase in a power as a result of diffusion can be expressed by $(1/\theta)\partial/\partial\theta(\theta D\, \partial P/\partial\theta)$, a term typical for radial diffusion in the cylindrical configuration. D is a coupling coefficient and is assumed to be independent of θ.

Therefore, Equation (5.38) is rewritten as follows:

$$\frac{\partial P}{\partial z} + \frac{1}{v_g}\frac{\partial P}{\partial t} = -A\theta^2 P + \frac{1}{\theta}\frac{\partial}{\partial\theta}\left(\theta D\,\frac{\partial P}{\partial\theta}\right) \qquad (5.38)$$

where the group velocity can be estimated from a simple ray picture that

$$v_g = \frac{c\cos\theta}{n} \simeq \frac{c}{n(1 + \theta^2/2)} \qquad (5.39)$$

Substituting Equation (5.39) into (5.38) gives

$$\frac{\partial P}{\partial z} = -A\theta^2 P - \frac{n}{2c}\theta^2\frac{\partial P}{\partial t} + \frac{1}{\theta}\frac{\partial}{\partial\theta}\left(\theta D\,\frac{\partial P}{\partial\theta}\right) \qquad (5.40)$$

where the delay that is common to all modes has been ignored. Now Equation (5.40) is multiplied by e^{-st} and integrated over t from $t = 0$ to $t = \infty$. With the help of the Laplace transform, e.g.,

$$p(\theta, z, s) = \int_0^\infty e^{-st} P(\theta, z, t)\, dt$$

the following is obtained:

$$\frac{\partial p}{\partial z} = -Ab^2\theta^2 p + \frac{1}{\theta}\frac{\partial}{\partial\theta}\left(\theta D \frac{\partial p}{\partial\theta}\right) \tag{5.41}$$

where

$$b = \left(1 + \frac{ns}{2cA}\right)^{1/2} \tag{5.42}$$

The result of Equation (5.41) is identical to the steady-state case as described by Equation (5.33), except the loss coefficient contains a product of Ab^2. This difference indicates that in the time-dependent case for an impulse response, $s \neq 0$, hence $b \neq 1$. Physically, this difference implies that the spatial spread of energy during pulse propagation is equivalent to the time spread of energy, which can lead to pulse narrowing in the time-dependent solution.

To obtain a closed-form solution of Equation (5.41), a solution of the following form is assumed:

$$p(\theta, z, s) = f(z, s) \exp\left(\frac{-\theta^2}{\Theta^2}\right) \tag{5.43}$$

where Θ represents the angular pulse width, which varies from Θ_0 to Θ_z as z increases. Substituting Equation (5.43) into (5.41) gives

$$\frac{\partial f(z, s)}{\partial z} + f(z, s)\frac{2\theta^2}{\Theta^3}\frac{\partial\Theta}{\partial z} = -Ab^2\theta^2 f(z, s) - Df(z, s)\left(\frac{4}{\Theta^2} - \frac{4\theta^2}{\Theta^4}\right)$$

By rearranging terms and separating the variables

$$\frac{1}{f}\frac{\partial f}{\partial z} + \frac{4D}{\Theta^2} = -\frac{2\theta^2}{\Theta^3}\left(\frac{\partial\Theta}{\partial z} + \frac{Ab^2\Theta^3}{2} - \frac{2D}{\Theta}\right) \tag{5.44}$$

Because this equation must hold true for all values of θ, two separate equations are obtained

$$\frac{\partial f(z, s)}{\partial z} = -\frac{4D}{\Theta^2} f(z, s) \tag{5.45}$$

and

$$\frac{\partial\Theta}{\partial z} = -\frac{A}{2}b^2\Theta^3 + \frac{2D}{\Theta} \tag{5.46}$$

For very large z, the quantity $\partial\Theta/\partial z$ in Equation (5.46) must vanish, so that

$$\Theta_\infty = \frac{(4D/A)^{1/4}}{\sqrt{b}} \tag{5.47}$$

Substituting Equation (5.47) into Equation (5.45) yields

$$f(z, s) = f(0, s) \exp(- \sqrt{b}\, \gamma_\infty z) \tag{5.48}$$

where γ_∞ is the steady-state attenuation coefficient for a very long fiber $(z \to \infty)$. The expression for γ_∞ in Equation (5.48) is given by

$$\gamma_\infty = \frac{4D}{\Theta_\infty^2 \sqrt{b}} = 2\sqrt{ADb} \tag{5.49}$$

For completeness the solutions of Equations (5.45) and (5.46) are given without the details, in the following:

$$\Theta^2(z, s) = \frac{\Theta_\infty^2}{b} \frac{b\Theta_0^2 + \Theta_\infty^2 \tanh(b\gamma_\infty z)}{\Theta_\infty^2 + b\Theta_0^2 \tanh(b\gamma_\infty z)} \tag{5.50}$$

and

$$f(z, s) = \frac{f(0, s)b\Theta_0^2}{\Theta_\infty^2 \sinh(b\gamma_\infty z) + b\Theta_0^2 \cosh(b\gamma_\infty z)} \tag{5.51}$$

where Θ_0 is the initial angular width at $z = 0$.

To find the time-dependent solutions, the inverse Laplace transform of the results above is needed. The closed-form Laplace transformation of Equation (5.43) exists only for the special cases when $z \ll 1/\gamma_\infty$ and $z \gg 1/\gamma_\infty$. In the former case of a short fiber, $\sinh b\gamma_\infty z$ and $\tanh b\gamma_\infty z$ are replaced by their argument $b\gamma_\infty z$ and set $\cosh b\gamma_\infty z = 1$. Equation (5.43) can be written with the help of Equations (5.42), and (5.47) to (5.51) as follows:

$$p(\theta, z, s) = \frac{f(0, s)}{1 + \gamma_\infty z} \exp\left[-\theta^2 \left(\frac{1}{\Theta_0^2} + \frac{nz}{2c} s \right) \right] \tag{5.52}$$

Equation (5.52) has the inverse Laplace transform

$$P(\theta, z, t) = F\left(0, t - \frac{n\theta^2 z}{2c} \right) (1 + \gamma_\infty z)^{-1} \exp\left(- \frac{\theta^2}{\Theta_0^2} \right) \tag{5.53}$$

The factor $(1 + \gamma_\infty z)^{-1}$ represents the loss in the short length of fiber; the expression $\exp(-\theta^2/\Theta_0^2)$, which describes the angular power distribution, is conserved under the transformation. The coefficient $F(0, t - n\theta^2 z/2c)$ shows that the portion of the input pulse propagating at an angle θ is delayed by $n\theta^2 z/2c$. The result of these calculations clearly shows that mode coupling has not affected pulse shape after propagating a very short distance.

By integrating Equation (5.52) over all angles the total output power for the case $z \ll 1/\gamma_\infty$ is

$$p(z, s) = 2\pi \int_0^\infty p(\theta, z. s)\theta d\theta = \frac{\pi f(0, s)\Theta_0^2}{(1 + \gamma_\infty z)(1 + n\Theta_0^2 zs/2c)}, \tag{5.54}$$

Now let $f(0, s) = 1$, which corresponds to an infinitesimally short input pulse of energy. The inverse Laplace transformation of Equation (5.54) yields an impulse response of the fiber:

$$P(z, t) = \frac{2\pi c}{nz(1 + \gamma_\infty z)} \exp\left(- \frac{2ct}{n\Theta_0^2 z}\right) \tag{5.55}$$

Equation (5.55) can be derived without the use of the power-flow equation, because in a very short fiber, mode coupling is not expected to play an important role. The fact that coupled-mode theory leads to these results indicates that the diffusion model is valid for predicting impulse response.

Now, investigate the impulse reponse after a short pulse has propagated over a very long distance (e.g., $z \gg 1/\gamma_\infty$). In this case $\tanh b\gamma_\infty z = 1$ and $\sinh b\gamma_\infty z = \cosh b\gamma_\infty z = 1/2 \exp(b\gamma_\infty z)$. Substituting these approximations into Equation (5.43) and assuming that $\Theta_0 \simeq \Theta_\infty$ gives

$$p(\theta, z, s) = \frac{2b}{1 + b} \exp\left[-b\left(\frac{\theta^2}{\Theta_0^2} + \gamma_\infty z\right)\right] \tag{5.56}$$

Upon integrating Equation (5.56) over all angles, the following results:

$$p(z, s) = \frac{2\pi\Theta_0^2}{1 + b} \exp(-b\gamma_\infty z) \tag{5.57}$$

Substituting Equation (5.42) for b and taking the inverse Laplace transform

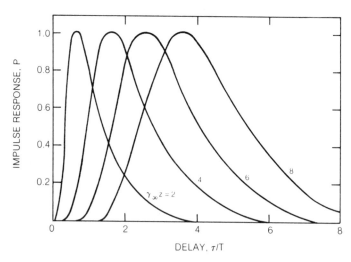

FIGURE 5.6 Normalized impulse response as a function of normalized time as a short pulse propagates in a multimode fiber over a normalized distance $\gamma_\infty z$. [From *The Bell System Technical Journal* (Ref. 5.3). Reprinted with permission of American Telephone and Telegraph Co., © 1973.]

of $p(z,s)$ gives

$$P(z, t) = \Theta_0^2 \sqrt{\frac{\pi}{Tt}} \left(\frac{1}{2} + \frac{t}{\gamma_\infty zT} \right)^{-1} \exp \left(-\frac{t}{T} - \frac{\gamma_\infty^2 z^2 T}{4t} \right) \quad (5.58)$$

where

$$T = \frac{n}{2cA} = \frac{n\Theta_0^2}{2c\gamma_\infty} \quad (5.59)$$

The results of Equation (5.58) for pulse shapes are plotted in Figure 5.6 as a function of the delay time t/T for various normalized lengths $\gamma_\infty z$. These impulse responses are normalized for equal peak value. Expression (5.58) is very different from Expression (5.55) for uncoupled modes. Equation (5.55) shows that when $t = T$, the pulse amplitude is reduced by a factor of e^{-1} if $z = 1/\gamma_\infty$. Therefore, the normalized length $1/\gamma_\infty$ is defined as the distance within which a 1-neper loss is incurred. This loss is equivalent to a total loss in decibels of

$$\alpha(\mathrm{dB}) = 4.35\gamma_\infty z \quad (5.60)$$

5.7 PULSE DISTORTION

For a Gaussian input pulse, the output remains Gaussian, but the width is subject to change and dependent on the fiber length. Previous sections discussed phenomena involving pulse delay and pulse broadening for two limiting cases. This section derives expressions for the mean pulse delay and mean pulse width as a function of propagation distance z in a multimode fiber, in which mode coupling predominates. In the application of the diffusion model, the nth moments of an impulse response are defined as

$$M_n = \int_0^\infty t^n P(z, t) \, dt \quad (5.61)$$

Because of the general relation between $P(z, t)$ and its Laplace transform $p(z, s)$, M_n can be expressed in terms of the nth derivatives of p as

$$M_n = (-1)^n \left. \frac{\partial^n p}{\partial s^n} \right|_{s=0} \quad (5.62)$$

The normalized moments, m_n, can be simply expressed in terms of the nth derivatives of $\ln p$ as given by

$$m_n = \frac{M_n}{\int_0^\infty P \, dt} = (-1)^n \frac{1}{p} \frac{\partial^n p}{\partial s^n} = (-1)^n \frac{\partial^n \ln p}{\partial s^n} \quad (5.63)$$

Physically, the first and second normalized moments of an impulse response about its mean values represent the mean delay τ_θ and the square of the half-width σ_θ^2 measured at the $1/e$ intensity point, respectively. Using Equation (5.43) and again assuming that $\Theta_0 \simeq \Theta_\infty$, τ_θ and σ_θ^2 can be calculated as follows:

$$\tau_\theta = - \left. \frac{\partial \ln p}{\partial s} \right|_{s=0} = \frac{T}{2} \left[\gamma_\infty z + \left(\frac{\theta^2}{\Theta_0^2} - \frac{1}{2} \right) (1 - e^{-2\gamma_\infty z}) \right] \quad (5.64)$$

and

$$\sigma_\theta^2 = \left. \frac{\partial^2 \ln p}{\partial s^2} \right|_{s=0}$$

$$= \frac{T^2}{4} \left[\gamma_\infty z + \left(\frac{\theta^2}{\Theta_0^2} - \frac{5}{4} \right) - 2\gamma_\infty z \left(2\frac{\theta^2}{\Theta_0^2} - 1 \right) e^{-2\gamma_\infty z} \right. \quad (5.65)$$

$$\left. + e^{-2\gamma_\infty z} + \left(\frac{\theta^2}{\Theta_0^2} - \frac{1}{4} \right) e^{-4\gamma_\infty z} \right]$$

where T is defined by Equation (5.59).

Figure 5.7 shows the variation of τ_θ and σ_θ as a function of $\gamma_\infty z$ for $\theta = 0$ and $\theta = \Theta_\infty$. For a very short length, the pulse propagates without broadening and merely suffers a mode-dependent delay of $n\theta^2/2c$, as expected. However, before the normalized length $1/\gamma_\infty$ is reached, the pulse in all modes begins to widen. Once $1/\gamma_\infty$ is passed, the pulse width in all

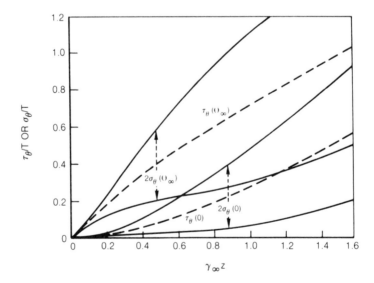

FIGURE 5.7 Time delay and pulse spreading as a function of fiber length. [From *The Bell System Technical Journal* (Ref. 5.3). Reprinted with permission of American Telephone and Telegraph Co., © 1973.]

modes increases essentially as $T(\gamma_\infty z)^{1/2}$. This result differs significantly from that when mode coupling is ignored. For $\gamma_\infty z \gg 1$, Equations (5.64) and (5.65) become

$$\tau_\theta = \frac{T}{2}\left(\gamma_\infty z - \frac{1}{2} + \frac{\theta^2}{\Theta_0^2}\right) \tag{5.66}$$

and

$$\sigma_\theta = \frac{T}{2}\left(\gamma_\infty z - \frac{5}{4} + \frac{\theta^2}{\Theta_0^2}\right)^{1/2} \tag{5.67}$$

To calculate the θ-independent delay and width, the first and second derivatives of $p(z, s)$ must be performed. Without going through the details, the results are

$$\tau = \frac{T}{2}\left[\gamma_\infty z + \frac{1}{2}(1 - e^{-2\gamma_\infty z})\right] \tag{5.68}$$

and

$$\sigma = \frac{T}{2}\left[\gamma_\infty z(1 - 2e^{-2\gamma_\infty z}) + \frac{3}{4} - e^{-2\gamma_\infty z} + \frac{1}{4}e^{-4\gamma_\infty z}\right]^{1/2} \tag{5.69}$$

The ratio σ/T is shown in Figure 5.8 as a function of the normalized length $\gamma_\infty z$. For $z \ll 1/\gamma_\infty$, the width σ approaches $T\gamma_\infty z$, as expected. At $z = 1/4\gamma_\infty$, σ begins to follow a new asymptote:

$$\sigma = \frac{T}{2}\sqrt{\gamma_\infty z} \tag{5.70}$$

The derivation from the $T\gamma_\infty z$ curve as shown in Figure 5.8 is an indication of mode coupling. The effect of mode coupling reduces the width by a factor of $\sqrt{4\gamma_\infty z}$ in exchange for an increase in the overall attenuation by $4.35\gamma_\infty$ dB/km. The physical contents of these results can best be summarized by defining a coupling length $L = 1/4\gamma_\infty$, at which the width of the impulse response changes from a linear to a square-root dependence of the length. Furthermore, if $\tau = \tau' + T\gamma_\infty z/2$ enters into Equation (5.58) with $\tau' \ll Y\gamma_\infty z/2$, the impulse response, for large z is

$$P(z, t') = \Theta_0^2 \sqrt{\frac{2\pi}{\gamma_\infty z}} \exp\left(-\gamma_\infty z - \frac{2\tau'^2}{T^2\gamma_\infty z}\right) \tag{5.71}$$

Equation (5.71) indicates that pulse shape changes from exponential to Gaussian in time with the variance $4\sigma^2$, as given by Equation (5.70). Beyond the coupling length L, the width of the impulse response increases only as the square root of the fiber length. Physically, the power carried by lower-order modes or small values of θ travels faster and tends to lead the pulse, but in

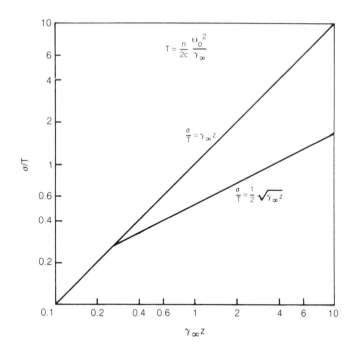

FIGURE 5.8 Asymptotic behavior of the width of an impulse response as a function of normalized length. [From *The Bell System Technical Journal* (Ref. 5.3). Reprinted with permission of American Telephone and Telegraph Co., © 1973.]

the meantime, feeds back continuously into higher-order modes by diffusion. The power carried by these higher-order modes tends to fall behind the main pulse but catches up with the main body of the pulse by diffusion. As a result, both the leading and trailing edges of the pulse are clipped to an extent that the characteristic width increases only as $\sqrt{L}$. Initially, a linear growth of the pulse width with length is only temporary while waiting for a redistribution of power as it approaches a steady-state condition.

5.8 SCATTERING AND ABSORPTION LOSSES

Because of the severity of optical power losses in fibers due to scattering and absorption, for a long time glass fibers were not seriously considered as candidates for use as optical transmission media. Only within very recent years, since techniques for glass material preparation and fiber manufacturing have been improved to minimize these losses, has the field of fiber optics grown tremendously through a concerted research and development effort that has increased the level of understanding of optical fiber transmission by several orders of magnitude. These loss mechanisms are discussed here and a discussion of fiber fabrication techniques is in Chapter 7. As evident from the discussion of material dispersion, a longer-wavelength transmitter in the region of 1.2 to 1.6 μm for an optical fiber system is a good choice. Another reason for choosing a longer wavelength is to minimize scattering loss. This phenomenon is inherent in all glasses, because all optical materials contain defects that scatter light as it propagates over a long distance. The amount of power scattered by those defects is dependent on the defect density and the scattering cross-section, C. The value of C is a measure of the scattered power P_s in a single scattering event for an incident light intensity I_0 (watts/cm^2). By definition

$$P_s = CI_0 \qquad (5.72)$$

In Equation (5.72), C has the dimensions of an area and is related to the geometrical cross-sectional area of the scatterer and the strength of the scattering interaction S as

$$C = \pi r^2 S \qquad (5.73)$$

where r is the average radius of the scatterer and S is the dimensionless factor, which depends on the ratio r/λ.

Consider that when a plane wave is incident on a spherical particle, the scattered wave in the far field and the field inside the sphere can be expanded in terms of spherical coordinates. The expansion coefficients can be determined by matching these fields at the boundary of the scatterer and also by satisfying the conditions at infinity. These coefficients are related to the scattering amplitudes, which, in general, are very complex functions and yield information concerning both the amplitudes and phase of the scattered field.

In all types of glass, scatterers are primarily impurities such as oxides and transition metal ions, with sizes typically much smaller than the wavelength (e.g., $r/\lambda \ll 1$). In this limit commonly known as Rayleigh scattering, expressions for scattering amplitudes are considerably simplified. Assuming that incident light is linearly polarized and that secondary waves are irradiated from microscopic scatterers in the form of an induced electric dipole, with its moment $|\mu|$ parallel to the polarization, the scattering cross-section in the far field can be expressed simply by

$$C = \frac{8\pi}{3\lambda^4} |\mu|^2 f(\theta) \qquad (5.74)$$

where $f(\theta)$ is an angular factor describing the radiation pattern of a dipole. Note that the most outstanding character in this expression is the $1/\lambda^4$ dependence, which has already been illustrated in Figure 1.4. For the case of a simple dipole, the perpendicular polarization scatters isotropically and the parallel polarization scatters as $\cos^2 \theta$, which yields an equal distribution in both the forward and backward directions, and zero at the right angle.

The total scattered power per unit length is a product of P_s and the number density of the scatterers in the fiber. Because the scatterers distribute randomly in glass, a density fluctuation exists over the length of the fiber. This problem can be traced back to thermal fluctuations that cause the impurities to diffuse in the form of Brownian motion in the molten state before solidification. The magnitude of this density fluctuation is proportional to the product of the softening point of the glass, T_s, and the isothermal compressibility, κ. In terms of these parameters, the scattering cross-section is (Ref. 5.4)

$$C = \frac{8\pi^3}{3\lambda^4} (n^2 - 1)kT_s\kappa \qquad (5.75)$$

where k is the Boltzmann constant. For fused silica with softening temperatures near 1500°C, Equation (5.75) yields a loss of 1.7 dB/km at 0.85 μm, which is consistent with the experimental results. According to Equation (5.75), a lower softening point may lead to a lower scattering loss. In reality, this situation is complicated by the fact that in materials with lower softening temperatures, a variety of impurities can easily be introduced. This detriment creates a composition fluctuation that can affect the refractive index much more severely than the density fluctuation. Consequently, the composition fluctuation in high-index glass can introduce higher losses than that in lower-index glass. In both cases the $1/\lambda^4$ dependence is the dominating factor. Other scattering processes exist for which the scattering cross-section is relatively independent of λ. This situation occurs when r/λ approaches unity at which point Mie scattering begins to dominate. In this case, the fibers are usually made with very poor quality because they contain very large scattering centers with scale sizes greater than λ. Consequently, loss mechanisms other than scattering can play an important role.

In addition to scattering losses, glass has absorption losses that arise from both the intrinsic structure of the material and impurity absorption. The intrinsic absorption originates from a charge transfer between various energy bands with characteristic spectra lying primarily in the ultraviolet region. However, these bands are sufficiently wide, with their spectral wings well extended into the near-infrared region; therefore, they could cause some absorption loss. However, recent measurements indicate that for wavelengths beyond 0.8 μm, band-edge absorption is almost certainly less than 1 dB/km. Impurity absorption, on the other hand, is caused by metal ions such as Fe, Cu, V, and Cr. At an impurity level of 10 ppb, the loss figure could run up to 20 dB/km. Highly purified silica glass is now routinely made without discerning the loss component due to impurities. However, in highly doped glass, losses due to metal ions are troublesome. Special care must be taken to reduce the impurity level in this glass.

Transition metal ions have incompletely filled inner electron shells which gives rise to their characteristic absorptions by inducing transitions between those levels. Unlike metals, the formation oxidation in glass leaves transition ions with unfilled levels. Even though transitions between different oxidation states are forbidden, the perturbation introduces a splitting in these levels that is responsible for the spectra observed. The coloration observed in heavily doped glass can be used to characterize different impurities under different conditions of oxidation. Proper balance between oxidation and reduction through control of the partial pressure of oxygen in the melt can reduce absorption loss to a minimum. In general, when glass is overly reduced, the absorption tends to increase at longer wavelengths. If the opposite is true, the absorption loss tends to increase at shorter wavelengths. The process of the reduction is to convert ion species from one type to another (e.g., from Fe^{3+} to Fe^{2+}, etc.). To accomplish such a chemical reaction, the partial pressure of oxygen may be varied by many orders of magnitude. Glass in the melt can be oxidized in many ways. For example, oxidation can be accomplished by adding oxides of arsenic and antimony, or by using bubbling gas such as CO and CO_2 to stir the melt. At present the production of high-purity glass is possible only by chemical vapor disposition technique for the growth of glass preforms that is discussed in Chapter 7.

Another mechanism responsible for absorption loss involves vibrational energy associated with some of the common bonds present in glass. In most cases the vibrational spectra of glass lie in the infrared region of 2 to 10 μm, in which the overtones of the fundamental stretching vibration of the hydroxyl ion (OH) play an important role. The fundamental is centered around 2.8 μm, with its first three overtones at 1.4, 0.97, and 0.75 μm, respectively. Attempts have been made by the Corning Glass Works and others to identify the OH overtones by matching individual absorption lines. However, the measured line shape deviates significantly from the expected Lorentzian profile. This type of measurement is extremely difficult unless the glass under investigation can be made without any loss other than that due to the OH stretching mode. To reduce OH content, a technique has been developed by which the oxide powders of glass are heated in an oven at 250°C for a few days. The glass is otherwise prepared in the usual way but under a controlled-humidity atmosphere, and the melt is bubbled with gases of varying dew points. From the measurement of absorption loss in this glass, which is about only 1 dB/km at 0.9 μm, the estimated OH content corresponds to only 1 ppm weight of water.

5.9 BENDING LOSSES

Optical losses associated with the cabling process can be introduced as a result of fiber bending. This problem becomes serious when the radius of curvature $R(z)$ of the bend is small but large compared with the radius of the fiber. Because of the stiffness of the fiber, $R(z)$ normally changes slowly and continuously with fiber strength. This type of loss is found most often in single mode or quasi single mode fiber systems, where very large bit rates are desired. In these systems, only the fundamental mode is excited. Due to the bending, the fundamental mode power is eventually lost through coupling to radiation modes.

Problems of this type are usually solved by using coupled mode theory, which involves a system of coupled equations that are usually solved by perturbation methods. Several methods have been introduced, with varying degrees of accuracy. Unfortunately, all these methods are rather tedious and involve first deducing the eigenvalue equations by using proper boundary conditions, including bending before calculating the coupling coefficient between the guided mode and leaky modes.

One method introduced by Marcuse (Ref. 5.5) derives the loss formula for single mode fibers with a constant radius of curvature $R(z)$ by expressing the field in the cladding of a curved fiber section in terms of a superposition of leaky waves. The field near the core is assumed to be the same as that of a straight fiber. The expansion coefficients used in the super position of fields outside the core are determined by matching the transverse field components to the field of the leaky modes along the fiber surface that is tangential to the curve guide. Let P_{in} and P_{out} be the input and output power through a curved single mode fiber with a bending path length L between two points A and B. The ratio of the output power to the input power can be expressed as

$$\frac{P_{out}}{P_{in}} = \left(\frac{P_{out}}{P_{in}}\right)^2_{\text{at A or B}} \exp(-2\alpha L) \tag{5.76}$$

where (P_{out}/P_{in}) at A or B are the transition losses at point A or B, which are assumed to be equal. The pure bending loss coefficient 2α for the fundamental mode is given by (Ref. 5.5)

$$2\alpha = \frac{\sqrt{\pi}\,(u/a)^2}{2(\gamma/a)^{3/2}V^2\sqrt{R}\,K_1^2(\gamma)} \exp\left(-\frac{2}{3}\frac{\gamma^2}{\beta^2 a^3}R\right) \tag{5.77}$$

Figure 5.9 shows the measured bending loss for a single mode step-index fiber, having a core diameter of 3.9 microns and a silica cladding of radius 48 microns. The refractive index difference between the core and cladding is 0.04, corresponding to an NA of 0.11. The measured 2α values normalized to a 10 cm bend length were obtained using a GaAlAs diode laser operating at a wavelength of 790 nm. The dashed curve shows the calculated values using Equation (5.77). The experimental values of 2α are slightly lower than the theoretical values and show oscillations that decrease in amplitude as the radius of the bend is reduced. This oscillatory behavior can be explained by the coupling between the fundamental mode and the whispering gallery

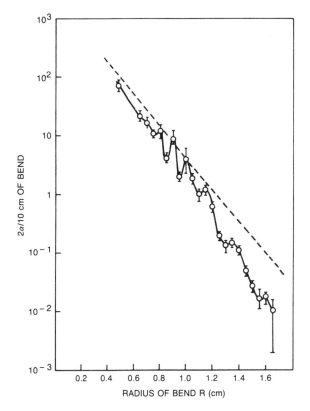

FIGURE 5.9   Bending loss 2α for 10 cm bend length as a function of the bend radius R for $\lambda = 790\,\text{nm}$. Dashed curve represents the calculated values. (Ref. 5.6. Reprinted with permission of IEEE.)

mode. The whispering gallery mode is formed from the light leaving the fundamental mode due to pure bending loss. The minima or troughs in the oscillations result from synchronized, e.g., in-phase, coupling of light from the whispering gallery mode to the fundamental mode, and the peaks of the oscillations correspond to asynchronous, e.g., out-of-phase coupling. The decrease in amplitude of the oscillations is attributed to the decrease in reflectivity of the cladding/coating interface as the bend radius is reduced. At longer wavelengths and/or larger bend radii, oscillations become more pronounced with peak values, which correspond to asynchronous coupling, coinciding with the theoretical bending loss.

The loss due to fiber deformations such as microbending along the fiber axis and random core diameter fluctuations is a rather complex problem. A common technique used to evaluate the micro-deformation losses is to make use of the results of the coupled mode theory, where the coupling strength between the guided mode of a single mode fiber and the cladding modes is used to evaluate the power loss coefficient. Because of the stochastic nature of the deformations, the coupling coefficients c_{ij} are independent of the fiber length. Usually an autocorrelation function is assumed, which characterizes the micro-deformations by the rms deviation and the correlation length of the random function. To a good approximation, the autocorrelation function is assumed to be Gaussian:

$$R(u) = \langle f(z)f(z + u)\rangle \qquad (5.78)$$
$$= \sigma^2 \exp(-u^2/L_c^2)$$

where $f(z)$ is the random deformation function, σ is the rms deviation of $f(z)$, and L_c is the correlation length. This assumption leads to an expression for the spatial power spectrum function of the above autocorrelation function as given by (Ref. 5.7)

$$\Phi(\beta_0 - \beta_{ij}) = \sqrt{\pi}\ \sigma^2 L_c \exp\left\{-\frac{1}{4}(\beta_0 - \beta_{ij})^2 L_c^2\right\} \qquad (5.79)$$

where β_0 and β_{ij} are the propagation constants of the fundamental mode and cladding modes, respectively.

The loss coefficient is defined by

$$\langle 2\alpha \rangle = \sum_{i=0}^{\infty} \sum_{j=1}^{n} c_{ij}^2 \Phi(\beta_0 - \beta_{ij}) \qquad (5.80)$$

where $\langle\ \rangle$ denotes the ensemble average, the summations are taken over all cladding modes, and c_{ij} are the coupling coefficients between the guided and cladding modes. It is given by

$$c_{ij} = \frac{k_0 \int_0^{\infty} \int_0^{2\pi} N(r, \phi)E_0E_{ij}r\ dr\ d\phi}{\left(\int_0^{\infty} \int_0^{2\pi} rE_0^2\ dr\ d\phi \int_0^{\infty} \int_0^{2\pi} rE_{ij}^2\ d\,r\ d\phi\right)^{1/2}} \qquad (5.81)$$

where N is the normalized refractive index distribution that characterizes micro-deformation of the fiber. Substituting Equations (5.79) and (5.81) into (5.80) and assuming that N can be expressed in terms of a sum of Dirac's δ-function for a distorted index distribution yields (Ref. 5.7)

$$\langle 2\alpha \rangle = \frac{1}{2} k_0^2 \sqrt{\pi}\ \sigma^2 L_c \sum_{j=1}^{n} \exp\left\{-\frac{1}{4}(\beta_0 - \beta_{ij})^2 L_c^2\right\} (n\alpha\,\Delta E_0E_j)^2 \qquad (5.82)$$

Figure 5.10 shows the calculated microbending losses (solid curves) for a step-index fiber with $a = 5$ microns, $\Delta = 0.003$, and $\sigma = 1$ nm for several values of L_c as a function of wavelength. Also shown in Figure 5.10 are calculated loss coefficients (dashed curves) for the same fiber having a ran-

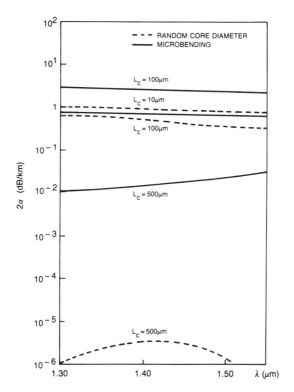

FIGURE 5.10 Micro-deformation losses as a function of wavelength for a step-index single mode fiber with $a = 5$ μm, $\Delta = 0.003$ and rms $\sigma = 1$ nm for several values of L_c. (Ref. 5.7. Reprinted with permission of IEEE.)

dom core diameter deformation. From these results (Figure 5.10), it is apparent that for small values of L_c (≤ 10 microns), the coupling loss coefficients for both microbending and random core diameter deformation are of the same order of magnitude. For large correlation length (≥ 500 microns), the difference between these two types of deformations becomes very large, as much as four orders of magnitude in dB/km.

PROBLEMS

5.1. Calculate the group delay between the fastest and the slowest mode in a 1-km-long step-index fiber with $n_0 = 1.5$ and $\Delta = 0.003$, using a light source at 0.9-μm wavelength and $2a = 10$ μm.

5.2. Derive Equation (5.26).

5.3. For a typical LED source emitting in the wavelength region of 0.8 to 0.9 μm, the spectral width is about 20 nm. Calculate the material dispersion of a silicate fiber with a LED source using the results of Figure 5.2, over a fiber length of 1 km.

5.4. Using the result of Equation (5.29), show that for $\alpha = \infty$,

$$\sigma_{\text{step}} = \frac{1}{2\sqrt{3}c} L(n - \lambda n')\Delta \times (1 + 3\Delta + \frac{24}{5}\Delta^2)$$

5.5. Using Equation (5.29), show that for $\alpha = 2$, and δ is equal to zero

$$\sigma_{\text{graded}} = \frac{1}{2\sqrt{6}c} L(n - \lambda n')\Delta^2$$

5.6. Assuming an effective numerical aperture NA of a fiber to be $n\theta_\infty$, derive an expression for the width of the impulse response for the case $z \gg 1/\gamma_\infty$ as a function of NA.

5.7. For a fixed fiber length L, much larger than $1/\gamma_\infty$, calculate the loss penalty for an increase in bandwidth $B = 1/4\sigma$ by a factor of 2.

5.8. Show that the pulse width σ is

$$\sigma = \begin{cases} T\gamma_\infty z & \text{for } \gamma_\infty z \ll 1 \\[2ex] \dfrac{T}{2}\sqrt{\gamma_\infty z} & \text{for } \gamma_\infty z \gg 1 \end{cases}$$

REFERENCES

5.1. D. Gloge, *Appl. Opt.*, *10*, 2252 (1971).

5.2. R. Olshansky and D. B. Keck, *Appl. Opt.*, *15*, 483 (1976).

5.3. D. Gloge, *Bell Syst. Tech. J.*, *6*, 801 (1973).

5.4. R. D. Maurer, *J. Chem. Phys.*, *25*, 1206 (1956).

5.5. D. Marcuse, *J. Opt. Soc. Am.*, *66*, 216 (1976).

5.6. A. J. Harris and P. F. Castle, *IEEE J. Lightwave Tech.*, *LT-4*, 34 (1986).

5.7. A. Bjarklev, *IEEE J. Lightwave Tech.*, *LT4*, 341 (1986).

6

Single
Mode Fibers

6.1 INTRODUCTION

The available bandwidth of graded-index multimode fibers tends to be limited by the modal dispersion arising from errors in the refractive index profile. Therefore, many high-data-rate transmission systems switch from graded-index multimode fibers to single mode fibers. This chapter first summarizes the results of previous chapters obtained for the HE_{11} or the LP_{01} mode, which is the only allowable mode in single mode fibers. In a perfect single mode fiber, the HE_{11} mode is circularly degenerate. However, this degeneracy can be removed by the birefringence or anisotropic properties of the fiber. As a result, a time delay exists between the two orthogonal modes with different propagation constants. This time delay introduces a distortion of transmitted signals. On the other hand, chromatic dispersion represents a major cause of signal distortion in single mode fibers. This chapter deals with these effects and formulates design criteria for fiber systems that can be the best candidates for wideband and coherent optical systems. With a good understanding of these effects, a dispersion-free and polarization preserving fiber can be designed that cannot only improve bandwidth and reduce loss, but also optimize optical fiber telecommunication systems.

6.2 DESCRIPTION OF THE HE₁₁ MODE

If the V value of a fiber is smaller than 2.405, only two degenerate HE_{11} modes are allowed to propagate. With a perfect circular symmetry and without any birefringence, these two modes are indistinguishable, and in principle, no modal dispersion is expected. In practice, this is often not the case. Equation (4.60) is a solution of the eigenvalue equation (4.48), and gives a functional relationship of u in terms of V for HE_{11} modes. The result is

$$u(V) = \frac{(1 + \sqrt{2})V}{1 + (4 + V^4)^{1/4}} \tag{6.1}$$

and is plotted in Figure 6.1. Table 6.1 gives the u and γ values for V up to 2.4. Above $V = 2.405$, other modes, namely TE_{01}, TM_{01}, and HE_{21} begin to emerge along with the HE_{11} mode.

The field components in a step-index fiber are given by Equations (4.35) to (4.43). For the HE_{11} mode the transverse field component is

$$E(r) = -\sqrt{\frac{2}{\pi}} \left(\frac{\epsilon_0}{\mu_0}\right)^{1/4} \frac{\gamma\sqrt{n_2 P}}{aVJ_1(u)} \begin{cases} J_0(ur/a) & (r < a) \\ \dfrac{J_0(u)}{K_0(\gamma)} K_0(\gamma r/a) & (r > a) \end{cases} \tag{6.2}$$

where ϵ_0 and μ_0 are the dielectric permittivity and magnetic susceptibility of free space, respectively. They are related to the velocity of light, c, as $c = 1/(\mu_0\epsilon_0)^{1/2}$. All other parameters have their usual meanings as defined in Chapter 4, with the exception that n_c is replaced with n_2 to represent the refractive index of the cladding. Figure 6.2 is a plot of E for $V = 1.2, 1.8,$

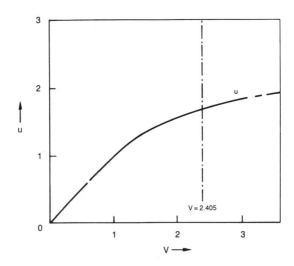

FIGURE 6.1 Transverse propagation constant of the HE_{11} mode as a function of normalized frequency V.

**TABLE 6.1   The Transverse Propagation
Constant u and Exponential Decay Constant γ
for HE₁₁ Mode As a Function of the
Normalized Frequency V**

V	u	γ
0.6	0.59997	0.0056
0.8	0.7974	0.0640
1.0	0.9793	0.2024
1.2	1.1341	0.3921
1.4	1.2618	0.6065
1.6	1.3670	0.8315
1.8	1.4545	1.0604
2.0	1.5282	1.2902
2.2	1.5911	1.5194
2.4	1.6453	1.7473
2.6	1.6926	1.9736
2.8	1.7342	2.1963
3.0	1.7711	2.4214
∞	2.405	∞

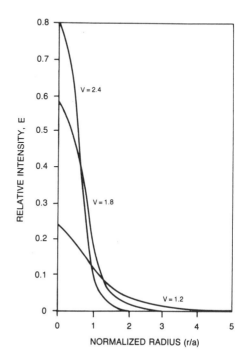

FIGURE 6.2 The transverse field $E(r)$
as a function of normalized radius r/a for
three different values of V.

and 2.4. In all cases, the field intensity extends into the cladding ($r/a > 1$). As V increases, the evanescent field decreases with respect to r. Therefore, in dealing with single mode fiber, the cladding material and structure are crucial parameters and must be carefully considered in the system design. Observe that the shapes of these curves are very similar to Gaussian distributions. To a good approximation, it is possible to replace the exact wave functions that are usually expressed in terms of Bessel functions to Gaussian functions of appropriate forms, which are more convenient to use, especially for calculating splicing and coupling losses. The field component is expressed in the form

$$\psi(r) = \left(\frac{4\sqrt{\mu_0/\epsilon_0}\, P}{\pi n_2 \omega_0^2}\right)^{1/2} \exp(-r^2/\omega_0^2) \tag{6.3}$$

where P is the power carried by the field, n_2 is the refractive index of the cladding, and ω_0 is the width of the Gaussian field, which can be empirically chosen to best fit these curves (Ref. 6.1) and written as

$$\frac{\omega_0}{a} = 0.65 + \frac{1.619}{V^{3/2}} + \frac{2.879}{V^6} \tag{6.4}$$

This equation gives the optimum width of the Gaussian profile that best approximates the actual field distribution of a step-index fiber for V values up to 2.8. The error introduced in this approximation by Equation (6.3), is within a fraction of 1% from the exact field profile. For smaller V values ($V < 1.2$), the mismatch between the Gaussian field and the exact field becomes more pronounced. More discussion can be found in Section 6.5.

The evanescent field far from the core-cladding interface ($\gamma r > 2a$) can be expressed by using the following asymptotic expansion

$$K_0(\gamma r/a) \simeq \sqrt{\frac{\pi a}{2\gamma r}} \exp(-\gamma r/a) \tag{6.5}$$

Substituting Equations (6.2) and (6.5) into (4.61) and (4.62) gives the expression for the fractional residual power P_r. The amount of power which resides above a given radius $r > a$ can be obtained by replacing the limits of the integral of Equation (4.62) from r to ∞, as given by

$$\frac{P_r}{P} = \frac{\pi}{2}\left(\frac{u}{\gamma V K_1(\gamma)}\right)^2 \exp(-2\gamma r/a) \tag{6.6}$$

Figure 6.3 is a plot of P_r/P_T as a function of V for a number of values of r/a from 1 to 10. It shows the fractional power carried by a HE_{11} mode which travels outside a given radius of the fiber as a function of V. It is clear from this figure that as V values decrease from 2.4, more and more modal power resides in the cladding. For example, at $V = 1.6$, about 40% of the power propagates in the region $r > a$. This V value represents the practical limit

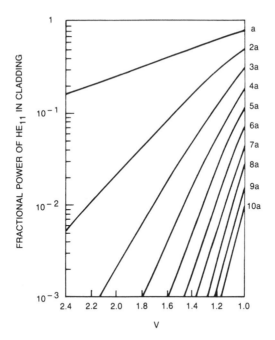

FIGURE 6.3   Fractional power of the HE_{11} mode as a function of the normalized frequency V at various multiples of the core radius a.

of operation below which bending losses become too excessive. For a given V, a corresponding u value can be calculated from Equation (6.1). Because a large fraction of modal power propagates in the cladding, special attention must be given to the cladding material and its index profile, which are discussed next.

6.3 DOUBLY-CLADDED STEP-INDEX FIBERS

Usually, a fiber is designed for a given wavelength by specifying the optimum radii for the core and cladding, and the difference in refractive indices between the core n_1 and cladding n_2. The wavelength of interest for all long-haul light wave systems is now centered on InGaAsP lasers emitting in the range from 1.3 to 1.7 microns. This wavelength range is considerably longer than the wavelength emitted by GaAs/GaAlAs lasers. For low-loss fibers, it is desirable to keep Δ as low as possible to minimize the loss introduced into the core upon appropriate doping. In addition to the loss consideration, material and waveguide dispersion at long wavelengths must be kept at a minimum. Because of all these requirements, new fiber designs have emerged. The new designs shown in Figure 6.4, involve doubly-cladded fibers. Figure 6.4(a) depicts a structure that consists of a germanium doped

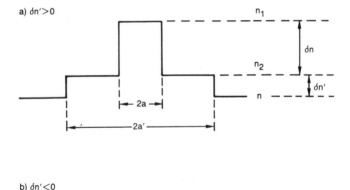

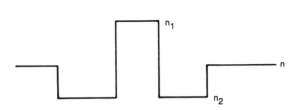

FIGURE 6.4   Refractive index variation along the diameter of a doubly-cladded and weakly guiding single mode fiber.

silica core, surrounded by a lightly phosphorous oxide doped silica cladding. This structure has two possible variations, namely, the first kind has a large core but a low Δ, and the second kind has a small core but a high Δ. Figure 6.4(b) depicts a W-type structure that is made by manipulating the material dispersion of the core and cladding compositions. Next, the effects of double cladding on propagation, cutoff, and dispersion properties will be examined. The exact solution for these fibers with double cladding is extremely complex. Only numerical (Ref. 6.2) solutions to Maxwell's equation exist. This section simply formulates the problem and presents the numerical results in graphical form.

Consider the structures shown in Figure 6.4 to be weakly-guided fibers having a core radius a and a core refractive index n_1. The inner cladding has a radius a' and a refractive index n_2. The refractive index of the surrounding medium or outer cladding is n. For this fiber, the V-parameters are defined as

$$V_1 = k_0 a (n_1^2 - n^2)^{1/2} = k_0 a [2n(\delta n + \delta n')]^{1/2}$$

$$V_2 = k_0 a' \mid n_2^2 - n^2 \mid^{1/2} = k_0 a' [2n \mid \delta n' \mid]^{1/2}$$

where

$$\delta n = n_1 - n_2 \qquad \delta n' = n_2 - n$$

The radial dependence of the field components can be separated into two sets of wave functions, each of which is a linear combination of Bessel and modified Bessel functions. For $\beta < k_0 n_2$

$$\psi(r) = \begin{cases} A_0 J_l(ur/a) & (r < a) \\ A_1 J_l(u'r/a') + A_2 Y_l(u'r/a') & (a < r < a') \\ A_3 K_l(\gamma r/a) & (r > a') \end{cases} \tag{6.7}$$

For $\beta > k_0 n_2$

$$\psi(r) = \begin{cases} A_0' J_l(ur/a) & (r < a) \\ A_1' I_l(\gamma'r/a') + A_2' K_l(\gamma'r/a') & (a < r < a') \\ A_3' K_l(\gamma r/a') & (r > a') \end{cases} \tag{6.8}$$

where the radial propagation parameters, u, u', γ, and γ' are defined as

$$\begin{aligned} u &= a(k_0^2 n_1^2 - \beta^2)^{1/2} \\ u' &= a'(k_0^2 n_2^2 - \beta^2)^{1/2} \\ \gamma &= a'(\beta^2 - k_0^2 n^2)^{1/2} \\ \gamma' &= a'(\beta^2 - k_0^2 n_2^2)^{1/2} \end{aligned} \tag{6.9}$$

By matching ψ and $d\psi/dr$ at two interfaces $r = a$ and $r = a'$, four homogeneous equations involving four coefficients, A_i for the case $\beta < k_0 n_2$ emerge:

$$J_l(u)A_0 - J_l\left(u'\frac{a}{a'}\right)A_1 - Y_l\left(u'\frac{a}{a'}\right)A_2 = 0$$

$$uJ_l'(u)A_0 - u'\frac{a}{a'}J_l'\left(u'\frac{a}{a'}\right)A_1 - u'\frac{a}{a'}Y_l'\left(u'\frac{a}{a'}\right)A_2 = 0$$

$$- J_l(u')A_1 - Y_l(u')A_2 + K_l(\gamma)A_3 = 0$$

$$- u'J_l'(u')A_1 - u'Y_l'(u')A_2 + \gamma K_l'(\gamma)A_3 = 0$$

A nontrivial solution is that the determinant of this set of homogeneous equations vanishes. This condition leads to the following eigenvalue equations for $\beta < k_0 n_2$:

$$\frac{[\hat{J}_l(u) - \hat{Y}_l(u'c)]\,[\hat{K}_l(\gamma) - \hat{J}_l(u')]}{[\hat{J}_l(u) - \hat{J}_l(u'c)]\,[\hat{K}_l(\gamma) - \hat{Y}_l(u')]} = \frac{J_{l+1}(u'c)Y_{l+1}(u)}{J_{l+1}(u)Y_{l+1}(u'c)} \tag{6.10}$$

Similarly, for $\beta > k_0 n_2$

$$\frac{[\hat{J}_l(u) - \hat{K}_l(\gamma'c)]\,[\hat{K}_l(\gamma) - \hat{I}_l(\gamma')]}{[\hat{J}_l(u) + \hat{I}_l(\gamma'c)]\,[\hat{K}_l(\gamma) - \hat{K}_l(\gamma')]} = \frac{I_{l+1}(\gamma'c)K_{l+1}(\gamma')}{I_{l+1}(\gamma')K_{l+1}(\gamma'c)} \tag{6.11}$$

where

$$\hat{J}_l(u) = \frac{J_l(u)}{uJ_{l+1}(u)}, \qquad \hat{K}_l(\gamma'c) = \frac{K_l(\gamma'c)}{\gamma'cK_{l+1}(\gamma'c)}, \text{ etc.} \tag{6.12}$$

and

$$c = \frac{a}{a'} \tag{6.13}$$

At cutoff ($\gamma = 0$), $\beta = k_0 n$. The radial propagation constants take the values

$$u = V_1 \quad (r < a)$$

$$u' = V_2 \text{ or } \gamma' = V_2 \, (a < r < a')$$

$$\gamma = 0 \quad (r > a')$$

Close to cutoff, $\gamma \to 0$, the LP_{01} mode is always guided for $\delta n' > 0$, but it may be leaky if $\delta n' < 0$. Using the asymptotic forms of the Bessel function, the cutoff condition for higher-order modes when $\delta n' > 0$ is

$$\frac{\hat{J}_l(V_1) - \hat{Y}_l(V_2 c)}{\hat{J}_l(V_1) - \hat{J}_l(V_2 c)} = \frac{J_{l+1}(V_2 c)Y_{l+1}(V_2)}{Y_{l+1}(V_2 c)J_{l-1}(V_2)} \tag{6.14}$$

and if $\delta n' < 0$

$$\frac{\hat{J}_l(V_1) - \hat{K}_l(V_2 c)}{\hat{J}_l(V_1) + \hat{I}_l(V_2 c)} = \frac{I_{l+1}(V_2 c)K_{l-1}(V_2)}{K_{l+1}(V_2 c)I_{l-1}(V_2)} \tag{6.15}$$

Near cutoff the following holds true:

$$\hat{K}_0(V_2 c) \simeq - \ln(V_2 c)$$

$$\hat{J}_0(V_1) \simeq 2V_1^{-2}$$

$$\hat{I}_0(V_2 c) \simeq 2(V_2 c)^{-2}$$

$$c^2 = \frac{I_1(V_2 c)K_1(V_2)}{K_1(V_2 c)I_1(V_2)}$$

Using the approximations in Equation (6.15) for $l = 0$ gives

$$c^2 + \left(\frac{V_1}{V_2}\right)^2 = 1$$

or

$$\frac{a'}{a} = \left(\frac{|\delta n'|}{\delta n}\right)^{-1/2} \tag{6.16}$$

Equation (6.16) is plotted in Figure 6.5. This curve shows the limit between guiding and leaking areas for the LP_{01} mode in the a'/a vs $\delta n'/\delta n$ diagram. Figure 6.6 shows the variation of the cutoff value of V for the first three modes, LP_{01}, LP_{11}, and LP_{02} as a function of $\delta n'/\delta n$ for $a'/a = 2$ and

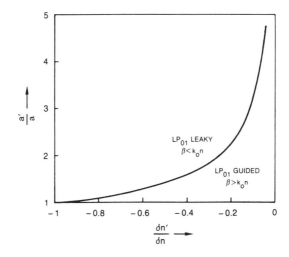

FIGURE 6.5   Cutoff limit for the LP_{01} mode. (Ref. 6.2. Reprinted with permission of IEEE.)

$a'/a = 5$. Figure 6.7 shows V_c for the LP_{01} and LP_{11} modes vs a'/a for various $\delta n'/\delta n$ values ranging from 0.1 to 0.75.

The advantage of using a doubly-cladded fiber is to obtain zero dispersion at wavelengths varying from 1.3 to 1.7 microns. Dispersion-free fibers can be made by choosing fiber parameters in such a way that material dispersion can be designed to cancel waveguide dispersion without intro-

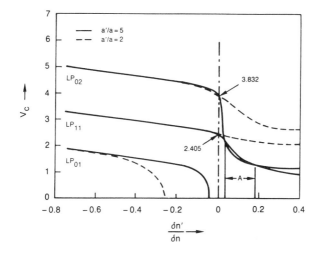

FIGURE 6.6   Normalized frequency at cutoff V_C versus $\delta n'/\delta n$ for $a'/a = 2$ and 5. Note that in Area A for $a'/a = 5$, the LP_{02} mode is guided, whereas the LP_{11} mode is leaky. (Ref. 6.2. Reprinted with permission of IEEE.)

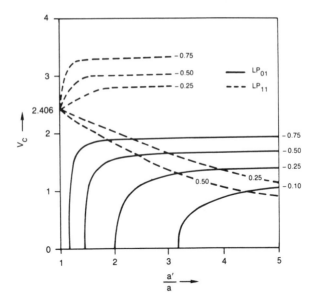

FIGURE 6.7 Normalized frequency at cutoff V_C as a function of a'/a for various $\delta n'/\delta n$ values. (Ref. 6.2. Reprinted with permission of IEEE.)

ducing excessive propagation loss. The group delay τ_g in a fiber as given by Equation (5.10) is

$$\tau_g = \frac{LN}{c}\left[1 + \Delta \frac{d}{dV}(Vb)\right] \qquad (6.17)$$

where N is called the group index and is defined by

$$N \equiv n - \lambda \frac{dn}{d\lambda} \qquad (6.18)$$

and b is the normalized propagation parameter defined by Equation (5.7). For weakly-guided doubly-cladded fibers

$$b = \frac{\beta/k_0 - n}{\delta n + \delta n'} \qquad (6.19)$$

and

$$\Delta = \frac{\delta n - \delta n'}{n} \qquad (6.20)$$

Substituting Equation (6.19) into (6.9) u, u', γ, and γ' can be expressed in terms of b, $\delta n/\delta n'$, and a'/a. From these relationships, the expression for b as a function of V for given values of $\delta n'/\delta n$ and a'/a emerges.

Because all light sources have a finite spectral width, chromatic dispersion of each mode depends on $d\tau_g/d\lambda$ (see Section 5.3). Differentiating Equation (6.17) with respect to λ yields

$$\frac{d\tau_g}{d\lambda} = \frac{L}{c}\frac{dN}{d\lambda}\left[1 + \Delta\frac{d}{dV}(Vb)\right] + \frac{LN}{c}\left[\frac{d}{dV}(Vb)\frac{d\Delta}{d\lambda} + \Delta\frac{dV}{d\lambda}\frac{d^2}{dV^2}(Vb)\right] \quad (6.21)$$

The right-hand side of Equation (6.21) can be regrouped into two terms as

$$\frac{d\tau_g}{d\lambda} = M\left[1 + \Delta\frac{d}{dV}(Vb)\right] - \frac{N\Delta}{\lambda c}\left[V\frac{d^2}{dV^2}(Vb) - p\frac{d}{dV}(Vb)\right] \quad (6.22)$$

where M is the material dispersion and $p = (\lambda/\Delta)\ d\Delta/d\lambda$ is the mean profile dispersion parameter. The values of M and p are usually known, e.g., for Ge doped fibers, $M = 22$ ps/nm/km and $p = 0.1$ at $\lambda = 1.55$ μm. With a doubly-cladded fiber design, it is possible to achieve zero dispersion, especially at longer wavelengths, by making the two terms in Equation (6.22) cancel each other. A single-cladded, step-index fiber cannot provide sufficient waveguide dispersion to compensate for the material dispersion unless the Δ-value increases significantly. In doing so, the fiber becomes very lossy. W-type fiber, on the other hand, can overcome this problem because it has a unique property that can be described qualitatively by using the β-ω diagram shown in Figure 6.8. At large V values, very little difference exists in β between these two types of fibers. As V decreases, the effect of outer

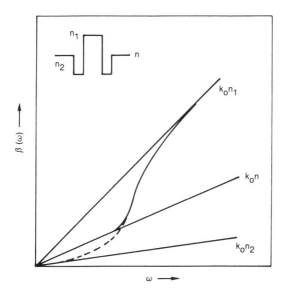

FIGURE 6.8 Behavior of the LP_{01} mode propagation constant in a depressed inner cladding fiber. The dashed curve represents expected behavior in the absence of outer cladding, whereas the solid curve corresponds to actual behavior. It indicates that the dispersion $d^2\beta/d\omega^2$ is strongly increased near the cutoff ($\beta \simeq k_0 n$).

cladding becomes dominant. As a result, the β vs ω curve tends to approach asymptotically the value β = kn, instead of β = kn_2. This behavior introduces a sharp curvature to the $b(V)$ curve, thus producing the stronger waveguide dispersion as shown in Figures 6.9 and 6.10.

Quantitatively, it is necessary to compute the $d(Vb)/dV$ and $Vd^2(Vb)/dV^2$ terms in Equation (6.22). First, $b(V)$ can be computed numerically for given values of $\delta n'/\delta n$ and a'/a. Then, $d(Vb)/dV$ and $Vd^2(Vb)/dV^2$ can be computed from $b(V)$ for three neighboring values of V. In this way, the results shown in Figures 6.9 and 6.10 are fairly accurate for all V values except at cutoff. At cutoff, it is expected that $d(Vb)/dV$ and $d^2(Vb)/dV^2$ for all guided modes approach zero as V approaches V_c. The results of numerical computations for $d(Vb)/dV$ and $d^2(Vb)/dV^2$ are plotted for two different

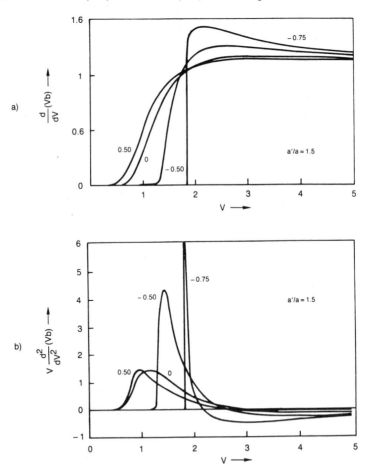

FIGURE 6.9 (a) Dispersion parameters $d(Vb)/dV$ and (b) $V\, d^2(Vb)/dV^2$ for $a'/a = 1.5$ and for various $\delta n'/\delta n$ values. (Ref. 6.2. Reprinted with permission of IEEE.).

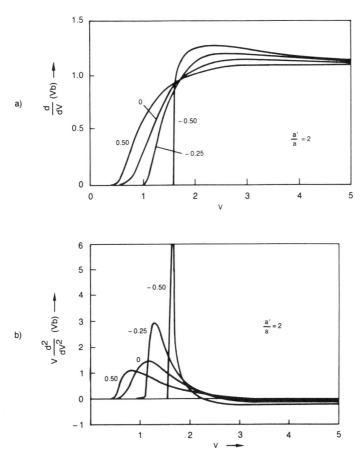

FIGURE 6.10   (a) Dispersion parameters $d(Vb)/dV$ and (b) $V\,d^2(Vb)/dV^2$ for $a'/a = 2$ and for various $\delta n'/\delta n$ values. (Ref. 6.2. Reprinted with permission of IEEE.)

values of a'/a, 1.5 and 2.0, respectively. Each quantity is computed for a set of $\delta n'/\delta n$ values varying from -0.7 to 0.5. Using these results, the three waveguides with the parameters shown in Figure 6.11 can be designed. Figure 6.11(a) shows singly-cladded, step-index fibers having $n_1 - n = 6.5 \times 10^{-3}$ and $2a = 6.5$ μm. Figure 6.11(b) shows a doubly-cladded fiber having $\delta n + \delta n' = 6.3 \times 10^{-3}$, $\delta n' = -6.3 \times 10^{-3}$, $a'/a = 1.5$, and $2a = 6.5$ μm. Figure 6.11(c) also shows a doubly-cladded fiber having $\delta n + \delta n' = 5.8 \times 10^{-3}$, $\delta n' = -1.9 \times 10^{-3}$, $a'/a = 2.0$, and $2a = 5.4$ μm. Also shown in Figure 6.11 is the chromatic dispersion, $d\tau_g/d\lambda$, for these three fibers. The zero dispersion for cases (a), (b), and (c) occurs at $\lambda = 1.3$, 1.55, and 1.7 μm, respectively.

It is possible to shift the zero dispersion for singly-cladded fibers to wavelengths longer than 1.3 μm. To accomplish this shift, it is necessary to

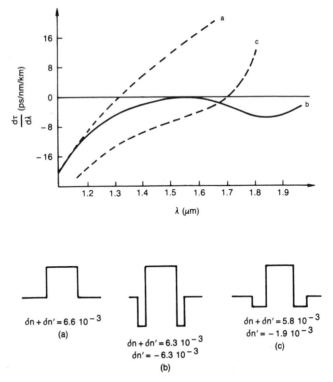

FIGURE 6.11 Three possible types of dispersion-free single mode fiber design: (a) The best singly cladded step-index fiber with a core diameter $2a = 6.5$ μm. (b) A doubly cladded fiber with $2a = 6.5$ μm, and $a'/a = 1.5$. (c) A doubly cladded fiber with $2a = 5.4$ μm and $a'/a = 2.0$. (Ref. 6.2. Reprinted with permission of IEEE.)

increase the index difference Δ between the core and cladding to a very large value (e.g., $\Delta \sim 10^{-2}$ at $\lambda = 1.55$ μm and $\Delta \sim 1.5 \times 10^{-2}$ at $\lambda = 1.7$ μm) for dispersion-free operation. To produce such a high Δ-value, a large amount of *Ge* dopants must be introduced into the core. This process introduces an excessive propagation loss that arises primarily from scattering by the *Ge* dopants residing in the core at a high concentration. With doubly-cladded fiber, it is possible to obtain dispersion-free propagation at long wavelengths by keeping *Ge* concentration at a level, which is considerably lower than that required for dispersion-free, singly-cladded fibers.

6.4 ELLIPTICAL CORE POLARIZATION-PRESERVING FIBERS

Due to either the lack of circular symmetry and/or internal stresses, a circular single mode fiber is usually birefringent. Intrinsic anisotropy can remove the two-fold degeneracy of the HE_{11} mode and consequently produces a group delay difference. The magnitude of internal birefringence is usually very small, however it could be severely magnified by environmental conditions such as tension, twists, and bending experienced after cable installation. Because extrinsic birefringence varies in an unpredictable manner, the polarization state of the output cannot be predetermined. This unknown factor results in an instability of the polarization state of the propagating mode, commonly known as polarization mode dispersion. It can be a serious problem to polarization sensitive sensors and phase coherent optical fiber communication systems. One solution to this problem is to excite the fundamental mode with properly oriented polarization in a birefringent-free, single mode fiber so that the polarization of the mode at the output is preserved. In practice, this situation is almost impossible to realize. Another solution is to maximize the modal birefringence of the fiber by increasing the internal birefringence to a level well above all that could be induced by external conditions.

The common way to produce a polarization-preserving fiber is to start with a single mode fiber having an elliptical core and subsequently subject this fiber to twist and spin. Even though the state of polarization in this fiber may change over a long propagation length and the degree of polarization may also deteriorate to some extent, the output polarization can always be recovered using a phase compensator and a polarizer. The maximum degree of linear polarization achieved after phase compensation, however, is limited by the depolarization process inherent in the fiber. To understand the polarization preserving effect in a strongly birefringent single mode elliptical fiber, the propagation properties of an elliptical fiber under stress are analyzed. These results will show that two eigen-polarization modes exist in this fiber, each of which is independent of the other, therefore, no mode-coupling between these two eigen-modes is expected.

Consider a fiber whose core or refractive index profile is elliptically deformed with a semi-major axis a and semi-minor axis b. The eccentricity e is defined as

$$e^2 = 1 - a^2/b^2 \qquad (6.23)$$

The modes propagating in an elliptical fiber with a step-index profile can be described in a series of Mathieu functions (Ref. 6.3). Difficulties arise when solving the eigenvalue equation involving infinite determinants. If the β value of a mode is required, it is possible to obtain a solution by carefully truncating the infinite determinants at the expense of introducing errors in the process. Similar difficulties are encountered in the numerical approach to solve the eigenvalue equation. The uncertainty in these approximations arise from the fact that $\delta\beta$ is a very small difference between two nearly degenerate modes.

Following a perturbation theory (Ref. 6.4), the expression for the phase difference $\delta\beta$ between the x and y polarization states of the fundamental mode in a weakly-guided elliptical core fiber is given by

$$\delta\beta = \frac{a\sqrt{2\Delta}}{2Vn_0^2 N} \iint \psi \left[\frac{\partial\psi}{\partial y} \frac{\partial n^2}{\partial y} - \frac{\partial\psi}{\partial x} \frac{\partial n^2}{\partial x} \right] dA \tag{6.24}$$

where

$$n^2(r, \theta) = n_0^2[1 - 2\Delta f(r, \theta)] \tag{6.25}$$

The function f in Equation (6.25) is completely arbitrary. In Equation (6.24), the normalization factor N is given by

$$N = \iint \psi^2 \, dA \tag{6.26}$$

where the integral extends over the elliptical cross-section, and ψ is the scalar field in an elliptical fiber. The above approach involves determining the propagating modes of an elliptical fiber in the limits of small ellipticity and weak guiding ($\Delta \ll 1$). In this approximation, the scalar wave equation can be solved in the elliptical coordinates and then the scalar propagation constant can be corrected for the presence of a slight birefringence that splits the state of polarization using first order perturbation theory. The modal field ψ for $1 = 0$ is obtained by using the corresponding field for a circular fiber as a perturbation. The field expression for the HE_{11} mode in the core ($r < a$) is given by (Ref. 6.5)

$$\psi = \frac{J_0(ur/a)}{J_0(u)} - \frac{e^2}{2J_0(u)} \left[gJ_0(ur/a) - \frac{ur}{2a} \frac{K_0^2(\gamma)}{K_1^2(\gamma)} J_1(ur/a) \right.$$

$$\left. - \frac{u^2}{4} \frac{K_0(\gamma)K_2(\gamma)}{K_1^2(\gamma)} J_2(ur/a) \cos 2\theta \right] \tag{6.27}$$

and in the cladding ($r > a$)

$$\psi = \frac{K_0(\gamma r/a)}{K_0(\gamma)} - \frac{e^2}{2K_0(\gamma)} \left[g' K_0(\gamma r/a) - \frac{\gamma r}{2a} \frac{J_0^2(u)}{J_1^2(u)} K_1(\gamma r/a) \right.$$

$$\left. - \frac{u\gamma}{4} \frac{K_0(\gamma)J_2(u)}{K_1(\gamma)J_1(u)} K_2(\gamma r/a) \cos 2\theta \right] \tag{6.28}$$

where g and g' are normalization factors related by

$$g - g' = V^2 J_0(u)/2uJ_1(u) \tag{6.29}$$

Substituting Equations (6.25) to (6.29) into (6.24) yields an explicit expression for the phase difference $\delta\beta$ between the two fundamental modes with orthogonal polarizations as given by $\delta\beta = (e^2/a)(2\Delta)^{3/2} \phi$, where

$$\phi = \frac{u^2\gamma^2}{8V^3} \left[1 + \frac{J_0^2(u)}{J_1^2(u)} \left(1 - \frac{\gamma^2}{u^2} + \frac{\gamma^2}{u} \frac{J_0(u)}{J_1(u)} \right) \right] \tag{6.30}$$

Figure 6.12 is a plot of the normalized phase difference ϕ as a function of V. After propagating over a length L, the group delay difference τ_g defined

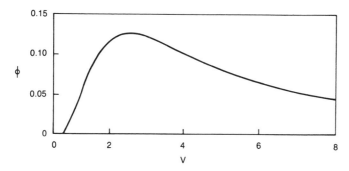

FIGURE 6.12 Normalized phase difference ϕ between the two orthogonal fundamental modes in a slightly elliptical fiber as a function of V. (Ref. 6.5. Reprinted with permission of IEE England.)

by Equation (5.6) is

$$\tau_g = -\frac{L\lambda^2}{2\pi c}\frac{d}{d\lambda}(\delta\beta) \tag{6.31}$$

If material dispersion is neglected, Equation (6.31) can be approximated as

$$\frac{\tau_g}{L} \simeq \frac{na(2\Delta)^2}{c}\frac{d\phi}{dV} \tag{6.32}$$

Figure 6.13 shows normalized group delay difference per unit length $d\phi/dV$ in p-sec/Km as a function of V. Note that a zero dispersion point occurs at $V = 2.47$. At this point, higher-order modes are very close to their cutoffs, and therefore are expected to be very lossy over a long propagation length. It is feasible to obtain dispersion-free operation at $V = 2.47$ where a minimum value for both the material dispersion and elliptical birefringence can

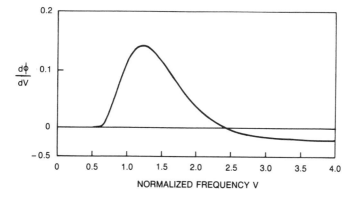

FIGURE 6.13 The normalized group delay difference $d\phi/dV$ as a function of V, showing the zero crossing at $V = 2.47$. (Ref. 6.6. Reprinted with permission of IEEE.)

be achieved simultaneously by restricting the wavelength to 1.3 microns. This operation is not possible for any other wavelength (see Section 5.4).

Figure 6.14 shows the group delay per unit length τ_g/L for fibers with 1% ellipticity and $V = 2.4$ as a function of λ. A family of curves is plotted for various values of Δ or the corresponding values for $2a$. The results of Figure 6.14 show that the delay difference between two polarizations of the HE_{11} mode remains below 0.1 psec/km in the wavelength range that varies from 0.9 to 1.6 microns provided that a core diameter greater than 10 microns is used. If the e value increases, τ_g/L increases as e^2. For a fiber of 100 Km the delay difference remains below 10 psec for $e = 1\%$, but increases rapidly to 1000 psec for $e = 10\%$. Because it is very likely that 10% ellipticity could be induced by, for example, thermal expansion through a residual mismatch between the core and the cladding of a poorly prepared fiber, it is extremely

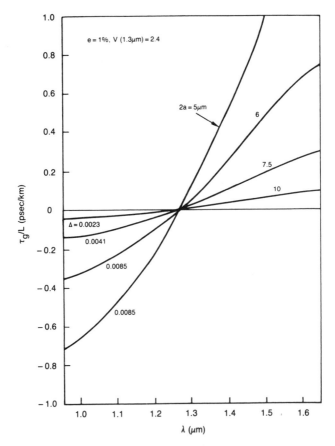

FIGURE 6.14   The group delay time τ_g per unit length as a function of λ for fibers with different core diameters but all having 1% ellipticity and for $V = 2.4$. (Ref. 6.6. Reprinted with permission of IEEE.)

important to use either a very low birefringent fiber with nearly perfect symmetry or a very strong birefringent fiber in which the state of polarization can be preserved.

The polarization preservation characteristics of a single mode fiber under elastic deformations can be understood by examining the dynamic behavior of the propagating modes in a twisted elliptical core fiber. The following will show that two eigen-polarization modes propagate without degradation in the state of polarization provided that a sufficiently large twist rate is maintained. Let E_x and E_y be the two orthogonally polarized field components of the HE_{11} mode in the unperturbed circular fiber. When this fiber is subjected to a strong twist at a twist rate φ_t per unit length, the shearing stress introduces an elastic deformation (ellipticity) that is represented by the anisotropy in the dielectric constant tensor involving photoelastic coefficients. This deformation causes these two orthogonal components to couple along the fiber. The coupled mode equations that govern the polarization evolution are

$$\frac{d}{dz} \begin{pmatrix} E_x \\ E_y \end{pmatrix} = -i \begin{pmatrix} c_{11} & c_{12} \\ c_{21} & c_{22} \end{pmatrix} \begin{pmatrix} E_x \\ E_y \end{pmatrix} \tag{6.33}$$

where

$$c_{11} = \beta + B_e \cos 2\varphi_t z$$

$$c_{12} = B_e \sin 2\varphi_t z - iB_t \tag{6.34}$$

$$c_{21} = c_{12}^*$$

$$c_{22} = \beta - B_e \cos 2\varphi_t z$$

Parameters B_e and B_t are birefringences caused by ellipitical core deformation and twist deformation, respectively. They can be expressed in terms of fiber structures and perturbation parameters as follows:

$$B_e = \frac{enk_0\Delta^2}{2} G(V) \tag{6.35}$$

$$B_t = \sigma p \varphi_t / n \tag{6.36}$$

where $\sigma = E/2(1 + \nu)$ is the modulus of rigidity, and E and ν denote Young's modulus and Poisson's ratio, respectively. p is the photoelastic coefficient. $G(V)$ is the normalized frequency dependence of the coupling coefficient caused by an elliptical core deformation and is given by (Ref. 6.7)

$$G(V) = \frac{\gamma^2}{V^4} \left[u^2 + (u^2 - \gamma^2) \frac{J_0^2(u)}{J_1^2(u)} + u\gamma^2 \frac{J_0^3(u)}{J_1^3(u)} \right] \tag{6.37}$$

Consider a rotating coordinate system (ξ, η) that rotates by an angle $\varphi_t z$, with respect to an (x, y) stationary coordinate system. The transformation relation is

$$\begin{pmatrix} E_x \\ E_y \end{pmatrix} = \begin{pmatrix} \cos \varphi_t z & -\sin \varphi_t z \\ \sin \varphi_t z & \cos \varphi_t z \end{pmatrix} \begin{pmatrix} E_\xi \\ E_\eta \end{pmatrix} \tag{6.38}$$

Substituting Equation (6.38) into (6.33) yields

$$\frac{d}{dz} \begin{pmatrix} E_\xi \\ E_\eta \end{pmatrix} = -i \begin{pmatrix} \beta + B_e & i(\varphi_t - B_t) \\ -i(\varphi_t - B_t) & \beta - B_e \end{pmatrix} \begin{pmatrix} E_\xi \\ E_\eta \end{pmatrix} \tag{6.39}$$

Now assuming

$$E_\xi = A \exp(-i\lambda z) \tag{6.40}$$
$$E_\eta = B \exp(-i\lambda z)$$

and substituting Equation (6.40) into (6.39) two homogeneous equations involving A and B coefficients are obtained. A nontrivial solution is that the determinant of the coefficients be zero. From this condition, two eigenvalues $\lambda_\pm$ are given by

$$\lambda_\pm = \beta \pm [B_e^2 + (\varphi_t - B_t)^2]^{1/2} \tag{6.41}$$

Substituting $\lambda_\pm$ into the homogeneous equations and solving B in terms of A yields the electric component ratio for the two eigen-polarization modes as

$$\frac{E_\eta}{E_\xi} = \frac{B}{A} = i \frac{\varphi_t - B_t}{B_e \pm [B_e^2 + (\varphi_t - B_t)^2]^{1/2}} \tag{6.42}$$

The fact that the ratio of these two components is an imaginary number implies that the principal axis of the polarization ellipse is always pointed to the fiber rotation axis ξ or η for any twist rate. The two eigen-polarization modes have the following properties: The major axes of the polarization ellipses are mutually perpendicular; their ellipticities are identical to each other; and, the end point loci for the electric vectors rotate in opposite directions. These two eigen-polarization mode vectors E_ξ and E_η, represented by the rotation coordinates, satisfy the decoupled mode equations as

$$\frac{d}{dz} \begin{pmatrix} E_\xi \\ E_\eta \end{pmatrix} = -i \begin{pmatrix} \lambda_+ & 0 \\ 0 & \lambda_- \end{pmatrix} \begin{pmatrix} E_\xi \\ E_\eta \end{pmatrix} \tag{6.43}$$

The decoupling of these two modes in these principal axes occurs because the off-diagonal elements in the above coupling matrix vanish. Therefore, a fiber with a sufficiently large twist rate retains a high degree of polarization not susceptible to environmental effects. This dynamic behavior can be described by the eigen-polarization states being circularly polarized for a sufficiently twisted fiber. The mode dispersion between these two circular polarizations approaches zero as the twist rate φ_t is sufficiently greater than the birefringences B_e and B_t of the elliptical core. Any incident polarization state is split into a pertinent combination of two eigen-polarization modes. The relative group delay difference $(L/c)\ d(\delta\lambda)/dk_0$ and power distribution $E_\eta^2(z)/E_\xi^2(z)$ remain at the same values that determine the degree of polarization. The principal axes of the eigen-polarization modes rotate according to the fiber twist angle as stated in Equation (6.42).

6.5 SPLICING AND COUPLING LOSSES

The typical core diameter of a single mode fiber is less than 10 microns. This dimension creates a difficulty in splicing and coupling. To gain a deeper appreciation of this problem, this section analyzes losses caused by the misalignment of two fibers in a splice, where two fibers have different diameters and are tilted or offset with respect to each other. Realizing that the mode profiles of single mode fibers are very nearly Gaussian in shape, the problem is treated using Gaussian beams with which the analysis is greatly simplified. Equations (6.3) and (6.4) can be used to compute the transmission losses between two misaligned Gaussian beams. The power transmission coefficient T for a cylindrical fiber is defined by

$$T = |\ c_0\ |^2 \qquad (6.44)$$

where

$$c_0 = \frac{1}{2P} \int_0^{2\pi} d\theta \int_0^\infty (\mathbf{E} \times \mathbf{H})_z\ rdr \qquad (6.45)$$

Taking the linearly polarized Gaussian beam given by Equation (6.3) as an input to a step-index fiber to excite a guided mode whose field component is described by Equation (6.2), makes it possible to compute T numerically using Equation (6.44). Clearly, the value of T depends on the width parameter ω_0. Equation (6.4) can be used to obtain optimum coupling between a Gaussian input laser beam and a HE_{11} mode of a step-index fiber with a near-unity transmission coefficient ($T \simeq 1$) for V values varying from 1.2 ($T = 0.946$) to 2.8 ($T = 0.998$). At $V = 2.4$, $T = 0.9965$. It has been shown (Ref. 6.1) that for a graded index fiber with a truncated parabolic index profile, the optimum beam width ω_0 is

$$\frac{\omega_0}{a} = \frac{1.405}{V} + \frac{0.23}{V^{3/2}} + \frac{18.01}{V^6} \qquad (6.46)$$

This equation gives the best fit for the width of a Gaussian beam to the mode of a parabolic index fiber.

Three types of splicing defects that commonly occur are shown in Figure 6.15. The power transmission coefficient T has been derived using Gaussian field distributions for all three cases (Ref. 6.8). Only the results

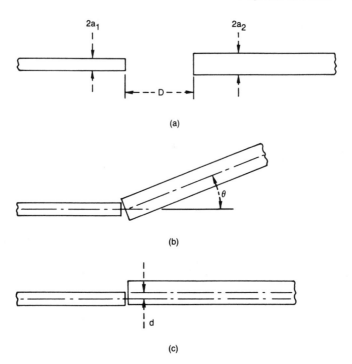

FIGURE 6.15 Three common types of splice imperfections.

are presented here. For splicing two fibers with different sizes and separated by a distance D shown in Figure 6.15(a), the power transmission coefficient is given by

$$T = 4\left(4Z^2 + \frac{\omega_1^2}{\omega_2^2}\right) \bigg/ \left[\left(4Z^2 + \frac{\omega_1^2 + \omega_2^2}{\omega_2^2}\right)^2 + 4Z^2\frac{\omega_2^2}{\omega_1^2}\right] \quad (6.47)$$

where ω_1 and ω_2 are the beam widths belonging to the fibers with radii a_1 and a_2, respectively. Z is the normalized fiber separation distance, and is defined by

$$Z = \frac{D}{n_2 k_0 \omega_1 \omega_2} \quad (6.48)$$

From the above result, if $D = 0$, the power transmission coefficient becomes

$$T_0 = \left(\frac{2\omega_1\omega_2}{\omega_1^2 + \omega_2^2}\right)^2 \quad (6.49)$$

Equation (6.49) shows that if two fibers have the same radius, e.g., $\omega_1 = \omega_2$, then the power transmission coefficient is unity for a perfect alignment.

If $\omega_1 \gg \omega_2$, then

$$T_0 = 4\left(\frac{\omega_2}{\omega_1}\right)^2 \Big/ \left[1 + \left(\frac{\omega_2}{\omega_1}\right)^2\right]^2 \simeq 4\left(\frac{\omega_2}{\omega_1}\right)^2 \tag{6.50}$$

The above equation indicates that transmission is greatly reduced and dependent on the square of the ratio of the beam sizes. If the separation distance D is very large an asymptotic expression for T in the limit that $D \to \infty$ is

$$T_\infty = \frac{1}{Z^2} \tag{6.51}$$

The above equation indicates that the power transmission coefficient decreases as $1/D^2$.

For the fiber splicing at a tilting angle θ shown in Figure 6.15(b), the power transmission coefficient is given by

$$T_\theta = T_0 \exp\left[-\frac{2(\pi n_2 \omega_1 \omega_2 \theta)^2}{(\omega_1^2 + \omega_2^2)\lambda^2}\right] \tag{6.52}$$

Equation (6.52) shows that when $T_\theta = T_0/e$, the tilting θ reaches the value θ_e given by

$$\theta_e = \left(\frac{\omega_1^2 + \omega_2^2}{2}\right)^{1/2} \frac{\lambda}{\pi n_2 \omega_1 \omega_2} \tag{6.53}$$

If two fibers with different sizes are offset by a distance d shown in Figure 6.15(c), the power transmission coefficient is given by

$$T_d = T_0 \exp\left[-\frac{2d^2}{\omega_1^2 + \omega_2^2}\right] \tag{6.54}$$

For a reduction of transmitted power by a factor of $1/e$, an offset distance d_e is

$$d_e = \left(\frac{\omega_1^2 + \omega_2^2}{2}\right)^{1/2} \tag{6.55}$$

Using the above results, a useful relation is obtained for splicing two identical fibers ($\omega_1 = \omega_2$):

$$d_e \theta_e = \frac{\lambda}{\pi n_2} \tag{6.56}$$

Equation (6.56) indicates that as one of the two defects becomes smaller, the other must become larger and vice versa. This relationship is mutually exclusive and is imposed by physical and geometrical conditions. The results of the above analysis indicate that the splicing loss is relatively less critical to longitudinal separation of two fibers than to tilting and/or offset. For example, consider again the case where $\omega_1 = \omega_2$. If $V = 2.4$, $\lambda = 1$ micron, and $\Delta = 2 \times 10^3$, a core radius $a = 4.15$ micron results for $n_2 = 1.457$. Equation (6.4) gives $\omega_1 = 4.56$ micron. For $\theta = 0$, the transmission coefficient is $T_d = 0.368$ if $d = \omega_1$ and $T_d = 0.9$ if $d = 1.5$ micron. For $d = 0$, $T_\theta = 0.368$ if $\theta = 2.7°$ and $T_\theta = 0.9$ if $\theta = 0.9°$.

PROBLEMS

6.1. Assume the wave function $\psi(r)$ takes on the form given by Equation (6.8). Derive the eigenvalue equation for the W-type doubly-cladded fiber ($\beta > k_0 n_2$).

6.2. Plot the index profile of a W-type single mode fiber having $2a = a'$ such that the LP_{01} mode is just below its cutoff.

6.3. Using the results of $d(Vb)/dV$ and $d^2(Vb)/dV^2$ vs V curves, show that at $\lambda = 1.55$ microns a Ge-doped W-type fiber with properly chosen parameters such as $\delta n'/\delta n$ and a'/a can have zero dispersion. In other words, $d\tau_g/d\lambda = 0$; and, the waveguide dispersion compensates for the Ge-doped silica fiber material dispersion.

6.4. Using the definition of ϕ given by Equation (6.30), calculate the normalized group delay difference for a single mode fiber $d\phi/dV$ per unit length. Show that $d\phi/dV = 0$ at $V = 2.47$.

6.5. In a twisted single mode fiber the two lowest-order degenerate modes are coupled and governed by Equation (6.39). Show that by rotating the axes of the ellipse at the angle given by Equation (6.42) it is possible to obtain two decoupled modes governed by Equation (6.43).

6.6. For $V = 2.8$, show that the matching between a Gaussian beam and the HE_{11} mode of step-index fiber is 99.8%.

REFERENCES

6.1. D. Marcuse, *Bell Systems Technical Journal*, 56, 703 (1977).

6.2. M. Monerie, *IEEE J. Quant. Elect. QE-18*, 535 (1982).

6.3. J. E. Lewis and G. Deshpande, *IEE J. Microwaves Opt. & Acoust.*, 3, 147 (1979).

6.4. A. W. Snyder and W. R. Young, *J. Opt. Soc. Am.*, 68, 297 (1978).

6.5. J. D. Love, R. A. Sammut, and A. W. Snyder, *Elect. Lett.*, 15, 615 (1979).

6.6. D. Marcuse and Chinlon Lin, *IEEE J. Quant. Elect.*, QE-17, 869 (1981).

6.7. J. I. Sakai and T. Kimura, *IEEE J. Quant. Elect.*, QE-17, 1041 (1981).

6.8. J. A. Arnaud, *Beam and Fiber Optics*, Academic Press, New York (1974).

7

Glass Materials, Fiber Fabrication, and Characterization Techniques

7.1 INTRODUCTION

Low-loss fibers with $\alpha < 5$ dB/km can now be routinely manufactured by using ultrapure glass materials and advanced fiber drawing techniques. To make good optical fibers, many requirements for the growth of glass materials with varying indices while keeping the fiber defect-free at a long length must be met simultaneously. This chapter first discusses the structure of glass and its physical properties. Methods by which different glass with desired properties can be formed in fibers are then introduced. The chapter includes a section on various measurement techniques for the characterization of fiber index profile, dispersion, and losses.

7.2 GLASS MATERIALS

It is interesting to note that the earth's crust is composed of approximately 62% oxygen and 21% silicon. These two elements are the major constituents of glass, with some minorities such as metal ions either substituting for silicon in the tetrahedral structure or coordinating themselves so as to form voids between the silicon tetrahedra. The silicon tetrahedra are arranged such that the oxygen atoms actually form the most closed packing. The structure of a typical silicate glass is shown in Figure 7.1. The exact composition of glass can vary tremendously, but it must contain predominately

(a)

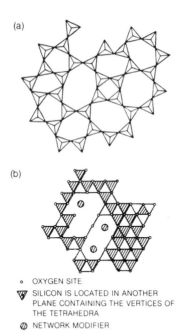

(b)

○ OXYGEN SITE

▽ SILICON IS LOCATED IN ANOTHER PLANE CONTAINING THE VERTICES OF THE TETRAHEDRA

⊘ NETWORK MODIFIER

FIGURE 7.1 (a) Lattice arrangement of SiO₄ tetrahedra in a glass; (b) two-dimensional network of Si—O group, including the effect of additional network modifiers.

oxygen, silicon, boron, sodium, and aluminum, the network-forming atoms listed in Table 7.1. Other oxides (see Table 7.1), called network modifiers, serve to change or modify the basic properties of glass, such as its index of refraction, thermal expansion, absorption coefficient, and melting point.

The strongest bond in glass is the Si—O bond in the silicon tetrahedra. The absence of symmetry in this structure allows the strength of the bond to vary from one tetrahedron to the next. The interatomic distances in each tetrahedron are virtually the same, 1.62 Å. Each silicon atom is tetrahedrally coordinated by four oxygen atoms, and each oxygen atom is bonded to two silicon atoms. If other oxides are added to the silica glass, the total number of oxygen atoms present in the glass is increased. Consequently, some of the oxygen atoms that are bonded to only one silicon atom must pick up additional bonds with other atoms present. For example, the addition of sodium tends to break up the Si—O network to form a sodium silicate glass. As a result, the structure is less tightly bonded and its melting point is lowered. On the other hand, metal ions can diffuse into glass and be distributed randomly among the voids in the network, as shown in Figure 7.1. Other types of glass exist as well; for example, where the network is formed by B_2O_2, Na_2O_3, and so on. These materials are frequently used in industrial products but are not suitable for optical fibers because of their optical quality.

TABLE 7.1   Some of the Most Common Types
of Glass and Their Composition

Network Former	Network Modifier
SiO_2	K_2O
B_2O_3	MgO
Al_2O_3	CaO
Na_2O_3	PbO

The distinctive property of all glass is that it undergoes a continuous decrease in viscosity when heated. Therefore, glass softens gradually instead of going through an abrupt melting stage as encountered in crystals. This property is, of course, unique to an amorphous solid. If a glass has been stressed, some type of preferred orientation effect takes place in the immediate vicinity of the induced strains and local anisotropy results. Therefore, the index of refraction depends on the previous thermal history of the glass.

The refractive index and the thermal expansion coefficient can also vary with the composition of a glass. These differences in makeup are the most common ways to control the differences in refractive indices between the core and cladding. By varying the concentration of the modifiers in a silicate network, the desired Δ can be obtained along an isothermal expansion curve. This variation is usually done with two additional oxides. For example, the sodium calcium silicate group is one such system. Another interesting system is the sodium borosilicate group, because it not only has the freedom to modify the index while holding the expansion coefficient constant, but also has a relatively low softening point, a matter of great importance for fiber manufacturing. However, great care must be taken in selecting the composition for this system because a relatively wide region exists in the phase diagram in which the glass product is found to be unstable and has a tendency to separate into two different glass groups, one with a silica network and the other with a boric oxide network. For a very large refractive index difference, the lead silicate group is the system to use. However, losses in lead silicate fibers have been found to be much greater than those in other silicate glass.

Of course, many possible combinations form glass. The groups mentioned above can be formed by the conventional melting technique of oxide powders. In this way, they can be manufactured in large quantities at relatively low cost. For high-purity, low-loss glass, material preparation must be modified from that of the melting method. The most successful technique has been Chemical Vapor Deposition (CVD), which is discussed in the following section.

7.3 PREFORM PRODUCTION

Many techniques are used for making glass fibers. For example, glass fibers can be drawn directly from the melt of the oxide powders in a crucible. This method is relatively simple and economical for making glass fibers, and in this case it is not necessary to prepare glass preforms. However, to produce fibers with extremely low-loss, a two-step process is often used: first, the production of an ultrapure glass preform, and second, drawing fibers from this preform. Preforms consist of glass rods with the desired index profile. Many techniques have been developed to make preforms. The most attractive ones are those that provide for rapid growth of solid layers of glass with high degrees of purity. The difficulty still exists when making a homogeneous preform of long length with a desired graded-index profile. The fundamental limitation is the finite length of the preform, which limits the total length of the fiber that can be drawn.

The Corning Glass Works used a simple technique to produce the first remarkably low-loss (~20 dB/km) fibers in the late 1960s. This technique, commonly known as the "soot" process, involved hydrolyzing a mixture of $SiCl_4$ and O_2 with an additive of either $TiCl_4$ or $GeCl_4$ vapors to produce a soot of either Ti-doped or Ge-doped SiO_2 material, deposited on pure SiO_2 glass. Because the refractive index of these doped materials was higher than that of pure silica, this deposition technique was also utilized to make the preform material for drawing optical fibers. The process involved proper injection of a stream of doped SiO_2 particles deposited on a pure silica tube. A layer of soot was formed on the inner surface of the tube. After accumulating a sufficient thickness, the tube was heated and collapsed to form a preform rod that had a core with a higher refractive index and a cladding with a lower index.

Since this early experiment by the Corning Glass Works, many modifications and improvements have been introduced by using a variety of dopants, forming the soots on the outside surface of a removable mandrel, and so on. It has been shown that using this soot process, preforms can be made at very rapid deposition rates and at reasonably high purity levels. This process can also be used to produce graded-index fibers with well-controlled profiles. In fact, all CVD processes offer this capability. CVD is a process in which chemical reactions take place at a relatively lower temperature and deposition is initiated from a heated surface. It is a very reliable process because pressure, flow rate, and temperature can be controlled very accurately. A typical CVD system is shown in Figure 7.2. A fused quartz tube is placed in an oven with differential temperature zoning. As the gas mixture flows into the heated region, reaction with the heated surface occurs and a glass layer is deposited at a rate governed by these parameters. A simpler system, which is frequently used, consists of rotating a tube heated by a multiple-burner torch. If the temperature is very high, rotating the tube

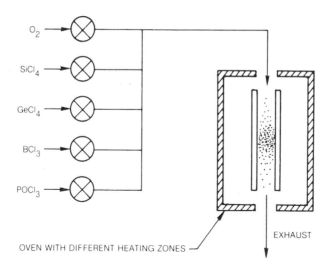

O₂

SiCl₄

GeCl₄

BCl₃

POCl₃

OVEN WITH DIFFERENT HEATING ZONES

EXHAUST

FIGURE 7.2 Schematic diagram of a CVD system for the growth of glass preforms.

is often needed to prevent the tube from sagging. At high temperatures, the distinction between the soot and CVD processes becomes less clear.

The products of chemical reactions for the mixtures described above are given in Table 7.2, together with a comparison of refractive indices. The exact value for the refractive index depends on the thermal history of the glass, because it undergoes an anomalous change in refractive index as a result of quenching. Because borosilicate glass has the lowest refractive index, it is often used as the cladding material. Table 7.3 shows a group of doped silica core fibers with borosilicate glass as the cladding material. Also shown in Table 7.3 are some typical values of Δn between the core and cladding. The exact Δn values for these fibers depends not only on their thermal history but also on the mole concentration of the doping.

The transmission loss of fibers in the near infrared is due primarily to OH absorption. The most significant reduction of OH ions can be accomplished by the preconsolidation treatment of the CVD-produced preform, using, for example, $SOCl_2$ at temperatures up to 1450°C. The dehydration effect of heat treatment of the preform in $SOCl_2$ vapor over a long period

TABLE 7.2   Glass Networks and Their Mixtures

Mixture	Network	Refractive Index
$SiCl_4$, O_2	SiO_2	n_0
$GeCl_4$, O_2	GeO_2	$n > n_0$
$POCl_3$, O_2	P_2O_5	$n > n_0$
BCl_3, O_2	B_2O_3	$n < n_0$

TABLE 7.3   Doped Silica Glass Fibers Using Borosilicate
Glass as Cladding

Core		Cladding		
Dopant	Network	Dopant	Network	Δn (%)
P_2O_5	SiO_2	B_2O_3	SiO_2	0.8
GeO_2	SiO_2	B_2O_3	SiO_2	1.2
GeO_2, B_2O_3	SiO_2	B_2O_3	SiO_2	1.3

of time (~5 h) reduces the OH ion content from a level of 30 ppm to below 0.1 ppm. Figure 7.3 shows a typical reduction of OH content as a function of temperature using $SOCl_2$ treatment. The OH residual content decreases rapidly to a level of 0.3 ppm in the vicinity of 700°C. Further reduction in the OH content occurs only very gradually with increasing temperature. A minimum level of OH residual of about 7 parts per billion has been obtained. This level of OH residual corresponds to an absorption loss of 0.45 dB/km at 1.39 μm.

To produce a graded-index preform with a smooth profile, it is necessary to deposit several hundred layers of doped silicate glass with slightly different mole concentrations. This requirement makes the CVD a rather tedious process to follow. Several methods have been introduced to assist the CVD process for rapid growth of a large number of layers. One of them, which makes use of plasma-augmented CVD to increase the reaction rate in the hot zone is described here. In a plasma-augmented CVD, an inert gas (Ar) is excited by an inductive radio-frequency (RF) circuit. The discharge must be sustained under the deposition environment. The introduction of endothermic materials such as oxygen and other oxides tends to quench the plasma and therefore requires higher sustaining RF power. Plasma-

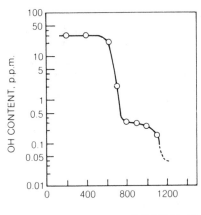

FIGURE 7.3   OH ion content as a function of dehydration temperature. [After T. Edahiro, M. Kawachi, S. Sudo, and H. Takata, *Electron. Lett.*, *15*, 482 (1979).]

augmented CVD cannot only provide a very rapid deposition rate, but can also initiate growth at relatively low temperature and eliminates the possibility of tube deformation.

To produce very long fibers, a method called vapor-phase axial deposition (VAD) has been introduced and has gained considerable popularity. This VAD method has been shown to offer an important advantage by avoiding the formation of cracks due to thermal mismatch between the core and cladding, a problem that often occurs in the conventional CVD process. Figure 7.4 illustrates schematically the essential features of the VAD method. Gaseous mixtures such as $SiCl_4$, $GeCl_4$, $POCl_3$, and O_2 are fed into an oxygen–hydrogen burner, which produces a stream of glass soot resulting from the flame hydrolysis. This stream of fine glass particles is directed toward one end of the starting rod, at which a porous glass rod is grown in the axial direction. The starting rod is rotated about its axis and moved upward at a speed consistent with the growth rate of the preform. Glass

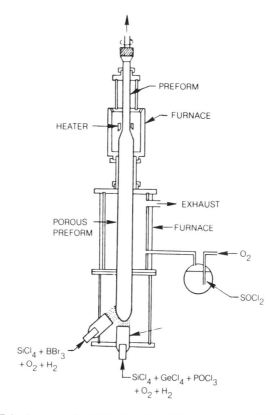

FIGURE 7.4   Apparatus for fabrication of low-OH-content optical fibers by VAD method. [After S. Sudo, M. Kawachi, T. Edahiro, T. Izawa, T. Shioda, and H. Gotok, *Electron., Lett., 14,* 534 (1978).]

particles of lower refractive index are deposited on the porous glass rod from another oxygen-hydrogen burner to form the cladding. The porous glass rod is then vitrified to a bubble-free transparent fiber preform in an electric resistance furnace at temperatures of approximately 1650°C. By varying the mixtures and adjusting the flow rates, it is possible to create a graded-index profile; however, the reproducibility and control of the index profile is more difficult to obtain by this method than by plasma-augmented CVD.

The reduction of OH ions is accomplished by using $SOCl_2$ gas to create a dehydration reaction with the OH ions and H_2O molecules contained in the porous glass rod. The dehydration process takes place in an electric furnace at a temperature of about 800°C. Under these conditions, OH ions and H_2O molecules diffuse to the surface and react with $SOCl_2$ to form HCl and SO_2 gases (dehydration process). The dehydrated preform produced by this method has been used to produce 20-km-long fibers with a measured loss coefficient of ~1 dB/km at 1.2 μm. The fiber had a 60-μm core diameter, a 150-μm cladding diameter, and a Δ value of 0.0014.

7.4 FIBER FABRICATION

Fibers can be drawn either from a preform or directly from melts of oxide powders. From the manufacturing cost point of view, it is more economical to produce high-purity preforms than to purify powders and maintain cleanliness. But at large production rates, manufacturers face the problem of frequent replacement of preforms, which can lead to a certain production waste and high manufacturing cost resulting from repeated shutdown and startup cycles. If the specification as to optical loss can be relaxed, a continuous process that offers very high output could easily be carried out using the double-crucible technique. A double-crucible configuration for fiber pulling from melts is shown in Figure 7.5. Two concentric crucibles, each with a specially designed nozzle, are configured with their axes in the vertical direction. The inner crucible is filled with a composite core material and the outer crucible is filled with cladding material. As molten glass flows through these nozzles under Poiseuille flow conditions, the ratio of core radius to cladding radius is determined by the simple expression

$$\frac{a_{core}}{a_{clad}} = \sqrt{\frac{Q_{core}}{Q_{clad}}} \tag{7.1}$$

The quantity Q in Equation (7.1) represents the volumetric flow and is given by

$$Q = \frac{\pi P r^4}{8\eta l} \tag{7.2}$$

where P is the pressure difference across the nozzle, η the viscosity, and r and l the radius and length of the nozzle, respectively.

With a double-crucible apparatus, fiber drawing can be very simple if powder materials can be fed directly into the crucibles. However, because the production of crude glass from powders usually requires several stages of processing, including melting, mixing, oxidation, or reduction, it is very difficult to obtain homogeneous and high-purity glass fibers directly from powders. Instead, premelted glass must be poured into the crucibles. An alternative method is to fill the crucible with small pieces of glass and subsequently melt the glass slowly, allowing plenty of time for gas bubbles to escape.

This double-crucible method has also been used to produce graded-index fibers. It is accomplished by selecting a glass pair that allows interdiffusion to occur. Thallium, in particular, is a suitable dopant for the core because it can greatly increase the refractive index and is easily diffusible. Diffusion occurs as soon as the core glass enters the molten cladding glass

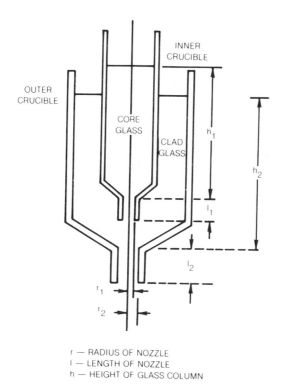

r — RADIUS OF NOZZLE
l — LENGTH OF NOZZLE
h — HEIGHT OF GLASS COLUMN

FIGURE 7.5 Cross-sectional view of a double crucible showing parameters important for controlling the dimension of the core and the cladding of a fiber.

in the vicinity of the nozzles, where both the core and cladding flow together during production. Using the simple diffusion equation with radial symmetry, the concentration of the diffusible species can be written as

$$N\left(\frac{r}{a}\right) = N_0 \int_0^\infty \exp\left(-\frac{Dtu^2}{a^2}\right) J_0\left(\frac{u}{a}r\right) J_1(ur)\, du \qquad (7.3)$$

where N_0 is the initial concentration, D the diffusion coefficient, t the transit time through the nozzle region, and a the radius of the core. Figure 7.6 is a plot of normalized concentration as a function of r/a for different values of the diffusion parameter Dt/a^2. Normalized concentration is directly related to the graded-index profile. The diffusion parameter is related to the volumetic flow rate Q_{core} by the expression

$$\frac{Dt}{a^2} = \frac{D\pi l}{Q_{core}} \qquad (7.4)$$

The model above is an oversimplification of the actual situation. Fortunately, the measured profile for a diffused double-crucible fiber matches very closely the α profile for an α value of approximately 2.4.

Drawing fiber directly from a preform is a simple and straightforward process. The index profile is dictated primarily by that of the preform, as discussed in Section 7.3. To obtain a uniform fiber, both the pulling speed and feeding speed of the preform must be controlled. The equation of continuity requires that

$$A_f V_f = A_p V_p \qquad (7.5)$$

where A_f and A_p, and V_f and V_p are the area and velocity of the fiber and the preform, respectively. A schematic for drawing fibers from a preform

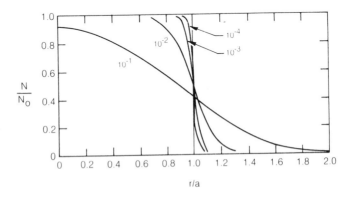

FIGURE 7.6 Index profiles calculated by a diffusion model for various values of Dt/a^2. [After K. B. Chan, P. J. B. Clarricoats, R. B. Dyott, G. R. News, and M. A. Sarva, *Electron. Lett.*, *6*, 748 (1970).]

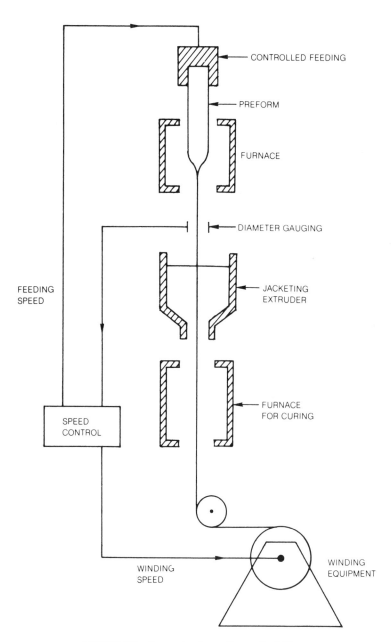

FIGURE 7.7 Apparatus for pulling optical fibers.

is shown in Figure 7.7. A detailed balance of Equation (7.5) can be accomplished by careful inspection of the fiber diameter with a sensing device as shown in Figure 7.7. A change in fiber diameter can be corrected by changing either the winding or the feeding speed, or both. It is assumed that a stable condition is maintained during the drawing process.

7.5 OPTICAL FIBER COUPLING

Before discussing various measurement techniques, it is essential to gain a good understanding of the power-transfer mechanism between a source and a fiber or between two fibers. The parameters involved in fiber coupling are the surface area, the emitting angle, the field of view, the numerical aperture of the emitter, and the receiver. If a lens is used, the parameters of the lens, such as the f-number and the magnification factor, must also be taken into account. For an emitting surface A_s, the intensity distribution is defined as

$$I = \int_{A_s} B \cos \theta \, dS \qquad (7.6)$$

where B is the radiance of the source and θ is the angle measured from the normal of the element dS. For most sources, B can be approximated by the expression

$$B(\theta) = B_0 \cos^n \theta \qquad (7.7)$$

where $n \geq 1$. In the case of a LED, $n \simeq 1$ and the distribution is commonly called Lambertian. A direct power transfer from A_s to a receiver surface A_r, as shown in Figure 7.8, can be computed by carrying out a double surface integral, as given by

$$P = \int_{A_r} \int_{A_s} B(\theta) \, dS \cos \theta_r \, dS_r / R^2 \qquad (7.8)$$

The geometric relationships as shown in Figure 7.8, yield

$$\cos \theta_r \, dS_r \simeq 2\pi R^2 \sin \theta_s \, d\theta_s$$

Therefore, a simplified expression for the power emitted from a source having a small emitting surface area A_s as compared with the magnitude of R and brightness B in Watts per cm²-sterad can be expressed as follows:

$$P_s = 2\pi A_s \int_0^{\theta_s} B(\theta) \sin \theta \, d\theta \qquad (7.9)$$

where θ_s is the emitting angle of the source. For an LED, $B(\theta) \simeq B_0 \cos \theta$ and the emitting angle θ_s is not restricted. The output power from this source

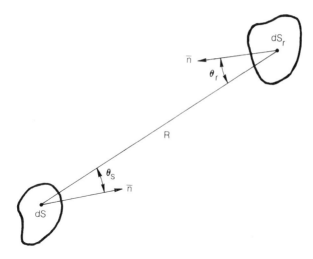

FIGURE 7.8 Geometric relationships between the element of an emitting surface and the element of a receiving surface.

emitted into the entire hemisphere has a simple form as given by

$$P_s = 2\pi A_s B_0 \int_0^{\pi/2} \cos\theta \sin\theta\, d\theta = \pi A_s B_0$$

The amount of power that can be coupled into a fiber or a fiber bundle from this source, in the case of a butt joint, is

$$P_f = 2\pi A_s f_p \int_0^{\theta_{NA}} B_0 \cos\theta \sin\theta\, d\theta$$

$$= \pi A_s f_p B_0 (1 - \cos^2\theta_{NA}) = P_s f_p (1 - \cos^2\theta_{NA})$$

where f_p is the packing fraction of the bundle. If θ_{NA} is small, $1 - \cos^2\theta_{NA} \approx \theta_{NA}^2$. Therefore, the power that the fiber can collect with a butt joint is simply proportional to the square of the numerical aperture of the fiber.

When a lens is used, the total power collected by the lens within the solid angle subtended by the lens (see Figure 7.9) can also be obtained from Equation (7.9) for a source having a constant radiance B_0, as given by:

$$P_{lens} = 2\pi B_0 A_s (1 - \cos\theta_l) \tag{7.10}$$

where θ_l is the collection angle of the lens and can be obtained by using the geometric relationships shown in Figure 7.9, as given by:

$$\theta_l = \tan^{-1}\left[\frac{M}{2(M+1)f^\#}\left(1 + \frac{d_s}{d_l}\right)\right] \tag{7.11}$$

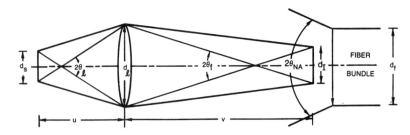

FIGURE 7.9 Geometric relationships for coupling between an emitter and a fiber by using a thin lens.

In Equation (7.11), the quantities M, $f^\#$, d_s and d_l are defined as:

$$M = \text{magnification power} \equiv \frac{v}{u} = \frac{d_I}{d_s}$$

$$d_l = \text{diameter of the lens}$$

$$d_s = \text{diameter of the source}$$

$$d_I = \text{diameter of the image}$$

$$f^\# = f\text{-number} \equiv \text{focal length}/d_l$$

The power emitted by the source into the full hemisphere is $P_T = 2\pi A_s B_0$. The power emitted by the source into a cone defined by $2\theta_s$ (see Figure 7.9) is

$$P_s = 2\pi A_s B_0 (1 - \cos \theta_s) \tag{7.12}$$

Substituting Equation (7.12) into (7.10) expresses the power collected by a lens in terms of the source power as

$$P_{\text{lens}} = P_{\text{source}} \frac{1 - \cos \theta_l}{1 - \cos \theta_s} \tag{7.13}$$

The total power received by the fiber through a lens can also be obtained by taking into account the mismatch between the angles of focusing θ_f and the numerical aperture of the fiber. It can be shown that

$$P_{\text{fiber}} = P_{\text{source}} \left(\frac{1 - \cos \theta_l}{1 - \cos \theta_s} \right) \left(\frac{1 - \cos \theta_{\text{NA}}}{1 - \cos \theta_f} \right) \tag{7.14}$$

where θ_f is also a property of the lens, and can also be expressed in terms of M, $f^\#$, d_s, and d_l by the same geometric relation as that obtained for θ_l:

$$\theta_f = \tan^{-1} \frac{1 + d_s/d_l}{2(M + 1)f^\#} \tag{7.15}$$

In the small-angle approximation, Equation (7.14) becomes

$$P_{\text{fiber}} \simeq P_{\text{source}} \left(\frac{\theta_{NA}}{\theta_s}\right)^2 \left(\frac{d_f}{d_s}\right)^2 \tag{7.16}$$

Equation (7.16) shows that optimum coupling occurs when $\theta_s = M\theta_{NA}$ and $d_f = Md_s$. The only advantage of using a lens is to provide proper matching of the angles and apertures. In most cases $\theta_{NA} < \theta_s$; therefore, an emitter with as small an aperture as possible must be selected to satisfy simultaneously all the requirements noted above.

The geometric optics treatment is valid only for the coupling of a LED to multimode fibers. Optical coupling to single mode fibers must be treated within the confines of elecromagnetic wave theory. Because single mode fibers only support one mode (HE_{11}), the radiation that is coupled into this fiber must be field-matched to this mode. Optical coupling between two single mode fibers has been treated in Chapter 6. If the LED source is partially coherent, it may be necessary to introduce a set of coherent modes to represent the source. The following treatment considers that the LED is an incoherent source. Let the source oriented perpendicular to the z-axis define the propagation axis of the fiber. The radiance of the source is of the form

$$B(x, y, \theta, \phi) = \frac{d^2P}{dA\ d\Omega} = B_0 N(x, y)F(\theta, \phi) \tag{7.17}$$

where $N(x, y)$ and $F(\theta, \phi)$ are the spatial and angular dependences of the source intensity with a maximum brightness B_0. Each unit area dA on the surface of the LED represents an emitting antenna. The power emitted by this source is given by

$$P_{\text{source}} = \iint B_0 N(x, y)F(\theta, \phi)\ dA\ d\Omega \tag{7.18}$$

where the double integrals extend over the full hemisphere and also the entire surface area of the source A_s. The power coupled to the fundamental mode from the LED is obtained by integrating the angular dependence over a solid angle subtended by the mode field, which can be approximated by a Gaussian beam profile as given by (Ref. 7.1)

$$P_{\text{fiber}} = T \frac{2\pi^2 n_0}{k_0^2 N_0} \left(\frac{\epsilon_0}{\mu_0}\right)^{1/2} \iint_{-\infty}^{\infty} dxdy\ N(x - x_0, y - y_0)E_T^2(x, y) \tag{7.19}$$

where $T = 4n_0/(1 + n_0)^2$ is the Fresnel transmission coefficient, (x_0, y_0) is the point at which the LED is centered, E_T and N_0 are related to the Gaussian field distribution as given by

$$E_T = \exp[-(x^2 + y^2)/r_0^2] \tag{7.20}$$

and

$$N_0 = \frac{\pi r_0^2 n_0}{2} \left(\frac{\epsilon_0}{\mu_0}\right)^{1/2} \tag{7.21}$$

where r_0 is the $1/e$ Gaussian mode radius. Hence, the coupling efficiency of the LED to the single mode fiber is

$$\eta = \frac{P_{\text{fiber}}}{P_{\text{source}}} \tag{7.22}$$

Consider a surface-emitting LED butt-coupled to a single mode fiber centered on the LED surface. The far-field radiation pattern is of the form $\cos^m \theta$, where m is a number close to unity. In this case

$$P_{\text{in}} = TB_0\lambda_0^2 \tag{7.23}$$

and

$$P_s = \frac{2\pi B_0 A_s}{m + 1} \tag{7.24}$$

Hence,

$$\eta = T\frac{m + 1}{2}\frac{\lambda_0^2}{\pi A_s} \tag{7.25}$$

where $m = 1$ and A_s is the area of the emitting surface of the LED.

Next, consider an edge-emitting LED. The spatial and angular dependences of the source can be approximated by

$$N = \exp\left[-\left(\frac{2x^2}{L^2} + \frac{2y^2}{W^2}\right)\right] \tag{7.26}$$

and

$$F(\theta, \phi) = \frac{\cos^{\mu + \nu}\theta}{\cos^\mu\theta\cos^2\phi + \cos^\nu\theta\sin^2\phi} \tag{7.27}$$

where μ and ν are the transverse and lateral power distribution coefficients, and $2W$ and $2L$ are the transverse and lateral $1/e^2$ full widths of the near-field, respectively. The fiber located at (x_0, y_0) centered on the LED junction plane, which is perpendicular to the y-axis. In this case

$$P_{\text{source}} = \frac{2\pi^2}{\mu + \nu + 2}B_0LW \tag{7.28}$$

$$P_{\text{fiber}} = TB_0LW\frac{\lambda_0^2 S_x S_y}{[(2r_0^2 + L^2)(2r_0^2 + W^2)]^{1/2}} \tag{7.29}$$

where

$$S_x = \exp\left[-2x_0^2/(2r_0^2 + L^2)\right] \tag{7.30}$$

$$S_y = \exp\left[-2y_0^2/(2r_0^2 + W^2)\right] \tag{7.31}$$

For the edge-emitting LED, the coupling efficiency can be improved with the help of a lens. Lens-assisted coupling is not useful for the surface-emitting LED because the dimensions of a typical surface-emitting LED are much greater than r_0.
As examples, consider the following cases:

1. For an edge-emitting LED operating at $\lambda_0 = 1.3$ microns, the $1/e^2$ full width in the parallel direction of the junction plane is $2L = 22$ microns; in the perpendicular direction, it is $2W = 0.95$ microns. The transverse and lateral power distribution coefficients, μ and ν, determined from far-field measurements, are 7 and 1, respectively. This source is coupled into a single mode fiber. The calculated coupling efficiency $C = 10 \log \eta = -18.6$ dB.

2. For a surface-emitting LED operating at $\lambda_0 = 1.3$ microns, having an effective aperture of 907 microns^2, the calculated coupling efficiency of this source to the same fiber used in case (1) is -34.8 dB, which represents the most significant loss-penalty to single-mode fiber communication systems choosing LEDs for light sources.

7.6 INDEX PROFILE MEASUREMENTS

The exact profile of either a step-index or a graded-index fiber is a very important parameter for determining the optical transmission properties of the fiber. Several methods for measuring index profile by either direct or indirect means exist. Direct methods involve measurements of either the reflected or transmitted optical power in the near and far fields. A more precise, but rather tedious method, is to measure fringes generated by the phase difference between the incident and refracted light from a fiber sample having polished planoparallel facets with a phase-contrast microscope. The difficulty of the latter technique is that very thin slices of fiber samples with planoparallel surfaces must be prepared at thicknesses no more than 50 μm. Attempts have been made to fabricate these samples by cladding the fiber in a glass tube filled with cement. A thin slice is cut from this rod and polished to optically flat surfaces. Another method for measuring the index profile involves dispersion measurements of time-resolved impulse response through a long fiber. This type of measurement not only gives information about the index profile, but also, and more important, the ultimate information-carrying capability or the bandwidth of the fiber. This section presents a detailed discussion of only one of the simplest index-measuring techniques and leaves pulse dispersion measurements to the next section.

The apparatus and experimental arrangement for measurements of the near-field power profile are shown in Figure 7.10. A LED can be used as the source, which provides uniform illumination across the face of the fiber using a microscope objective lens. The fiber sample is prepared with a fixed length of approximately 1 m. Two ends of the fiber must be optically polished. The detector with a pinhole is mounted on an X-Y scanner and used to trace out the near-field output profile.

The relation between the index profile of a graded-index fiber and the power distribution in the near-field can easily be established if the effects of nonuniform excitation, loss differences, and coupling among modes are ignored. If the fiber is uniformly illuminated, the power per unit solid angle at any point in the cross-section is constant. For a step-index fiber, the numerical aperture is given by $(n_0^2 - n_c^2)^{1/2}$. For a graded-index fiber, a local numerical aperture $A(r)$ at a point r in the cross-section is defined as

$$A(r) = [n^2(r) - n_c^2]^{1/2} \qquad (7.32)$$

The ratio of the power at a point r to the average power P_0 received by a fiber having a constant numerical aperture A_0 can be expressed as

$$\frac{P(r)}{P_0} = \frac{A^2(r)}{A_0^2} = \frac{n^2(r) - n_c^2}{n_0^2 - n_c^2} \qquad (7.33)$$

If the fiber length is very short, it is reasonable to assume that all modes are attenuated equally and propagating through the fiber without coupling. Therefore, the same power distribution should hold for transmitted power.

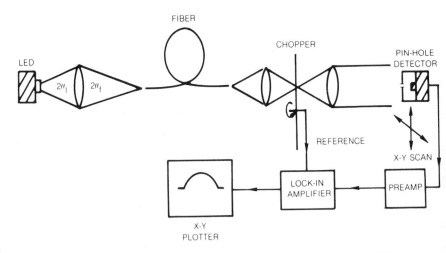

FIGURE 7.10   Experimental arrangement for measurements of the near-field output power as a function of normalized core radius.

Substituting Equation (1.2) into (7.33) for $n(r)$ yields

$$P(r) = P_0 \left[1 - \left(\frac{r}{a}\right)^x \right] \tag{7.34}$$

The power profile described by Equation (7.34) resembles the index profile and is exactly the same for small Δ values. Figure 7.11 is a plot of the near-field power distribution for various values of x.

In practice, it is necessary to correct the near-field profile to account for the contribution from leaky modes because they are usually present in graded-index fibers, and they can introduce errors in the profile measurement. The output power expression, therefore, must include a correction factor $C(r, z)$ due to leaky modes:

$$\frac{P(r)}{P_0} = \frac{n^2(r) - n_c^2}{n_0^2 - n_c^2} \, C(r, z) \tag{7.35}$$

where

$$C(r, z) = 1 + \frac{4}{\pi[n^2(r) - n_c^2]} \int_0^{\pi/2} d\phi \int_0^{\pi/2} \cos\theta \sin\theta \, d\theta \, \exp\left(-\frac{\alpha}{a} z\right) \tag{7.36}$$

is obtained from the following considerations: The second term on the right-hand side of Equation (7.36) represents the contribution due to leaky modes. The factor $\cos\theta$ represents the angular distribution of a Lambertian source, and the factor $\exp(-\alpha z/a)$ represents a leakage of power with an attenuation coefficient $\alpha(z)$ from the end face of the fiber. The integration is taken over the emission angle θ and the projection angle ϕ.

To convert Equation (7.36) into mode description, the following

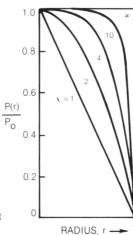

FIGURE 7.11 Plot of near-field output power profiles for various values of a.

changes of variables are made. Snell's law yields

$$\sin\theta = n(r)\sin\theta_g$$

or

$$\cos\theta\, d\theta = n(r)\cos\theta_g\, d\theta_g$$

where θ_g is the corresponding angle inside the guide. Now using the relations, e.g., Equation (4.120) among the components of the k vector:

$$\cos\theta_g = \frac{\beta}{k_0 n(r)}$$

and

$$\sin\phi = \frac{l/r}{\sqrt{k_0^2 n^2(r) - \beta^2}}$$

Substituting these changes of variables into Equation (7.36), the correction factor can be written in the form

$$C(r, z) = 1$$
$$+ \frac{4}{\pi[n^2(r) - n_c^2]} \int_0^l \frac{dl}{a^2 k_0^3 r} \int_{u_l}^{u_u} \frac{\exp(-\alpha z/a)u\, du}{\sqrt{n^2(r) - n_c^2 + u^2/k_0^2 a^3 - l^3/k_0^2 r^2}} \quad (7.37)$$

where

$$u^2 = a^2[k_0^2 n^2(0) - \beta^2], \qquad V^2 = a^2 k_0^2[n^2(0) - n_c^2]$$

and the upper limit $u_u = \sqrt{(V^2 + l^2)}$ if the separation between guided modes and leaky modes occurs at $r = a$. The lower limit u_l, for a parabolic profile, is

$$u_l = \sqrt{V^2\left(\frac{r}{a}\right)^2 + \left(\frac{a}{r}\right)^2 l^2} \quad (r > 0) \quad (7.38)$$

where l values for leaky modes can vary from 0 to V. The minimum value for u_l is obviously $V(l = 0)$. For small values of l, u_l values increase rapidly near the center of the core. To carry out the integral numerically, a value for r must be chosen at which the correction factor is required. Then, u_l must be computed for each value of l.

Before performing numerical integration, the normalized attenuation coefficient α/a must be calculated. The calculation has been carried out by Adams et al. (Ref. 7.2) using the result of WKB approximation. This method is not only very accurate but offers some physical insight into the problem. Figure 7.12 defines the region that separates the oscillatory and evanescent fields. The oscillatory fields, which represent guided modes, are bound within caustic points r_1 and r_2. The evanescent fields, which represent leaky

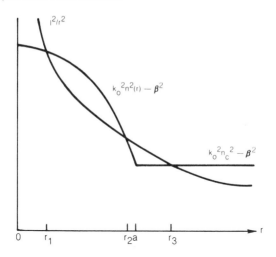

FIGURE 7.12 Graphical solutions of the points r_1, r_2, and r_3 that define the regions for guided and evanescent fields.

modes, are confined within the region bound by r_2 and r_3. Using the concept of quantum mechanical tunneling, a tunneling coefficient, T, which represents the probability of power leakage in the region bounded by r_2 and r_3 is defined as the ratio of the squares of the field amplitudes in these regions. By using the expression Equation (4.117)] for the phase factor of these fields, it can be shown that

$$T = \exp\left[-2 \int_{r_2}^{r_3} \sqrt{\frac{l^2}{r^2} - k_0^2 n^2(r) + \beta^2} \, dr \right] \qquad (7.39)$$

The attenuation coefficient $\alpha(r, z)$, normalized to the core radius a, can be expressed in terms of T as

$$\alpha(r, z) \, \Delta z = aT \qquad (7.40)$$

where Δz is the distance between two internal reflections. The result of WKB approximation provides

$$\Delta z = 2 \int_{r_1}^{r_2} \frac{\beta \, dr}{\sqrt{k_0^2 n^2(r) - \beta^2 - l^2/r^2}} \qquad (7.41)$$

Equations (7.39) and (7.41) can be integrated for a specific index profile. For simplication, the following assumptions are made:

1. The true profile will not deviate much from the perfect paraboloid profile.
2. The integral of Equation (7.39) will be evaluated only from a to r_3, because the contribution from r_2 to a is negligible.

With these assumptions, a closed-form solution can be obtained. Substituting the parabolic index profile into Equation (7.41) gives an arcsine solution for Δz as given by

$$\Delta z = \frac{\beta a^2}{V^2} \arcsin \left[\frac{u^2 - 2V^2 r^2/a^2}{\sqrt{u^4 - 4V^2 l^2}} \right]_{r_1}^{r_2} \tag{7.42}$$

where the integral limits can be obtained by solving the intersections of two curves as shown in Figure 7.12. At the caustic:

$$\beta^2 = k_0^2 n^2(r) - \frac{l^2}{r^2} \tag{7.43}$$

Substituting the parabolic index profile again into Equation (7.43) and solving the quadratic equation yields

$$r_{1,2}^2 = \frac{u^2 \pm \sqrt{u^2 - 4V^2 l^2}}{2V^2} a^2 \tag{7.44}$$

Substituting the results of Equation (7.44) into (7.42) and letting $r_2 = a$, the following simple result for Δz emerges:

$$\Delta z = \frac{\beta a^2}{V} \pi \tag{7.45}$$

An analytic solution can also be obtained for T if a parabolic profile is introduced into Equation (7.39). In the case that $r_2 = a$, a simple expression for T is given by (Ref. 7.2)

$$T = \left[\frac{u^2 - V^2}{(l + \zeta)^2} \right]^l \exp(2\zeta) \tag{7.46}$$

where

$$\zeta = \sqrt{l^2 - u^2 + V^2} \tag{7.47}$$

With the results of Equations (7.45), (7.46), (7.40), and (7.37), it is possible to evaluate numerically the correction factor C as a function of fiber length z and the V parameter. As expected, C approaches unity as z approaches infinity. The expression given by Equation (7.37) is rather complex and can be simplified by introducing a normalization parameter $X = (1/V) \ln(z/a)$. As a result, a single set of curves $C(r, z)$ can be obtained to specify the near-field intensity profile completely for a length of fully excited fiber having an arbitrary index profile. Most important is the fact that the same set of curves can be used to correct the measured intensity distribution to give the true refractive index profile of the fiber. Figure 7.13 gives a set of curves for $C(r, z)$ as a function of the normalized radius r/a for several values of X. These curves represent a good approximation not only for all X values

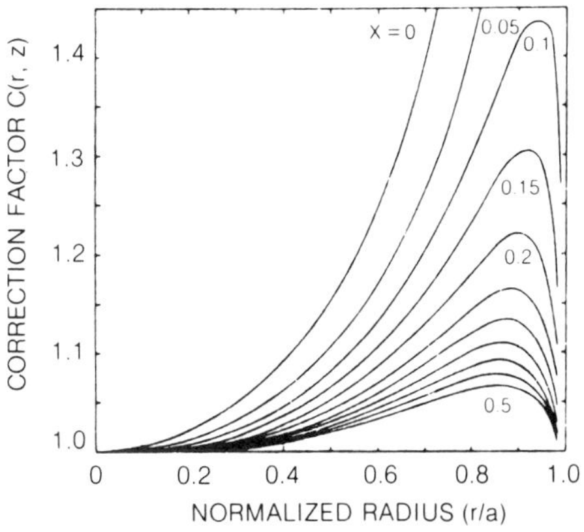

FIGURE 7.13 The correction factor for near-field output power measurements as a function of the radius of the core for various values of $X = \ln(z/a)/V$. (From Ref. 7.2. Reprinted with permission of North-Holland Publishing Company, Amsterdam.)

normally encountered, but also for a wide range of near-parabolic and power-law variations.

For example, take a 1-m-long graded-index fiber having a core distribution of 80 μm and a numerical aperture of 0.18. If λ = 0.9 μm, the X value is 0.2. These curves show that a correction to the index profile is only about 8% greater than the value measured in the near-field for a normalized radius of 0.6 and rises to 20% for r/a to be 0.85. The result of this approximation is proven to be good to within 2% for a wide range of index profiles, provided that $z/a > 10^3$.

7.7 DISPERSION MEASUREMENTS

Modal and material dispersion of a multimode fiber can be determined by measuring either the impulse response in the time domain or the spectral response in the frequency domain of the output. In the former case, pulse broadening can be measured either by sampling the waveform in real time with very high-resolution electronics or by analyzing its Fourier transform in slow time with a computer. Frequency-response measurements, on the other hand, are relatively simple to perform; however, they require the use of a very broadband optical modulator. Because these techniques are very useful in modern research, these methods will be described in some detail. Mechanisms responsible for dispersion in fiber were discussed in Chapter 5. Because all fibers are dispersive, it is expected that a short pulse after propagating through a fiber will be broadened in width and distorted in shape to a certain extent, depending on the fiber parameters. For a step-index fiber, optical loss usually increases with increasing mode order. The impulse response, illustrated in Figure 7.14(a), can therefore be expressed as

$$P(t) = P_0 \sum_{k=1}^{N} \delta(t - \tau_k L)e^{-\alpha_k L} \qquad (7.48)$$

where $\delta(t)$ is the Dirac delta function, P_0 a normalizing constant, and τ_k and α_k the group delay ($nskm^{-1}$) and attenuation coefficient of the kth mode in the step-index fiber.

For a graded-index fiber with no mode coupling, the impulse response is illustrated in Figure 7.14(b). The dip in the response curve shown in Figure 7.14(b) often occurs in practice because of imperfections in index grading. As a result, a short input pulse could excite two distinct groups of modes. Figure 7.14(c) shows the impulse response of a fiber involving strong mode coupling along its length ($\gamma_\infty z \gg 1$). In this case the expression for the impulse response is a Gaussian-like function [see Equation (5.71)] with width σ proportional to $\sqrt{L}$.

If the fiber is assumed to be a linear filter, which is a valid assumption

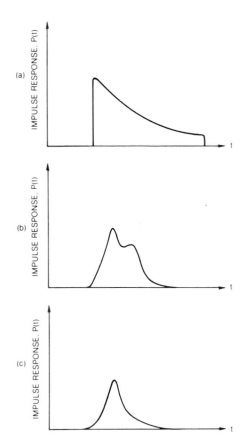

(a)

(b)

(c)

IMPULSE RESPONSE. P(t)

FIGURE 7.14 Waveforms of various possible impulse responses.

under normal circumstances when the optical power in a fiber is very low, the input and output relationship can be written in terms of a convolution integral given by

$$P_{out}(t) = \int h_f(t - \tau)P_{in}(\tau) \, d\tau \tag{7.49}$$

Equation (7.49) can sometimes be expressed by the symbol $h_f(t)*P_{in}(t)$. Physically, this integral represents the sum of impulse responses at the observation time t to impulses of strength $P_{in}(\tau) \, d\tau$ occurring at τ. The time delay $t - \tau$ at the instant of observation is a result of dispersion. It must be realized that the output waveform distortion is not only caused by the fiber but also by the measuring instrument with an impulse response function $h_i(t)$. Therefore, the output is a series of convolution integrals given by

$$P_{out}(t) = P_{in}(t)*h_f(t)*h_i(t) \tag{7.50}$$

To eliminate the effect of the instrument, the test fiber must be replaced with a short "strap" fiber. The output through the strap is simply

$$P'_{out}(t) = P_{in}(t)*h_i(t) \tag{7.51}$$

Now Fourier transforms of Equations (7.50) and (7.51) are taken and the two results are divided. The net result is the frequency response of the fiber alone, because a convolution in time domain transforms to a multiplication in the frequency domain.

The impulse function $P(t)$ can be expressed in terms of its power-transfer function $F(\omega)$ by the Fourier transform as

$$P(t) = \frac{1}{2\pi} \int_{-\infty}^{\infty} e^{-i\omega t} F(\omega) \, d\omega \tag{7.52}$$

Now, examine the convolution integral

$$P_2(t) = \int d\tau \, P_1(t - \tau)P_0(\tau) \tag{7.53}$$

where P_1 and P_0 have the corresponding transfer functions $F_1(\omega)$ and $F_0(\omega)$ as indicated by Equation (7.52). In terms of F_1 and F_0, Equation (7.53) can be written as

$$P_2(t) = \left(\frac{1}{2\pi}\right)^2 \int d\tau \int d\omega' e^{-i\omega'(t - \tau)} F_1(\omega') \int d\omega e^{-i\omega\tau} F_0(\omega)$$

$$= \left(\frac{1}{2\pi}\right)^2 \int d\tau e^{i\tau(\omega' - \omega)} \int d\omega' e^{-i\omega' t} F_1(\omega') \int d\omega F_0(\omega) \tag{7.54}$$

because

$$\int d\tau e^{i\tau(\omega' - \omega)} = 2\pi\delta(\omega' - \omega) \tag{7.55}$$

Substituting Equation (7.55) into (7.54) yields

$$P_2(t) = \frac{1}{2\pi} \int d\omega \int d\omega' \delta(\omega' - \omega) F_1(\omega') F_0(\omega) e^{-i\omega't}$$

$$= \frac{1}{2\pi} \int d\omega e^{-i\omega t} F_1(\omega) F_0(\omega) \tag{7.56}$$

Equation (7.56) indicates that the transfer function of a convolution integral $P_2(t)$ is the product of the transfer functions $F_1(\omega)$ and $F_0(\omega)$. That is,

$$F_2(\omega) = F_1(\omega) F_0(\omega) \tag{7.57}$$

Therefore the impulse response for the fiber can be obtained by taking the inverse Fourier transform of the frequency response. The experimental procedure in this case involves first taking measurements of $P_{out}(t)$ and $P'_{out}(t)$ and processing these pulse shape data with a minicomputer using a fast Fourier transform routine (FFT). The division yields the frequency response of the fiber. If necessary, a deconvolution process can be executed to obtain the impulse response.

It is desirable to observe the pulse shape in real time; however, in a time scale on the order of a few picoseconds, most optical detectors are inadequate to resolve the waveform. It is necessary to use an ultrafast optical shutter such as a Kerr cell, or alternatively, a streak camera, to reach the necessary time resolution. Figure 7.15 shows an experimental arrangement for real-time measurements of impulse response. In this setup, a 0.53-μm mode-locked Nd-doped glass laser with a pulse width of about 7 ps at the second harmonics of the laser ($\lambda = 1.06$ μm) is used as the source. The key component in this measuring apparatus is the optical shutter, which is made of an optically active carbon disulfide CS_2 gas cell. When the laser pulse at its fundamental wavelength with a very high peak power is passing through the cell, the optical field induces an instantaneous birefringence in the CS_2 gas for a duration comparable to the pulse width. This phenomenon is commonly known as the Kerr effect. The Kerr cell is placed between two crossed polarizers. In the absence of a 1.06-μm laser pulse, the signal pulse at 0.53 μm is totally extinguished by the crossed polarizers. During the brief opening period of the optical shutter, the signal pulse is allowed to transmit and to be recorded by a sensitive photomultiplier. The signal pulse at 0.53 μm is generated in a nonlinear KDP crystal that is responsible for the frequency doubling. After propagating through the fiber, this pulse is delayed by a variable optical delay line, which consists of a moving prism retroreflector. By means of this varying delay, the waveform of a distorted signal pulse can be sampled across its entire width with the optical shutter. A reference detector is placed in front of the shutter to provide a normalization factor for canceling any fluctuation in the sampled signals through the cell. This optical shutter is considered to be one of the fastest signal-processing techniques available at present.

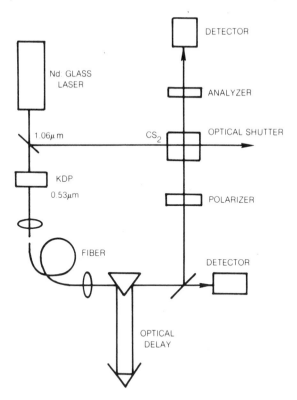

FIGURE 7.15   Apparatus for real-time measurements of ultra-short pulse waveform by using an optical shutter. (From Ref. 7.3. Reprinted with permission of
the Institute of Electrical and Electronics Engineers, Inc., © 1972.)

Figure 7.16 shows a log plot of the profile of an undistorted signal pulse
obtained by this sampling procedure. The sampled profile exhibits a width
of about 24 ps at two $1/e$ points, indicating that significant broadening has
already occurred as a result of this sampling procedure. This broadening is
expected, because at $\lambda = 0.53$ μm, both material dispersion and scattering
loss can be significant and can contribute to broadening and pulse-shape
distortion in addition to the effect of modal dispersion.

The experimental arrangement for making a detailed analysis of the
output waveform of an impulse response with a fast Fourier transform technique is shown in Figure 7.17. The output of a diode laser with a typical
pulse width of ≥200 ps is coupled into the testing fiber using a lens. The
output waveform is detected by a fast avalanche photodiode APD, and displayed on a sampling oscilloscope synchronized by delay trigger pulses. The
sampled waveform is then digitized and processed using a minicomputer.
The processed results of the averaged pulse response and its Fourier transform can be displayed with a printout. The transfer function of the fiber is

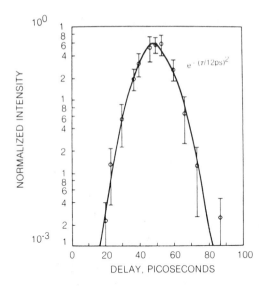

FIGURE 7.16   Time resolution of a CS_2 Kerr cell activated by a 1.06-μm Nd:glass laser with a pulse width of 10 ps. The input pulse width is 7 ps. (From Ref. 7.3. Reprinted with permission of the Institute of Electrical and Electronics Engineers, Inc., © 1972.)

obtained by dividing the Fourier transform of the averaged pulse response of the fiber by that of the strap.

By using a long-wavelength diode laser ($\lambda \geq 1.2$ μm), material dispersion can be reduced. Some major limitations on accuracy and reproducibility of the method above are detector nonlinearity, jitter in the delay trigger, laser instability, and spectral width. Factors such as nonlinearity and jitter can produce large errors in the deconvolution process. Using a

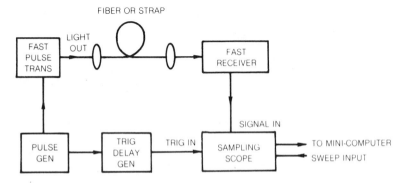

FIGURE 7.17   Apparatus for the measurement of the impulse response of a fiber by the Fourier transform technique.

normalization process with a strap, the jitter effect can be removed, and with sufficient optical attenuation, system nonlinearity can be minimized. As an alternative to time-domain measurements, transfer functions can be measured directly in the frequency domain. Figure 7.18 shows the experimental apparatus required for making frequency-response measurements of a fiber. In this case an incoherent light source can be used, but its spectral width must be filtered by a monochromator. Therefore, a monochromatic source is generated and passed through an electro-optic modulator. The output from the modulator is coupled into the fiber or strap. The output from the fiber is captured by a broadband receiver, whose output is fed to a spectrum analyzer through a calibrated attenuator. As modulation frequency is varied over a wide range, the spectrum analyzer records receiver output as a function of modulation frequency. By dividing the two sets of measurements obtained for the fiber and the strap, the transfer function of the fiber alone is obtained. This method is clearly simpler than the other two methods in the time domain. Furthermore, this method is insensitive to problems associated with system nonlinearity as for the case of the pulse sampling methods. On the other hand, the method in the frequency domain requires an external optical modulator and a monochromator, both of which are essential. The spectral purity of the source must be kept at a very high level to maintain a phase synchronization of the incoming waves entering the modulator. Otherwise, a chirping effect occurs and can lead to large dispersion.

If a source has a very broad spectral width, material dispersion becomes significant. This effect must be separated from the broadening due to modal dispersion. A simple way to measure the material dispersion of a fiber is to observe the delay through the fiber using narrow pulses from lasers at two different wavelengths. In this way material dispersion can be obtained by dividing the delay difference due to the wavelength difference of the lasers and normalizing to the fiber length.

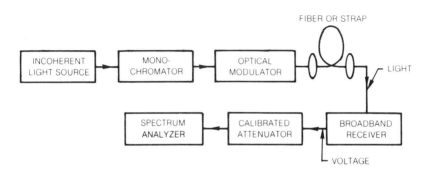

FIGURE 7.18   Apparatus for the measurement of frequency response of a fiber.

7.8 FIBER CONNECTION AND SPLICING

One practical problem associated with fiber optics applications is the difficulty of connecting and disconnecting fibers. This problem is due to the fact that the small cross-sections of the fibers require extremely high precision in alignment. A severe loss penalty can result if lateral displacement between two parallel fiber end faces exceeds a small fraction of their radius. More serious problems occur when the end faces are rough and not parallel. Clearly, this problem is much more severe for single mode fibers than for multimode fibers because the diameter of a single mode fiber is only a few micrometers. This section will treat the case of a single mode fiber and introduce techniques that have been developed for splicing single mode fibers at relatively low loss. These techniques can readily be extended to multimode fibers to yield negligible coupling losses.

Figure 7.19 illustrates two types of misalignment that can cause losses in fiber connections. Rigorous calculation of the joint loss in terms of the HE_{11} mode is difficult. If the HE_{11} field is replaced by a Gaussian function as an approximation, the calculation can be considerably simplified and has been carried out in Chapter 6. When two identical single mode fibers are displaced by a normalized distance $d = x/a$, where x is the offset from fiber axis and a is the fiber radius, as shown in Figure 7.19(a), the loss α_d in decibels is given with a reasonable degree of accuracy by the simple expression (Ref. 7.4)

$$\alpha_d = 2.17 \left(\frac{d}{\omega}\right)^2 \qquad \text{decibels} \qquad (7.58)$$

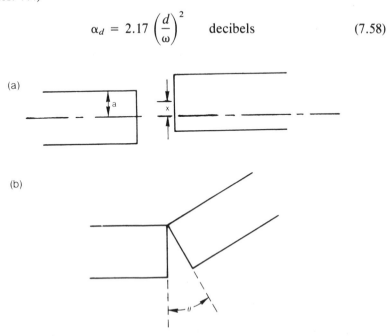

FIGURE 7.19   Two basic types of fiber coupling misalignment.

where ω, the spot size of the HE_{11} mode normalized to a, is given by Equation (6.4) as

$$\omega = 0.65 + 1.62V^{-1.5} + 2.88V^{-6} \qquad (7.59)$$

When the end faces of two fibers are tilted at an angle θ, as shown in Figure 7.19(b), the loss α_θ in decibels is (Ref. 7.4)

$$\alpha_\theta = 2.17 \left(\frac{\theta \omega n V}{NA} \right)^2 \quad \text{decibels} \qquad (7.60)$$

If the joint between two fibers is misaligned with both lateral and angular displacements, the total loss α_T, to a good approximation, is given by the expression (Ref. 7.4)

$$\alpha_T = 3.6d^2 + \left(\frac{n\theta}{NA} \right)^2 (7.6 + 7d^2) \qquad (7.61)$$

Most connectors are designed to produce a butt-joint, placing the fiber ends as close as possible. To maintain the joint within an acceptable tolerance, several techniques have been developed, as illustrated in Figure 7.20. Figure 7.20(a), (b), and (c) illustrate the main features of these demountable connectors that are widely used in multimode fiber design. They provide reasonably reproducible results with typical joint losses of about 0.2 to 3 dB, depending on the precision in making these adaptors and preparing the fiber end faces. In all cases, the primary goal is to assure that the lateral and angular misalignment of two fibers is minimized.

Figure 7.20(c) shows a typical design of demountable butt joint connectors that generally consist of a ferrule for each fiber end and a precision sleeve into which the ferrule fits. To increase the mechanical strength of the cable, Kevlar braids are crimped to the ferrule, which in turn securely attaches to SMA type screw-on connectors. In this case, the axial and angular alignments are determined by the precision and best fit of the ferrules into the tubular sleeve. In particular, the connector shown in Figure 7.20(d) has been used in the field very successfully (Ref. 7.5). For connecting single mode fibers, lenses and x, y, z positioners are often used. A quick demountable connector for single mode fibers is a very difficult task, however, significant progress has been made in recent years.

The best method for making a permanent connection between two fibers is to fuse the two prealigned ends with a short burst of electric arc. Practical fusion-splicing equipment is now available for use with multimode fibers in the field, because most optical fiber systems in service today utilize multimode fibers. This equipment usually consists of a microscope and an electric arc. The microscope is used for inspecting fiber alignment when the ends of two fibers are placed between two electrodes. A battery-operated power supply that can provide high voltage typically on the order of 10 kV and currents in the 10-mA range is needed to energize the electrodes to

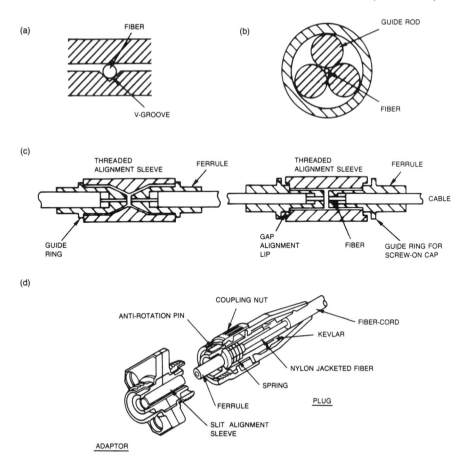

FIGURE 7.20   Examples of demountable couplers.

generate an electric arc. The splicing loss with this technique is usually very small (0.1 to 0.3 dB). Permanent splicing can also be made by thermal fusion to obtain similar results. Thermal fusion can be used to make a star coupler that joins together as many as 100 fibers. Fibers are first twisted together to form a ropelike joint. Tension is then applied to the fibers before fusion. For better results it is necessary to form a biconical taper at the joint. Using a star coupler, optical power can be fed into N fibers from any one of the fiber bundles on the opposite side of the taper joint. An insertion loss of less than 0.6 dB has been achieved with such a coupler.

An interesting application of a biconical tapered coupler is the formation of a data distribution network that contains N parallel terminals, each of which has a transmitter and a receiver and is arranged in accordance with Figure 7.21. The ratio of the received power P_R by one of the receivers to the input power P_I is determined by the total system losses. For this

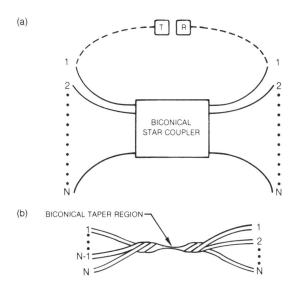

FIGURE 7.21   (a) Data distribution network using a biconical star coupler; (b) schematic diagram showing a biconical taper joint.

system the losses are (1) fiber-to-transmitter and fiber-to-receiver coupling losses, (2) the insertion loss of the star coupler, and (3) the power distribution loss through the star coupler, which is equal to 10 log N (dB). The first two loss factors are independent of the total number of terminals N, but the last loss factor is a function of N. An interesting feature of the parallel distribution system is that as N increases, the total system loss saturates rapidly. As a rule, a parallel distribution system has a considerable advantage over a series distribution network for $N > 10$.

One of the most commonly used techniques to obtain a reasonably good end face is to cleave the fiber under tension with a sharp diamond-tipped or tungsten carbide knife edge. A straight scratch on the surface produces a stress concentration and can lead to a very clear cleavage. If the initial scratch is not very sharp, a faulty cleavage often occurs at which one or more long tails are found to attach to the otherwise perfect cleavage plane.

Another method that yields a consistently good cleavage plane is to cut the fiber with a focused carbon dioxide (CO_2) laser beam. This method is not very convenient to use in field service, but it is certainly the most reliable method to use in a manufacturing facility.

The best method for measuring joint loss is to measure the power transmitted through the joint between two short fibers normalized to the same length as that of a fiber without the joint. Other methods, such as pulse echo and time-domain reflectometry (TDR) could be used to estimate splicing losses and to locate imperfections in a long length of fiber. Echo and reflection measurement techniques have obvious limitations, which can arise from many factors, such as end reflection, scattering, absorption, dispersion, and imperfections other than splicing. Therefore, results are often difficult to interpret. The distance between most repeater stations is in the 10km range, so the typical insertion loss of the fiber can easily exceed 50 dB. To detect a return signal by these methods, a very large dynamic range (>120 dB) for detection sensitivity is required.

PROBLEMS

7.1. A LED source has an output of 1 mW and a maximum emitting angle of 90°. The diameter of its emitting surface is 100 μm. Calculate its radiance in units of $W \; cm^{-2} \; sr^{-1}$.

7.2. Calculate the percent power that can be butt-coupled into a fiber directly from a surface emitting LED source that emits into a full hemisphere. The fiber has a numerical aperture $\theta_{NA} = 0.3$, and $d_f \simeq d_s$.

7.3. Derive Equation (7.11), assuming that the simple lens formula applies.

7.4. If a lens is used, repeat Problem 7.2 and compute the improvement factor for using a lens over the butt-joint.

7.5. Calculate the near-field power profile emitting from a fiber with a parabolic index profile.

7.6. Calculate caustic points r_1 and r_2 for a fiber with a parabolic index profile, in terms of its parameters V, a, u, and l.

7.7. For a 1-m-long graded-index fiber having a value of 50 for the parameter V, determine the correction factor for the near-field power profile at three points, corresponding to $r/a = 0.2$, 0.5, and 0.8.

7.8. Calculate the joint loss of two single-mode fibers ($V = 2.4$) having NA = 0.1 and $n = 1.5$, if only the angle between the two fibers is misaligned by 2°. What would be the joint loss if these two fibers were displaced by $\frac{1}{2}\omega$, where ω is the spot size of the HE_{11} mode?

REFERENCES

7.1. D. N. Christodoulides, L. A. Reith, and M. A. Saifi: *IEEE J. Lightwave Tech.*, *LT-5*, 1623 (1987).

7.2. M. J. Adams, D. N. Payne, and F. M. E. Sladen, *Opt. Commun. 17*, 204 (1976).

7.3. D. Gloge, A. R. Tynes, M. A. Duguag, and J. W. Hanson, *IEEE J. Quantum Electron.*, *QE-8*, 217 (1972).

7.4. W. A. Gambling, H. Matsumura, and C. M. Ragdale, *Electron. Lett.*, *14*, 618 (1978).

7.5. J. Minowa, M. Saruwatari, and N. Suzuki, *IEEE J. Quantum Electron.*, *QE-18*, 705 (1982).

8

Light Emission Processes in Semiconductors

8.1 INTRODUCTION

The process in which electrical energy is converted to light in solids is called electroluminescence. Light emission occurs during the recombination of an electron in the conduction band with a hole in the valence band or an electron at a donor with a hole at an acceptor. The radiative transitions for all light-emitting systems involves spontaneous or stimulated emission and absorption of photons. For semiconductor light sources the selection rules that determine the radiative transition probability are relaxed significantly by the continuous nature of energy bands. As a result, the spectrum is considerably broadened. Semiconductors can be categorized into two groups, one of which is associated with direct bandgap materials and the other with indirect bandgap materials. The radiative transition probability is usually very high for direct bandgap materials. However, complications can arise when a material is heavily doped or is subjected to a high injection of current. Detailed theoretical treatments on these subjects is beyond the scope of this text.

This chapter briefly reviews the fundamentals of the quantum theory of semiconductors without getting into too much mathematical detail. It discusses the mechanisms and circumstances by which various radiative transitions occur. In particular, it treats, in some detail, the probability of electron–hole transitions between the conduction and valence band of a direct bandgap material. The probability for radiative recombination is directly proportional to the recombination rate, and therefore, is related to the radiative lifetime for the emission process. Calculations of transition probabilities for spontaneous and stimulated emissions are difficult, however, relationships among the rates of these emission and absorption processes can be established. From these relationships, it is possible to obtain emission rates from the measurements of absorption spectra. These results allow us to estimate the quantum efficiency, threshold current density, and switching rate for semiconductor light-emitting devices. This chapter also includes an introduction to laser oscillation in a Fabry Perot optical cavity and a discussion on the limitations of the laser spectral width.

8.2 QUANTUM MECHANICAL DESCRIPTION OF SEMICONDUCTORS

The simple model commonly used to describe the motion of an electron in a semiconductor is to consider an electron in a periodic potential with a period determined by the lattice of the semiconductor. This potential is caused by the periodic charge distribution associated with the ion cores situated on the lattice sites plus a constant term due to the contribution of all other free electrons in the crystal. The wave function for the single electron in this potential can be obtained by solving Schrödinger's equation. The solution for a single electron then provides a set of states that can be shared by all the electrons in the crystal subject to the limitations of the Pauli exclusion principle, which states that no two electrons may occupy the same quantum state specified by a given quantum number, unless the state is degenerate. In an infinite one-dimensional crystal, Schrödinger's equation is of the form

$$\frac{d^2\psi}{dx^2} + \frac{2m}{\hbar^2}[E - V(x)]\psi = 0 \tag{8.1}$$

where $\hbar(h/2\pi)$ is Planck's constant, m the mass of a free electron, E the energy associated with the electron, and $V(x)$ the periodic potential. By assuming a proper periodic potential function, the one electron wavefunction which is called Bloch function, can be calculated. Results indicate that the allowed electronic energies occur in bands of permitted states separated by forbidden energy regions, commonly called bandgaps. The difference between a semiconductor and an insulator is that in semiconductors all bands are filled, except for one or two bands that are slightly filled or slightly empty. In actual crystals, the potential function experienced by the electron due to the ion cores is usually very complicated. Therefore, it is customary to approximate the function best suited for the situation. In general, there are two opposite approaches: (1) the free electron approximation, and (2) the tight-binding approximation. In the free electron approximation, the total energy of the electron is assumed to be large compared to the periodic potential energy. Under this assumption, the allowed bands are very broad and the forbidden energy bandgaps are very narrow. In the tight-binding approximation, the potential energy of the electron is assumed to account for nearly all of the total energy. In this case, the allowed bands are very narrow in comparison with the forbidden bands. The exact treatment is beyond the scope of this book. To understand the properties of bandgaps the Kronig-Penney model (Ref. 8.1) is introduced, which assumes a simple periodic square-well potential. Even though such an idealized potential is only a crude representation of actual crystals, it is possible to obtain an exact solution of the Schrödinger equation, which illustrates in a most explicit way many of the important features of quantum mechanical behavior of electrons in periodic lattices.

The one-dimensional periodic square-well $V(x)$ with the well spacing a and width b can be expressed by:

$$V(x) = \begin{cases} V_0 & a \le x < a + b,\ 2a + b \le x \le 2(a + b),\ \text{etc.} \\ 0 & \text{elsewhere} \end{cases} \quad (8.2)$$

In the region where $V = 0$, the wavefunction representing a free electron can be expressed by a linear combination of two opposite traveling-plane waves as

$$\psi(x) = Ae^{ikx} + Be^{-ikx} \quad (8.3)$$

with a corresponding energy E of a free particle as given by:

$$E = \hbar^2 k^2 / 2m \quad (8.4)$$

Within the potential well where $V = V_0$, the solution is of the form:

$$\psi(x) = Ce^{\alpha x} + De^{-\alpha x} \quad (8.5)$$

where

$$\alpha = \frac{1}{\hbar} \sqrt{2m(V_0 - E)}$$

The constants A, B, C and D are chosen so that ψ and $d\psi/dx$ are continuous at $x = 0$ and $x = a$. These requirements lead to the following equations:

$$A + B - C - D = 0$$

$$ikA - ikB - \alpha C + \alpha D = 0$$

$$Ae^{ika} + Be^{ika} - Ce^{\alpha a} - De^{-\alpha a} = 0$$

$$ikAe^{ika} - ikBe^{-ika} - Ce^{\alpha a} + De^{-\alpha a} = 0$$

The above set of homogeneous equations has a solution only if the determinant of the coefficients of A, B, C and D vanishes. Using this requirement, the following equation can be obtained after some algebraic manipulations:

$$\frac{(\alpha^2 - k^2)}{2\alpha k} \sin \alpha b \sin ka + \cos \alpha b \cos ka = \cos k'(a + b)$$

where k' is an additional wave number which has been introduced in order to satisfy the other boundary conditions, e.g. ψ and $d\psi/dx$ at $x = a + b$, etc. must be equal to those values at $x = a$, expect that the phase has been advanced by a factor $\exp ik'(a + b)$. This is a consequence of the Bloch Theorem. The wave-number k' is associated with the Bloch function $U_{k'}(x)$, which has the property that $U_{k'}(x) = U_{k'}(x + a)$.

The above equation can be simplified considerably if the shape of the square potential well is changed by a periodic delta function which becomes infinite in the limit as $b \rightarrow 0$, but the value of $\alpha^2 ab/2$ remains to be a finite quantity. With this modification, the above equation reduces to

$$\frac{P}{ka} \sin ka + \cos ka = \cos k'a$$

where $P = \alpha^2 ab/2$.

Figure 8.1(a) is a plot of the left-hand side of the above equation as a function of ka, for the case $P = 3\pi/2$. From Figure 8.1(a), it is clear that the allowed values of the energy E are given by those ranges of $k = (2mE/\hbar^2)^{1/2}$ for which the function lies between ± 1. The corresponding energy $E(k')$ in the unit of $\hbar^2\pi^2/2ma^2$ is plotted in Figure 8.1(b) as a function of $k'a$. At the lattice sites, $k'a = n\pi$, there is a series of energy gaps at which solutions of the Schrödinger equation do not exist. Physically, the electron wave suffers a Bragg reflection when the Bragg condition, $k'a = n\pi$, is satisfied. This reflection arises because the incident wave interferes constructively with the reflected wave at the lattice sites, or the Brillouin zone of the lattice.

This problem can be treated in the free electron approximation by using a more realistic potential function. A periodic potential, which is invariant under a crystal lattice translation, can be expressed in terms of the reciprocal lattice vector $2n\pi/a$ as a Fourier series:

$$V(x) = -\gamma \frac{\hbar^2}{2m} \sum_{n=-\infty}^{\infty} C_n e^{-i2\pi nx/a}$$

$$= \sum_{-\infty}^{\infty} V_n e^{-i2\pi nx/a} \tag{8.6}$$

In this approximation $V(x)$ is valid for small values of γ and can be tailored to fit a variety of periodic potentials of real material systems, provided that the Fourier coefficients are properly determined.

According to the Bloch Theorem, the wave function can be written in

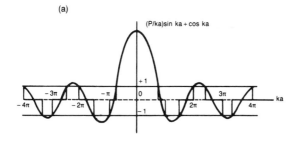

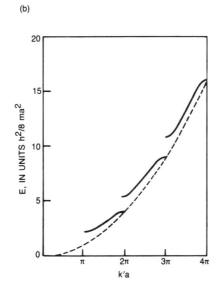

FIGURE 8.1   (a) A plot of the expression $(3\pi/2ka)\sin ka + \cos ka$ versus ka.
(b) A plot of the energy values as a function of $k'a$.

the form, $\psi(x) = e^{ikx}U(x)$, where $U(x)$ is a periodic function of period a, which is the lattice constant. Note that the wave function in the Bloch form differs from Equation (8.3) for a free particle, by a periodic function $U(x)$ with a period of the crystal lattice. It can be shown that for $\gamma = 0$, the wave function reduces to that for a free particle.

For $\gamma \neq 0$, Equation (8.1) can be solved in the vicinity of band edges by using the free-particle approximation. Under the assumption that the electron energy is large compared to the periodic potential energy, E is replaced by $\hbar^2 k_0^2/2m$ in Equation (8.1). At lattice sites or band edges, $k = n\pi/a$, where $n = 0, 1, 2 \ldots$.

In the vicinity of a band edge, the following wave function is assumed of the form

$$\psi(x) = B_0 e^{ikx} + \gamma e^{ikx} U(x)$$

where $U(x)$ is the Bloch function which can also be expressed in terms of a Fourier series:

$$U(x) = \sum_{n \neq 0} B_n e^{-i2\pi nx/a}$$

Substituting the above wave function in which E has been replaced by $\hbar^2 k_0^2/2m$ and neglecting γ^2 terms, we obtain

$$B_0(k_0^2 - k^2)e^{ikx} + \gamma \sum_{n \neq 0} [(k_0^2 - k^2)B_n + B_0 C_n]e^{ik_n x} = 0$$

where $k_n = k - 2\pi n/a$. Now multiplying the above equation by $e^{-ik_n x}$ and integrating over the unit cell from $x = 0$ to $x = a$ yields

$$B_0(k_0^2 - k^2)\int_0^a e^{i2\pi mx/a}\,dx + \gamma \sum_{n \neq 0}[(k_0^2 - k_n^2)B_n + B_0 C_n]\int_0^a e^{i2\pi(m-n)x/a} = 0$$

If $m = 0$, the second integral vanishes for all values of n in the sum. Therefore, the first term leads to

$$B_0(k_0^2 - k^2)a = 0 \quad \text{or} \quad k_0 = k$$

which shows that to this order of approximation ($\gamma^2 = 0$) there is no change in electron energy from that of a free particle under the influence of the periodic potential. If $m \neq 0$, the first integral vanishes. The second integral also vanishes, except for $n = m$. In this case,

$$\gamma[(k_0^2 - k_n^2)B_n + B_0 C_n]a = 0$$

or

$$B_n = \frac{B_0 C_n}{k_0^2 - k_n^2} = \frac{B_0 C_n}{k^2 - k_n^2}$$

Therefore, the wave function can be written as

$$\psi(x) = B_0 e^{ikx}\left[1 + \gamma \sum_{n \neq 0} \frac{C_n}{k^2 - k_n^2} e^{-i2\pi nx/a}\right]$$

The above treatment is good as long as k^2 does not approach one of the k_n^2 values. If $k^2 = k_n^2$, the above wave formation is not a good approximation. At the lattice sites, the wave function of a different form must be assumed. Since the electron undergoes Bragg reflection at the band edges ($k = \pm n\pi/a$), the wave function can be expressed as a superposition of two waves propagating in two opposite directions, and thus it has the character of a standing wave. Therefore, for the case $k = -k_n$

$$\psi(x) = e^{ikx}(B_0 + \gamma B_n e^{-i2\pi nx/a}) \tag{8.8}$$
$$= B_0 e^{ikx} + \gamma B_n e^{ik_n x}$$

and all other terms in the Fourier series are negligible compared to the γB_n term.

Substituting Equation (8.8) into Equation (8.1) in which E has been replaced by $\hbar^2 k_0^2/2m$ and retaining the γ^2 terms yields

$$B_0(k_0^2 - k^2)e^{ikx} + \gamma B_n(k_0^2 - k_n^2)e^{i(k-2\pi n/a)x}$$
$$+ \gamma B_0 \sum_{n' \neq 0} C_{n'} e^{i(k-2\pi n'/a)x} + \gamma^2 B_n \sum_{n' \neq 0} C_{n'} e^{i(k-2\pi n/a-2\pi n'/a)} = 0$$

Multiplying the above equation by e^{-ikx} and integrating from $x = 0$ to $x = a$ gives an equation that came from the first and last terms of the above equation when n' was equal to $-n$, of the form

$$(k_0^2 - k^2)B_0 + \gamma^2 C_n B_n = 0$$

Similarly, multiplying the above equation by $e^{-ik_n x}$ and integrating from $x = 0$ to $x = a$ provides another equation:

$$C_n B_0 + (k_0^2 - k_n^2)B_n = 0$$

The foregoing two homogeneous equations involving B_0 and B_n have non-trivial solutions only if

$$(k_0^2 - k^2)(k_0^2 - k_n^2) - \gamma^2 C_n^2 = 0$$

or

$$k_0^2 = \tfrac{1}{2}\left[(k^2 + k_n^2) \pm \sqrt{(k^2 - k_n^2 + 4\gamma^2 C_n^2)} \right]$$

Therefore, the $E(k)$ relationship is given as

$$E(k) = \frac{\hbar^2}{4m}\left\{ k^2 + \left(k - \frac{2\pi n}{a} \right)^2 \right.$$
$$\left. \pm \sqrt{\left[k^2 - \left(k - \frac{2\pi n}{a} \right)^2 \right]^2 + \left(\frac{4m\,|\,V_n\,|}{\hbar^2} \right)^2} \right\} \tag{8.9}$$

These results shows that at the band edge $k = \pm n\pi/a$, internal Bragg reflection occurs and is accomplished by a discontinuity at the energy gap in the E versus k curve (Figure 8.1). The width of the bandgap is given by $2\,|\,V_n\,|$, where V_n is the nth Fourier coefficient in the series expansion of the periodic lattice potential as given by Equation (8.7). Therefore, at the band edge

$$E = \frac{\hbar^2}{2m}\left(\frac{n\pi}{a} \right)^2 \pm |\,V_n\,|$$

Because the number of atoms in the crystal is very large, the allowed values for E and k, although discrete, are very close together and, for practical purposes, can be regarded as quasi-continuous bands of allowed values. This description applies also to a three-dimensional crystal, except that in this case more than two values of k are allowed for each eigenvalue of E. The k values can be both real and imaginary. More details can be found in Ref. 8.2.

As E increases, the allowed bands become very broad, whereas the forbidden bands become very narrow. These features are shown in Figure 8.2 and are quite general in the sense that they are relatively independent of the precise form of the potential function. Note that the electrical conductivity can only arise from bands that are partially filled. This phenomenon occurs in the proximity of a *pn* junction in a semiconductor, which is discussed in Chapter 9. If the bandgap energy E_g extending across the forbidden zone is small, an appreciable number of electrons will be excited thermally from the top of the valence band across the gap to the bottom of the conduction band.

Band theory also shows that the masses of electrons and holes in the bands generally have values different from those of a free electron. An electron in an energy band can have positive or negative effective mass. The states of positive effective mass occur near the bottom of a band because

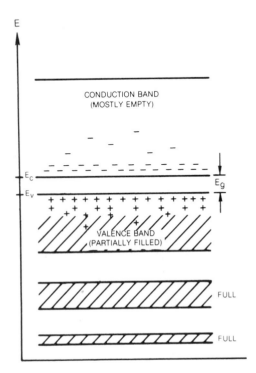

FIGURE 8.2 Energy-band diagram of a semiconductor.

the band has an upward curvature (see Figure 8.1b). The states of negative effective mass occur near the top of the band. These features can be understood by considering the momentum of an electron as it changes from the state k to the state $k + \Delta k$. Here the group velocity v_g of these particles must be introduced. This consequence is directly related to the fact that the energy of these particles is a function of k and therefore the group velocities v_g of these particles must be dealt with, as defined by

$$v_g = \frac{d\omega}{dk} = \frac{2\pi}{\hbar} \frac{dE}{dk}$$

where Planck's relation $E = \hbar\omega$ applies. Note from Figure 8.1 that $E(k)$ always has zero slope at the edges of the allowed bands, $k = \pm n\pi/a$. This result implies that at these points, v_g is zero, indicating that the particle is at rest. The electron or hole can be regarded as undergoing an internal reflection by the lattice potential. In that instance, the effective masses of these particles may be considered to be infinitely large. Therefore, when dealing with the dynamic behavior of electrons and holes in the bands, an effective mass m_e for electrons and m_h for holes are introduced. These masses can be significantly different from the mass m of a free particle. When the values of E become very large, the function $E(k)$ approaches the free energy of the particle as given by Equation (8.5).

Using DeBroglie's relation, which states that the momentum of a particle p is equal to $\hbar k$, where k is the wave number associated with a wave packet, it can be shown that an electron's effective mass m_e associated with the wave packet is

$$m_e = \frac{\hbar^2}{d^2E/dk^2} \qquad (8.10)$$

Equation (8.10) indicates that the sign of effective mass depends on the curvature d^2E/dk^2 of the E versus k curve. Fortunately, the relation between E and k is almost always parabolic or nearly parabolic over the range of energies of interest. What is important is that the effective mass be essentially constant over a sufficient energy interval δE in which a collision process can occur within a time of the order of the radiative recombination lifetime. This amount of energy typically is of the order of a few kT ($kT \simeq 0.025$ eV at 300°K), which is very small compared with the bandgap energy E_g (1 to 2 eV). In other words, in the interval between successive collision events, an electron is usually confined to a short segment of the $E(k)$ curve, which can be regarded to a good approximation as parabolic. For k values near the edge of the conduction band [e.g., $k = (n\pi/a) + k'$, where $k' \ll \pi/a$], $E(k)$ deviates from the free electron energy curve by a small amount, as shown in Figure 8.1.

In the neighborhood of the band edge, $E(k)$ is expressed in terms of

k' as

$$E(k') = \frac{\hbar^2}{2m} \left[\left(\frac{n\pi}{a}\right)^2 + k'^2 + \frac{2m|V_n|}{\hbar^2} \sqrt{1 + 4k'^2 \left(\frac{n\pi}{a}\right)^2 \left(\frac{\hbar^2}{2m|V_n|}\right)^2} \right] \quad (8.11)$$

The free-particle energy and bandgap energy are defined by the expressions

$$E_n = \frac{\hbar^2}{2m} \left(\frac{n\pi}{a}\right)^2 \quad \text{and} \quad E_g = 2\,|V_n|$$

Substituting E_n and E_g into Equation (8.11) and expanding the radical provides an approximation for $E(k')$ as

$$E(k') = E_n + \frac{E_g}{2} + \frac{\hbar^2 k'^2}{2m} \left(1 + \frac{4E_n}{E_g}\right) \quad (8.12)$$

Differentiating Equation (8.12) twice with respect to k' yields

$$m_e = \frac{m}{(1 + 4E_n/E_g)} \quad (8.13)$$

Table 8.1 gives the effective mass of electrons and holes for a number of binary compounds commonly used in making heterostructure light-emitting devices. Also listed in this table are values for the electron affinity χ, which is the energy required to liberate an electron from the conduction band edge to free space, and is useful for the calculation of the energy difference of the conduction bands ΔE_c between two dissimilar materials. The values for the dielectric constant ϵ, the refractive index n, the thermal conductivity σ, and the bandgap energy at 300°K are also included. For GaAs, $E_g = 1.42$ eV and $E_n = \chi + E_g/2 = 4.75$ eV; therefore, a value for the effective mass of an electron is approximately $0.067m$. In all direct bandgap materials, the effective mass of a hole in the valence band is approximately one order of magnitude larger than that of an electron. For GaAs, $m_h \approx 0.48m$.

TABLE 8.1   Constants for Selected III–V Binary Compounds

Compound	E_g (eV)	χ (eV)	m_e/m	m_h/m	ϵ	n	σ (W/cm-deg)
AlAs	2.163		0.15	0.79	10.1	3.178	9.91
AlSb	1.58	3.64	0.12	0.98	14.4	3.4	0.57
GaP	2.261	4.0	0.82	0.60	11.1	3.452	0.77
GaAs	1.424	4.05	0.067	0.48	13.1	3.655	0.44
GaSb	0.726	4.03	0.042	0.44	15.7	3.82	0.33
InP	1.351	4.4	0.077	0.64	12.4	3.450	0.68
InAs	0.360	4.45	0.023	0.40	14.6	3.52	0.27
InSb	0.172	4.59	0.0145	0.40	17.7	4.0	0.17

8.3 CARRIER DISTRIBUTION AND CONCENTRATION

To extend the analysis for a single electron or a single hole to a distribution of electrons in the conduction band or holes in the valence band, the term $\rho(E)$, which represents the density of states at any particular energy E, and $f(E)$, which represents the fractional occupation of states are introduced. This description is analogous to that used for describing the mode distribution in a fiber. In the case of a semiconductor, energy states are modes of the electronic wave functions, which are analogous to optical modes. The major distinction between the two cases is that in a quantum mechanical system the electron distribution is determined by Fermi–Dirac statistics, which are a direct consequence of the Pauli exclusion principle that prevents more than one particle from occupying an identical state. Fermi–Dirac statistics describe the statistical behavior of a system of particles such as conduction electrons in semiconductors.

The probability that an electronic state at a particular energy E is occupied by an electron is given by the Fermi–Dirac distribution:

$$f(E) = \frac{1}{\exp(E - F)/kT + 1} \qquad (8.14)$$

where F is the Fermi energy level, k the Boltzmann constant, and T the absolute temperature. Figure 8.3 is a plot of this distribution as a function of E for several values of temperature. At absolute zero, this distribution function is simply a step function for which

$$f(E) = \begin{cases} 1 & \text{for } E < F \\ 0 & \text{for } E > F \end{cases}$$

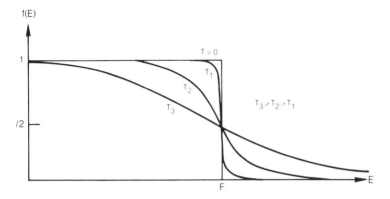

FIGURE 8.3 Temperature dependence of Fermi–Dirac distribution function $f(E)$.

As temperature increases, the edges of the step start to round off and the distribution function loses its steplike character and varies much more slowly with energy. It can be shown that in the high-temperature limit the Fermi–Dirac distribution approaches the classical Maxwell–Boltzmann distribution. From Equation (8.14), $E = F$ and

$$f(F) = \tfrac{1}{2}$$

This result implies that a quantum state at the Fermi level has a probability of occupation of $\tfrac{1}{2}$.

The following discussion introduces the concept of the density of states per unit energy for electrons and holes in a semiconductor. Consider a system of free particles of spin $\tfrac{1}{2}$ like electrons, each of which has two allowed momentum states: one has a spin oriented upward, the other oriented downward for each lattice point in the momentum space. From deBroglie's relation, the momentum vector is $\mathbf{p} = \hbar\mathbf{k}$, where $\mathbf{k}$ is the wave vector in the direction of propagation. In the momentum space with coordinates (p_x, p_y, p_z), each lattice point representing allowed momentum values occupies a unit cell with dimensions $(h/L_x, h/L_y, h/L_z)$, where the product $L_xL_yL_z$ is the geometric volume V of the unit cell. The volume in the momentum space V_p, which corresponds to the space that occupies a single quantum state of energy E, is simply $h^3/2V$, where a factor of 2 has been introduced to account for the two spin states. The volume in the momentum space between two spheres with radii, p and $p + dp$ is

$$dV_p = 4\pi p^2 \, dp$$

Because $p^2 = 2mE$

$$dV_p = 4\pi m^{3/2}(2E)^{1/2} \, dE$$

The density of states, which is the number of quantum states contained in this volume of momentum spaces, is simply equal to dV_p/V_p. Therefore

$$\rho(E) \, dE = \frac{8\pi\sqrt{2} \, m^{3/2}E^{1/2}}{h^3} \, dE \qquad (8.15)$$

where the geometric volume V of the unit cell is assumed to be unity.

Using the expression for an electron in the conduction band, e.g., $E - E_c = \hbar^2k^2/2m_c$, the density of states in the conduction band can be expressed as

$$\rho_c(E) \, dE = \frac{8\sqrt{2}}{h^3} \, \pi m_e^{3/2}\sqrt{E - E_c} \, dE \qquad (8.16)$$

where E_c and m_e are the band edge and effective mass of the electron in the conduction band, respectively. Similarly, the density of states in the valence band is

$$\rho_v(E) \, dE = \frac{8\sqrt{2}}{h^3} \pi m_h^{3/2} \sqrt{E_v - E} \, dE \tag{8.17}$$

where E_v and m_h are the band edge and effective mass of the hole in the valence band. Equations (8.16) and (8.17) indicate that the density of states is proportional to $m_{eff}^{3/2}$. The difference in effective mass for electrons and holes shows that the density of states in the valence band is almost 25 times greater than that in the conduction band.

When the system is in thermal equilibrium, a single Fermi energy level exists that uniquely determines the distribution of electrons and holes in these bands over the entire range of energy. As excess electrons are injected into one or both bands, the equilibrium condition is disturbed and it is necessary to establish two distinct distributions, each of which corresponds to a quasi-Fermi level. If F_c and F_v represent the quasi-Fermi levels of the conduction and valence bands, respectively, the occupation factors f_c and f_v for the two bands, can be written as

$$f_c = \frac{1}{\exp(E - F_c)/kT + 1} \tag{8.18}$$

and

$$f_v = \frac{1}{\exp(E - F_v)/kT + 1} \tag{8.19}$$

Equations (8.16) to (8.19) can be used to calculate the total concentration of electrons and holes by integrating the product of the density of states and the occupational factor over the entire energy range:

$$n = \int \rho_c(E - E_c) f_c \, dE \tag{8.20}$$

and

$$p = \int \rho_v(E_v - E)(1 - f_v) \, dE \tag{8.21}$$

The carrier concentration in the conduction band as given by Equation (8.20) can be written explicitly as

$$n = \frac{4\pi}{h^3} (2m_e)^{3/2} \int_{E_c}^{\infty} \frac{(E - E_c)^{1/2} \, dE}{\exp[(E - F_c)/kT] + 1} \tag{8.22}$$

Similarly, the expression for the hole concentration can be written as

$$p = \frac{4\pi}{h^3} (2m_h)^{3/2} \int_{-\infty}^{E_v} \frac{(E_v - E)^{1/2} \, dE}{\exp[-(E - F_v)/kT] + 1} \tag{8.23}$$

To carry out these integrals, it is necessary to determine F_c and F_v. For very lightly doped material, the Fermi level lies outside the band (see Figure 8.4). If F_c is at least $3kT$ below the conduction band edge in n-type material, and F_v is at least $3kT$ above the valence-band edge in p-type material, the exponential term in the denominators of Equations (8.22) and (8.23) is large compared to unity. Thus an approximate value for n can be obtained. By letting $x = (E - F_c)/kT$ and changing the limits of integration, a simple expression for the electron density is obtained from Equation (8.22):

$$n = \frac{4\pi}{h^3} (2m_e)^{3/2} \exp\left(\frac{F_c - E_c}{kT}\right) \int_0^\infty x^{1/2} \exp(-x)\, dx$$

Because the value of the integral is $\sqrt{\pi}/2$, the above result is reduced to

$$n = N_c \exp\left(\frac{F_c - E_c}{kT}\right) \qquad (8.24)$$

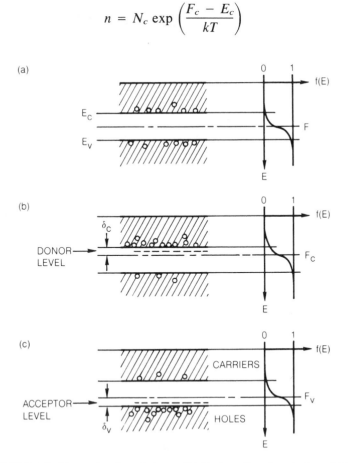

FIGURE 8.4 Energy levels and charge distribution of semiconductors: (a) intrinsic, (b) n-type; and (c) p-type materials.

where

$$N_c = 2 \left(\frac{2\pi m_e kT}{h^2} \right)^{3/2} \tag{8.25}$$

Similarly

$$p = N_v \exp \left(\frac{E_v - F_v}{kT} \right) \tag{8.26}$$

where

$$N_v = 2 \left(\frac{2\pi m_h kT}{h^2} \right)^{3/2} \tag{8.27}$$

At 300°K, the N_c and N_v values in GaAs are 4.35×10^{17} cm^{-3} and 8.87×10^{18} cm^{-3}, respectively.

Equations (8.20) and (8.21) are useful only if the impurity concentration is very low. However, most semiconductor light-emitting devices are either made of heavily doped materials or are operated under high injection currents. Under these conditions, it is necessary to modify the density of states by adding various types of tails at the edges of the conduction and valence bands. These smearing effects on the band structure caused by high density of impurities or electrons are difficult to treat theoretically. One common approach is to modify the normal density-of-state functions by an exponential tail state function with an empirical parameter that governs the effective depth of the tail. The parameter must be determined experimentally from the measurements of the emission and absorption spectra. More details can be found in the book by Thompson (Ref. 8.3).

8.4 EFFECTS OF DOPING

In an intrinsic material, the densities of electrons and holes are nearly equal. This condition occurs when $F_c = F_v$, as shown in Figure 8.4(a). Substituting $F = F_c = F_v$ into Equations (8.24) and (8.26) yields

$$n = N_c \exp \left(\frac{F - E_c}{kT} \right)$$

$$= N_v \exp \left(\frac{E_v - F}{kT} \right) = p$$

Because the product of $np = n_i^2$, an expression for intrinsic carrier concentrations can be written as

$$n_i = \sqrt{N_c N_v} \exp \left(- \frac{E_g}{2kT} \right) \tag{8.28}$$

which can be described simply by Boltzmann statistics.

Figure 8.4(b) shows an n-type material in which the Fermi level is close to the conduction band. This shift is primarily a result of the ionization of the donors by the amount

$$\delta_c = E_c - F_c \tag{8.29}$$

Figure 8.4(c) shows p-type material in which the Fermi level is close to the valence band. This result is due to the capture of electrons by the acceptors, thus creating holes in the valence band. Physically, consider the bonding structure within a solid. In the case of silicon and germanium, each atom has four valence electrons, and each electron is shared with a nearest neighbor. Ideally, no free electrons exist at 0°K. As temperature increases, electrons are excited into the conduction band with a concentration proportional to Boltzmann's factor, and, in the meantime, holes are created in the valence band. Under an applied electric field electrons and holes are moving in opposite directions at different velocities depending on their effective masses. In GaAs nearest-neighbor atoms have unequal numbers of valence electrons. Five valence electrons are associated with the As atom and three are as-

TABLE 8.2 Ionization Energies of Dopants in GaAs

Type	Element	E_i (eV)
Simple donors	S	0.006
	Se	0.006
	Te	0.006
	Sn	0.006
	C	0.006
	Ge	0.006
	Si	0.006
Simple acceptors	Cd	0.03
	Zu	0.03
	Mg	0.03
	Be	0.03
	C	0.03
	Su	0.2
	Pb	0.12
	Ge	0.038
	Si	0.035
Complex centers	Ge (acceptor)	0.08
	Si (acceptor)	0.1
Transition metals	Cr	0.8
	Mn	0.1
	Fe	0.2–0.5
	Co	0.1–0.5
	Cu	0.15
	Ag	0.24

Source: Data compiled by H. Kressel and J. K. Butler, *Semiconductor Lasers and Heterojunction LEDs*, Academic Press, Inc., New York, 1977.

sociated with the Ga atom. But the total number remains eight. Thus no net charge difference exists between the neighboring atoms.

The introduction of dopants into a semiconductor causes changes in carrier density and in ionization energy E_i. Consequently, the intensity and wavelength of the light emitted from the material will be affected. In Zn-doped GaAs, where the Ga atom is replaced by a Zn atom, the impurity acts as an acceptor. In Te-doped GaAs, the As atom is replaced by a Te atom and forms a donor. Various dopants in GaAs fall into three major categories: (1) simple donors, (2) simple acceptors, and (3) complex centers. They can be introduced into GaAs by various techniques involving epitaxial growth, which are discussed in Chapter 9. Table 8.2 lists various types of dopants and their ionization energies in GaAs.

From band theory, the effect of doping is described by forming the bandtail states as shown in Figure 8.5. These states become significant when the impurity concentration approaches N_c and N_v. The net effect of bandtail states is to reduce the separation between the valence and conduction band edges. Therefore, optical transitions involving emission and absorption occur at energies less than the bandgap energy of the undoped material. Another consequence of doping is that radiative efficiency increases with increasing electron or hole concentration in n- or p-type material, provided that nonradiative lifetime remains constant.

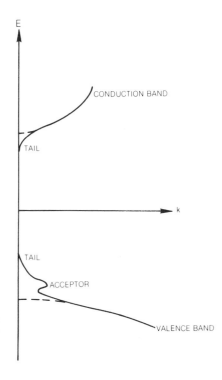

FIGURE 8.5 Effect of p-type impurities on density-of-state distribution for conduction and valence bands by introducing the bandtails to $E(k)$ curves.

8.5 RADIATIVE TRANSITIONS AND RECOMBINATION RATES

In thermal equilibrium the conduction band usually contains only a few filled states and the valence band has only a few empty states. Therefore, the radiative transitions through spontaneous emission are insignificant. One of the most efficient ways to produce electroluminescence is to inject currents through a *pn* junction by applying forward-bias to the junction. More details on *pn* junctions are discussed in Chapter 9. During the recombination of injected electrons with holes, light is emitted from the junction. The efficiency and rate of recombination are very different in direct versus indirect bandgap materials. For direct bandgap semiconductors such as GaAs, InP, GaSb, and so on, a band-to-band radiative recombination has the highest probability, because this process involves the most direct interaction between an electron, a hole, and a photon. As shown in Figure 8.6(a), the conduction band edge, which is the minimum, occurs at the same k value as for the maximum (band edge) of the valence band. In this case, the electrons and holes have a large overlapping range of k values to satisfy the momentum conservation requirement for the recombination process. This requirement is commonly referred to as the k selection rule. In these materials, the radiation emitted can also be absorbed by exciting an electron in the valence band back to the conduction band. For this case, the density of the emitted photons is an important factor and will be treated in the next section.

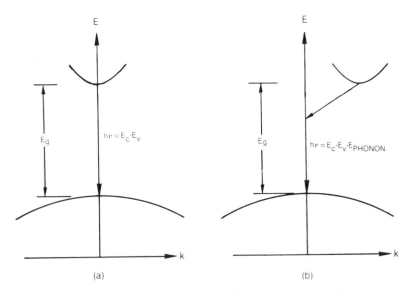

FIGURE 8.6 Processes occurring during a radiative transition in (a) direct bandgap and (b) indirect bandgap materials.

For indirect bandgap materials such as AlAs, GaP, Si, and so on, light emission involves a multistep process in which the recombination of an electron and a hole can occur only through collision with lattice vibrations or phonons as illustrated in Figure 8.6(b). Therefore, the efficiency and rate of recombination are considerably reduced for these materials. A direct bandgap material can be converted to an indirect material by alloy processing through which a binary is made into a ternary or a quaternary. For example, a certain amount of aluminum can be introduced into GaAs to form the $Al_x Ga_{1-x}As$ compound, which becomes an indirect material when the composition factor x reaches 0.45. This process, which has been used widely in making heterostructures, is discussed in detail in Chapter 9. For heavily doped indirect materials however, radiative transitions between deep impurity levels can be very efficient, because wave functions of spatially confined impurity states can extend quite far in momentum space, and consequently, relax the k selection rule.

The probability for radiative recombination, P_r, which is directly proportional to the recombination rate R, is related to the radiative lifetime τ_r as

$$P_r = \frac{1}{\tau_r} \tag{8.30}$$

If P_{nr} is the probability for nonradiative transitions, the quantum efficiency for luminescence can be expressed as

$$\eta = \frac{P_r}{P_r + P_{nr}} = \frac{1}{1 + \tau_r/\tau_{nr}} \tag{8.31}$$

Equation (8.31) represents the ultimate limiting value, because for actual devices the quantum efficiency is usually reduced by other losses, such as absorption and reflection. The efficiency for a band-to-band transition in direct bandgap semiconductors such as GaAs, InP, and GaSb, as well as the ternary and quaternary alloys of these compounds, is very high and can approach 100%. Therefore, these materials are commonly used for making highly efficient LEDs and LDs for optical fiber system applications.

Both the quantum efficiency and transition rate are determined by radiative lifetime. This value also determines the switching rate of the source. This section will introduce various possible radiative transitions and derive expressions for the rates of these transitions. For simplicity, this section deals first with transitions between two discrete levels and then extends the treatment by introducing the density of states and proper selection rules to make it suitable for an energy band.

Let E_1 and E_2 be the energy of a state within the valence and conduction bands, respectively. There are, in general, three types of radiative transitions: spontaneous emission, stimulated emission, and absorption. Each transition is associated with a probability coefficient: A_{21}, B_{21}, and B_{12}, respectively. The order of appearance for the subscripts indicates the direction in which these processes occur. For example, emission processes occur only from state 2 to state 1 and absorption from 1 to 2. The rate of a transition is related to the product of the occupation factor of the initial state and the nonoccupation factor of the terminal state with the appropriate probability coefficient. Therefore the spontaneous emission rate R_{sp} is written as

$$R_{sp} = A_{21}f_2(1 - f_1) \tag{8.32}$$

where f_1 and f_2 can be obtained by using Equations (8.18) and (8.19), which are

$$f_1 = \frac{1}{\exp(E_1 - F_v)/kT + 1} \tag{8.33}$$

$$f_2 = \frac{1}{\exp(E_2 - F_c)/kT + 1} \tag{8.34}$$

The transition rates for stimulated emission and absorption, on the other hand, depend not only on these occupation factors but also on the density of photons $\rho_p(\mathscr{E})$, where $\mathscr{E} = h\nu$. Therefore, the absorption transition rate can be written as

$$R_{ab} = B_{12}f_1(1 - f_2)\rho_p(\mathscr{E}) \tag{8.35}$$

and the downward stimulated emission rate can be written as

$$R_{st} = B_{21}f_2(1 - f_1)\rho_p(\mathscr{E}) \tag{8.36}$$

The spectral density of photons $\rho_p(\mathscr{E})$ can be determined in a manner similar to the concentration of electrons and holes. However, the distribution law that governs photons is different from that for charged particles, because photons are not subject to the Pauli exclusion principle, and they are regarded as a system of indistinguishable particles having integer spin. The Bose–Einstein distribution law applies to such a system. The average number of photons in a single quantum state is given by the expression

$$\langle N \rangle = \left[\exp\left(\frac{\mathscr{E}}{kT}\right) - 1 \right]^{-1} \tag{8.37}$$

Therefore, the photon density distribution per unit volume and unity energy $\mathscr{E}$ can be obtained by multiplying $\langle N \rangle$ by the density of states, $\rho(\mathscr{E})$. Note that $k = 2\pi n_0/\lambda = 2\pi n_0\mathscr{E}/hc$ and $dk = (2\pi/hc)(n_0 + \mathscr{E}\,dn_0/d\mathscr{E})\,d\mathscr{E}$, thus

$$\rho_p(\mathscr{E})\,d\mathscr{E} = \frac{k^2}{\pi^2}\,dk = \frac{8\pi n_0^2\mathscr{E}^2}{h^3c^3}\langle N \rangle\left(n_0 + \mathscr{E}\frac{dn_0}{d\mathscr{E}}\right)d\mathscr{E} \tag{8.38}$$

where n_0 is the refractive index of the material. Using Equations (8.37) and (8.38) and ignoring material dispersion yields

$$\rho_p(\mathscr{E}) = \frac{8\pi n_0^3 \mathscr{E}^2}{h^3 c^3} \left[\exp\left(\frac{\mathscr{E}}{kT}\right) - 1 \right]^{-1} \tag{8.39}$$

The relationships between the three probability coefficients can now be established by using detailed balancing between the upward and downward transitions. At thermal equilibrium the following relations apply

$$R_{ab} = R_{st} + R_{sp} \tag{8.40}$$

and by definition

$$F_c = F_v \tag{8.41}$$

Substituting Equations (8.32), (8.35), and (8.36) into (8.40) gives

$$\rho_p(\mathscr{E}) = \frac{A_{21} f_2 (1 - f_1)}{B_{12} f_1 (1 - f_2) - B_{21} f_2 (1 - f_1)} \tag{8.42}$$

The results of Equations (8.39) and (8.41) allows the separation of Equation (8.42) into temperature-dependent and temperature-independent terms as follows:

$$\exp\left(\frac{\mathscr{E}}{kT}\right)\left(\frac{8\pi n_0^3 \mathscr{E}^2}{h^3 c^3} B_{12} - A_{21}\right) = \left(\frac{8\pi n_0^3 \mathscr{E}^2}{h^3 c^3} B_{21} - A_{21}\right) \tag{8.43}$$

For all values of T, the quantities inside the parentheses on both sides of Equation (8.43) must vanish. Hence, the following relations:

$$A_{21} = \frac{8\pi n_0^3 \mathscr{E}^2}{h^3 c^3} B_{21} \tag{8.44}$$

and

$$B_{21} = B_{12} \tag{8.45}$$

Even though the analysis above is carried out for a system in thermal equilibrium, the probability coefficients are independent of the nature of the system, and can be used for systems that deviate from thermal equilibrium. In this case assume that field intensity changes only slowly with frequency and is essentially constant over the entire range of the emission spectrum.

8.6 POPULATION INVERSION

In thermal equilibrium, as shown by Figure 8.7(a), a photon with energy $\mathscr{E} = h\nu \geq E_g$ will be absorbed by the system and will excite a valence electron into the conduction band. This additional electron will again return by recombination to the valence band after an average time τ_{sp} that defines the spontaneous emission lifetime. Under nonequilibrium conditions, as shown by Figure 8.7(b), the photon can induce a downward event called stimulated

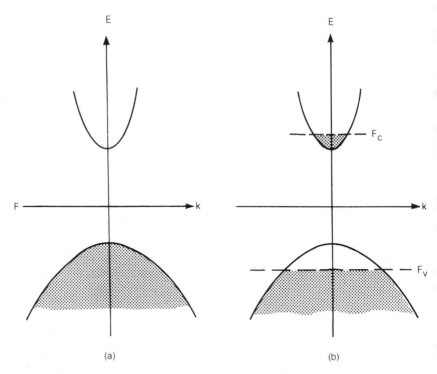

FIGURE 8.7 Electron energy diagram as a function of density of states for systems (a) in thermal equilibrium and (b) in nonequilibrium.

emission. In this case population inversion is established by injecting electrons over potential barriers at heterojunctions, as discussed in Chapter 9. Because an equal number of holes is generated in the valence band to maintain charge neutrality, the states in the valence band up to F_v are empty. Because the conduction band is filled up to F_c, photon energies greater than E_g but less than $F_c - F_v$ cannot be absorbed, but these photons can induce downward electronic transitions by recombining electrons from the conduction band with holes in the empty valence band. Therefore, the requirement for stimulated emission is

$$F_c - F_v > h\nu$$

In other words, to sustain stimulated emission, it is necessary to satisfy the condition

$$R_{st} > R_{ab} \qquad (8.46)$$

Substituting Equations (8.35) and (8.36) into (8.46) provides

$$f_2(1 - f_1) > f_1(1 - f_2)$$

This condition is equivalent to

$$\exp[(F_c - F_v)kT] > \exp[(E_2 - E_1)/kT]$$

which implies that

$$F_c - F_v > E_2 - E_1 = h\nu \tag{8.47}$$

Equation (8.47) is the requirement for stimulated emission as previously stated.

To extend the foregoing analysis for transition rates between two levels E_1 and E_2 to the case involving a band-to-band transition, it is necessary to take into account the density of states as given by Equations (8.16) and (8.17) and to integrate the rate expressions over the entire energy range allowed by appropriate selection rules. For example, the absorption rate as given by Equation (8.35) depends on the density of the filled states in the valence band, $\rho_v(E_v - E_1)f_1$, and also on the density of empty states in the conduction band, $\rho_c(E_2 - E_c)(1 - f_2)$. The selection rule in this case is limited to the condition that the amount of energy separation between the two bands is exactly the same as that given by $h\nu$. With these modifications Equation (8.35) becomes

$$R_{ab} = \int_{-\infty}^{\infty} B_{12}\rho_v(E_v - E_1)\rho_c(E_2 - E_c)f_1(1 - f_2)\rho_p(\mathscr{E})\delta(E_2 - E_1 - \mathscr{E})\,dE$$

$$\tag{8.48}$$

Equation (8.48) implies that the transition from the valence band to the conduction band is allowed only between states where ρ_v and ρ_c are separated exactly by $h\nu$. Under this circumstance, k selection rules can be ignored.

Similarly for spontaneous emission, the rate given by Equation (8.32) depends on the density of the filled conduction band states, $\rho_c f_2$, and the density of the empty valence band states, $\rho_v(1 - f_1)$. Equation (8.32) can be rewritten as

$$R_{sp} = \int_{-\infty}^{\infty} A_{21}\rho_c\rho_v f_2(1 - f_1)\delta(E_2 - E_1 - \mathscr{E})\,dE \tag{8.49}$$

The same modification must be made to the stimulated emission rate given by Equation (8.36). To evaluate these rates, one of the two probability coefficients A_{21} or B_{12} must be determined. The calculation of these coefficients requires extensive quantum mechanical knowledge, which involves the use of time-dependent perturbation theory, as described in most texts on quantum mechanics. Basically, the matrix element for the interaction Hamiltonian, H, involving an electron and the radiation field between the two states in the semiconductor must be calculated. This treatment is omitted here because of the difficulty in obtaining the appropriate wave functions associated with a highly doped semiconductor whose band structure deviates significantly from the simple parabolic model. Several models have been introduced to modify the smearing effect of band edges by adding bandtails in the structure, as illustrated in Figure 8.5. The density of states deviates significantly near the band edges from the usual parabolic dependence for a semiconductor with an intermediate doping level. The hole impurity band structure, which lies fairly deep in the valence band, is characteristic of p-type doping at levels around 10^{18} cm^{-3}. In n-type materials impurity bands are much less conspicuous than those shown in Figure 8.5, because the donor band is relatively shallower and merges with the conduction band quickly at levels greater than 10^{16} cm^{-3}.

It is possible to relate the three radiative transition rates as defined by Equations (8.32), (8.35) and (8.36) to the absorption coefficient $\alpha(\mathscr{E})$ under excitation. Consider a beam of photons with an occupation number N_{12} between two states in a given mode interacting with a semiconductor. The rate of change in this number can be written as

$$\frac{dN_{12}}{dt} = -R_{ab}N_{12} + R_{st}N_{12} + R_{sp} \tag{8.50}$$

where the first term is the decrease in mode occupation number due to absorption and the next two terms describe the increase in occupation number due to stimulated and spontaneous emission. As the radiative processes proceed, a steady-state is reached at which the rates of stimulated and spontaneous emission are nearly equal. At this point, $\langle N_{12} \rangle$ becomes the steady-state occupation number so that $dN_{12}/dt = 0$. Equation (8.50) indicates that

$$\langle N_{12} \rangle = \frac{R_{sp}}{R_{ab} - R_{st}} \tag{8.51}$$

Assuming that near the lasing threshold

$$R_{sp} = R_{st} \tag{8.52}$$

and substituting Equations (8.35) and (8.36) into (8.51) gives

$$\langle N_{12} \rangle = \frac{f_2(1 - f_1)}{f_1 - f_2} = [\exp(\mathscr{E} - \Delta F)/kT - 1]^{-1} \tag{8.53}$$

where ΔF is the separation in quasi-Fermi levels. If the numerator and denominator of Equation (8.51) are divided by R_{st} and compared with Equation (8.53) the following results:

$$R_{ab} = R_{sp} \exp(\mathscr{E} - \Delta F)/kT$$

or

$$R_{sp} = R_{ab} \exp(\Delta F - \mathscr{E})/kT \tag{8.54}$$

The result of Equation (8.54) is obtained by assuming that a steady-state has been reached at which $R_{sp} = R_{st}$. These results can be obtained without making any assumption about calculating the steady-state occupation number $\langle N_{12} \rangle$ using a statistical mechanical approach (Ref. 8.3). More details can be found in Refs. 8.4 and 8.5.

Equation (8.54) describes the rate of spontaneous emission into a single mode. The total number of photons emitted spontaneously per unit volume per unit energy is obtained by multiplying R_{sp} by the mode density $\rho_p(\mathscr{E}) = (k^2/\pi^2) \, dk/d\mathscr{E} = 2n_0^2 \mathscr{E}^2/\pi h^3 c^2 v_g$. In terms of the absorption coefficient, $\alpha(\mathscr{E})$ which is related to R_{ab} by

$$\alpha(\mathscr{E}) = \frac{1}{v_g} R_{ab} \qquad (8.55)$$

Equation (8.54) can be written by using Equation (8.55) and by accounting for the mode density:

$$R_{sp} = \alpha(\mathscr{E}) \frac{2\mathscr{E}^2 n_0^2}{\pi h^3 c^2} \exp(\Delta F - \mathscr{E})/kT \qquad (8.56)$$

To relate the absorption coefficient with the measured spontaneous emission intensity $I(\mathscr{E}, \Delta F)$, which is proportional to R_{sp}, Equation (8.56) is rewritten as

$$I(\mathscr{E}, \Delta F) \propto \mathscr{E}^2 \alpha(\mathscr{E}) \exp(\Delta F - \mathscr{E})/kT \qquad (8.57)$$

Equation (8.57) shows ΔF is a more fundamental parameter than the injection current. ΔF represents the separation of quasi-Fermi levels. Each excitation current has a corresponding value of ΔF.

Figure 8.8 shows the measured emission spectra as a function of $\mathscr{E}$ for a 0.2-micron thick active $Al_{0.08}Ga_{0.92}As$ layer clad on either side with 1.5-micron thick $Al_{0.36}Ga_{0.64}As$ layers. For this heterostructure, the laser threshold current is 56 mA. Two vertical lines are drawn in Figure 8.8, indicating the laser line energy hv_L and the upper bound hv_u beyond which the absorption coefficient $\alpha(\mathscr{E})$ is nearly independent of the excitation current. Equation (8.57) gives the expression of the ratios of intensities as

$$\frac{I(\mathscr{E}, \Delta F)}{I(\mathscr{E}, \Delta F_L)} = \frac{\alpha}{\alpha_L} \exp(\Delta F - \Delta F_L)/kT \qquad (8.58)$$

where ΔF_L is the forward-bias energy at laser threshold. Carrier injection modifies the absorption edges, but for photon energies sufficiently above the absorption edge, the absorption coefficient is independent of carrier injection. Therefore, for $\mathscr{E} > hv_u$, Equation (8.58) simply reduces to

$$\frac{I(\mathscr{E}, \Delta F)}{I(\mathscr{E}, \Delta F_L)} = \exp(\Delta F - \Delta F_L)/kT \qquad (8.59)$$

The absolute value of ΔF_L can be determined from the requirement that at the lasing condition, $\Delta F = \Delta F_L$, the gain spectrum (see Section 8.8)

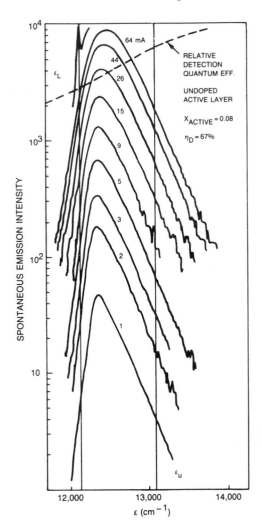

FIGURE 8.8 Measured spontaneous emission spectra. The laser threshold current is 56 mA. ϵ_L is the laser line energy and ϵ_u is an upper limiting value beyond which the absorption coefficient is independent of the injected current (Ref. 8.4. Reprinted with permission of J. Appl. Phys.).

has a maximum at $\mathscr{E} = \mathscr{E}_L$. The absorption coefficient can be expressed in absolute terms by determining α_L, the absorption coefficient at the laser line energy under lasing conditions. Equation (8.57) provides

$$\alpha = \alpha_L \frac{I}{I_L} \exp(\mathscr{E} - \mathscr{E}_L)/kT \exp(\Delta F_L - \Delta F)/kT \qquad (8.60)$$

Once α_L is determined, Equation (8.60) specifies $\alpha(\mathscr{E}, \Delta F)$. Equation (8.59) implies that the high energy tail of the spontaneous emission spectra differ from that at the laser line energy by a constant factor $\exp(\Delta F - \Delta F_L/kT)$. Now, consider two extreme cases, e.g., $I = 1$ mA and $I = 64$ mA, in Figure 8.8 and shift the highest spectrum curve at $I = 64$ mA downward so that

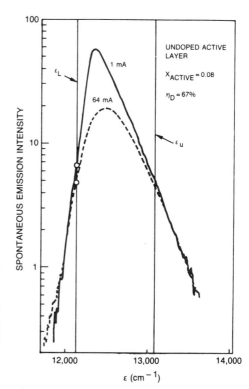

FIGURE 8.9 Spontaneous emission spectra at 1 mA and 64 mA. The dashed curve corresponds to that shown in Fig. 8.8, which has been shifted downward to coincide with the solid curve at $\mathscr{E}_u$. The ratio of these two curves is equal to the ratio of the absorption coefficients. (Ref. 8.4. Reprinted with permission of J. Appl. Phys.)

the tails of the two curves are brought into coincidence, as shown in Figure 8.9. The ratios of the luminescence intensities in Figure 8.9 are equal to the ratio of the absorption coefficients. At $h\nu_u$, the absorption coefficients differ only by 10%. This ratio increases to about 3.6 at the peak of the 1 mA spectrum and then decreases again. At the laser line energy $h\nu_L$, the ratio is only about 1.4 and goes down to unity where the two spectra cross at about 115 cm^{-1} below the laser line energy.

8.7 CARRIER LIFETIME

From the previous analysis of spontaneous and net stimulated transitions, it is possible to evaluate the radiative lifetime as limited by a band-to-band recombination. As defined by Equation (8.30), the recombination rate resulting from a specific transition is related to the carrier lifetime as

$$\tau_r = \frac{n}{R} \tag{8.61}$$

where n is the carrier concentration given by Equation (8.20) or (8.21) and

R is the corresponding transition rate. The measured values of α at various injection levels give the corresponding τ_r values. First consider the case involving either spontaneous emission or low injection level. The spontaneous recombination rate R_{sp} can be expressed by Equation (8.49) for the spontaneous emission rate. In terms of both electron and hole concentrations and by making use of Equations (8.20) and (8.21):

$$R_{sp} = \frac{8\pi n_0^3 \mathscr{E}^2}{h^3 c^3} \int_0^n \int_0^p B(n, p) \, dn \, dp \qquad (8.62)$$

where B is the rate coefficient for spontaneous emission. This form is useful especially in the case of very low injection, which results in a relatively undisturbed distribution of electrons and holes in the noninverted condition. To a good approximation, it is possible to take $B(n, p)$ outside the integral, and write a simple expression for the rate as

$$R_{sp} = B' np \qquad (8.63)$$

where B' is related to the rate constant and is equal to $(8\pi n_0^3 \mathscr{E}^2/h^3 c^3)B$. For the case of a very high injection level, the k value of at least one of the carrier types becomes very large, so that the k selection rule is relaxed to the extent that B is a constant over most of the integral. Therefore, the recombination rate can again be written in the simple form given by Equation (8.63). This simple relationship is no longer true when the Fermi levels move from within the bandgap into the bands. In this case, the value of B or the matrix element is strongly dependent on the energy of the initial and final states.

Under conditions of thermal equilibrium, the product np is equal to n^2. Hence the noninverted recombination rate is defined as $R_{sp}^0 = B'n^2$. Under nonequilibrium conditions, additional carriers $\Delta N = \Delta P$ are injected into the material. Therefore, the total recombination rate must be modified by the expression

$$R_{sp} = R_{sp}^0 + R_{sp}^{ex} \qquad (8.64)$$

where the recombination rate of the injected excess carrier is given by

$$R_{sp}^{ex} = R_{sp} - R_{sp}^0 = B'(n + \Delta N)(p + \Delta P) - B'np \qquad (8.65)$$
$$= B'\Delta N (n + p + \Delta N)$$

The radiative lifetime for the excess carriers is defined by

$$\tau_r = \frac{\Delta N}{R_{sp}^{ex}} = \frac{1}{B'(n + p + \Delta N)} \qquad (8.66)$$

At high currents (e.g., $\Delta N > n + p$) the radiative lifetime can be approximated by the expression

$$\tau_r \simeq (B'\Delta N)^{-1} \qquad (8.67)$$

Table 8.3 gives the values of B' and τ_r for three cases of p-type GaAs. The values are obtained from the absorption measurements under low injection levels. For p-type material, the radiative lifetime τ_r can be expressed in terms of $B'p_0$ or

$$\tau_r = \frac{1}{B'p_0} \qquad (8.68)$$

Also included in Table 8.3 are the calculated values of B' and τ_r based on the bandtails model. The calculated and measured values differ by only a factor of about 2. Figure 8.10 shows the radiative lifetime as a function of the hole concentration p_0 at 300°F. The data points represented by the squares are obtained by measuring the diffusion length in GaAs. According to simple diffusion theory the diffusion length Λ is

$$\Lambda = (D\tau_r)^{1/2} \qquad (8.69)$$

where D is the diffusion coefficient and is related to the mobility μ as

$$D = \frac{\mu kT}{e} \qquad (8.70)$$

At hole concentration levels greater than 10^{18} cm^{-3}, agreement exists among these lifetime data. Below this concentration level, the lifetime is dominated by nonradiative recombinations, which sharply degrades quantum efficiency. At levels above 10^{18} cm^{-3} quantum efficiency approaches 100% for most direct bandgap materials, and a recombination lifetime shorter than 1 ns is expected.

Besides spontaneous lifetime, two other lifetimes are relevant to the operation of a semiconductor laser: (1) stimulated lifetime and (2) photon lifetime. The values for the former are on the order of 10^{-11} s and decrease

TABLE 8.3   Radiative Recombination Rates, Lifetimes, and Other Parameters for p-type GaAs at 297°K

p_0 (cm^{-3})	E_g (eV)	μ (cm^2/V-s)	B' (10^{-10} cm^3/s) Exptl.	Theor.[a]	τ_r (ns) Exptl.	Theor.[a]
1.2×10^{18}	1.408	162	3.2	1.9	2.6	4.4
2.4×10^{18}	1.403	102	2.8	1.7	1.5	2.4
1.6×10^{19}	1.381	67	1.7	0.9	0.37	0.7

[a] From R. E. Fern and A. Onton, *J. Appl. Phys.*, 42, 3499 (1971).

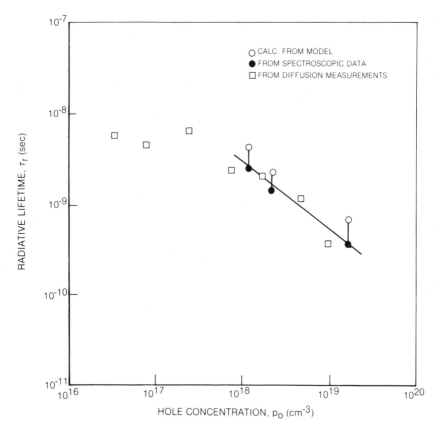

FIGURE 8.10   Radiative lifetime t_r as a function of hole concentration at 300°K. (From Ref. 8.6. Reprinted with permission of Academic Press, Inc., New York.)

with increasing light intensity. The latter has values on the order of 10^{-12} s for GaAs diode lasers and decreases with increasing cavity losses.

8.8 GAIN AND CURRENT RELATIONS

The optical gain in a semiconductor laser is related to the net rate between the stimulated emission and the absorption. The relationship given by Equations (8.52) and (8.54) provides

$$R_{\text{gain}} = R_{\text{st}} - R_{\text{ab}} = R_{\text{sp}}[1 - \exp(\mathscr{E} - \Delta F)/kT] \qquad (8.71)$$

The gain coefficient per unit length of a semiconductor laser, $g(\mathscr{E})$, is related to R_{gain} by an expression similar to Equation (8.55)

$$g(\mathscr{E}) = \frac{1}{v_g} R_{\text{gain}} \qquad (8.72)$$

Therefore, $g(\mathcal{E})$ and $\alpha(\mathcal{E})$ can be related through Equations (8.72), (8.71), (8.55), and (8.54) as

$$g(\mathcal{E}) = \alpha(\mathcal{E})[\exp(\Delta F - \mathcal{E})/kT - 1] \tag{8.73}$$

Equations (8.73), (8.60), and (8.59) can be used to convert the spontaneous emission curves in Figure 8.9 into the gain spectra shown in Figure 8.11. Now, the value of ΔF can be determined from the requirement that at the lasing condition $\Delta F = \Delta F_L$ the gain has a maximum of $\mathcal{E} = \mathcal{E}_L$. Substituting Equation (8.57) into (8.73) yields

$$g(\mathcal{E}, \Delta F) \propto I(\mathcal{E}, \Delta F)[1 - \exp(\mathcal{E} - \Delta F)/kT] \tag{8.74}$$

where $I(\mathcal{E}, \Delta F)$ is the measured spontaneous emission intensity. Now differentiating $g(\mathcal{E}, \Delta F)$ with respect to $\mathcal{E}$ and equating $dg/d\mathcal{E}$ to zero at $\mathcal{E} = \mathcal{E}_L$ yields

$$\exp(\Delta F_L - \mathcal{E}_L)/kT = 1 + I(\mathcal{E}_L, \Delta F_L)/kT(dI/d\mathcal{E})_{\mathcal{E}_L} \tag{8.75}$$

Equation (8.75) determines the value of ΔF_L for each current. Note that the calculated value of ΔF_L results in a maximum $g(\mathcal{E}_L)$ value at the laser line

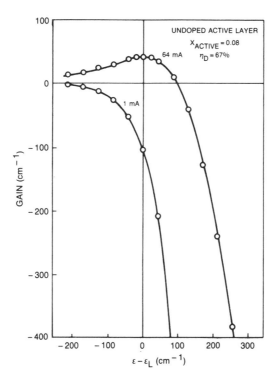

FIGURE 8.11   Gain spectra deduced from the spontaneous emission spectra in Figure 8.9. (Ref. 8.4. Reprinted with permission of J. Appl. Phys.)

energy. Also note that the loss at the laser line energy for the case of the lowest current 1 mA is only about 100 cm^{-1} for this laser.

To relate the gain to the minority carrier density n, ΔF must vary with n in a known fashion. This relationship can be derived by calculating the quasi-Fermi levels F_c and F_v from the following expressions for the carrier densities:

$$n = N_c \frac{2}{\pi} \int_0^\infty E^{1/2}[\exp(E + E_c - F_c)/kT + 1]^{-1} \, dE$$

and

$$p = N_v \frac{2}{\pi} \int_0^\infty E^{1/2}[\exp(F_v - E_v - E)/kT + 1]^{-1} \, dE$$

where the characteristic values of GaAs are: $N_c = 4.3 \times 10^{17}$ cm^{-3} and $N_v = 8.9 \times 10^{18}$ cm^{-3}. For undoped GaAs semiconductors, $n = p$, and for n-type materials, $n = p + N_D$. For p-type materials, $p = n + N_A$ where N_D and N_A are the concentrations of the donors and acceptors, respectively. The appropriate values for the minority carrier densities, $E_c - F_c$ and $F_v - E_v$ can thus be determined. The separation in the quasi-Fermi level ΔF can then be dertermined by the following expression:

$$\Delta F = E_g - (E_c - F_c) - (F_v - E_v) \tag{8.76}$$

The calculated minority carrier density as a function of bias energy difference $\Delta F - \Delta F_L$ is shown in Figure 8.12. The results of Figure 8.12 and the

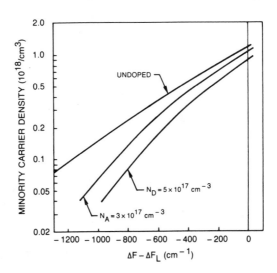

FIGURE 8.12 Calculations of minority carrier density versus bias energy change for undoped, slightly doped n-type and p-type GaAs active layers. (Ref. 8.4. Reprinted with permission of J. Appl. Phys.)

measured gain, deduced from Figure 8.8, can be plotted as a function of carrier density as shown in Figure 8.13 for an undoped GaAs active layer without aluminum. The absorption coefficients are typically in the 150–200 cm^{-1} range as the minority carrier density goes to zero. The gain versus carrier density curves for undoped and n-type GaAs and GaAlAs active layers usually exhibit a super-linear dependence with increasing carrier density. In contrast to this behavior, the gain versus carrier density relation for the p-type active layer usually appears to be slightly sub-linear with increasing minority carrier density.

In the linear regime, the gain coefficient can be written as

$$g(\mathscr{E}, N) = G(\mathscr{E})(N - N_0) \tag{8.77}$$

where $G(\mathscr{E})$ is the gain slope parameter, which is temperature-dependent. N and N_0 are the injected and nominal carrier densities, respectively. The typical values of G and N_0 for GaAs at $T = 300°K$ are 4×10^{-16} cm^2 and 7×10^{17} cm^{-3}, respectively. It is also useful to express the gain coefficient in terms of the nominal injection current density $J_{nom}(A/cm^2)$, which is related to the total radiative recombination rate R as

$$J_{nom} = eR \tag{8.78}$$

The actual current density $J(A/cm^2)$ is related to j_{nom} by

$$J = J_{nom} \, d/\eta \tag{8.79}$$

where d is the active-layer thickness and η is the quantum efficiency as-

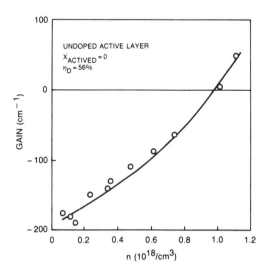

FIGURE 8.13   Gain versus carrier density at ϵ_L for an undoped GaAs active layer. (Ref. 8.4. Reprinted with permission of J. Appl. Phys.)

sociated with the transition. Equation (8.77) now becomes

$$g = \beta(J_{nom} - J_{nom}^{th}) \tag{8.80}$$

where J_{nom}^{th} is the nominal threshold current density at which the gain is equal to the absorption. For GaAs at room temperature, $J_{nom}^{th} = 4100$ A/$cm^2 - \mu m$ and $\beta = 0.044$ cm $- \mu m$/A. At low injection levels, the stimulated recombination process is negligibly small and R_{sp} can account for the entire luminescent process. In this case, R in Equation (8.78) can be related to the rate of spontaneous emission R_{sp} given by Equation (8.56) and can be written as

$$J_{nom} = \frac{8\pi e n_0^2}{h^3 c^2} \int \mathscr{E}^2 \gamma(\mathscr{E}, \Delta F) g(\mathscr{E}) \, d\mathscr{E} \tag{8.81}$$

where

$$\gamma = [1 - \exp(\mathscr{E} - \Delta F)/kT]^{-1} \tag{8.82}$$

The solution of this integral is not straightforward because of the difficulty in obtaining a satisfactory relationship for the density of state function and the precise dependence of g on $\mathscr{E}$. In practice it is useful to estimate the current density at $g(\mathscr{E}_m)$ where $\mathscr{E}_m$ is the photon energy corresponding to the maximum gain value lying between E_g and ΔF. Therefore, the value of J_{nom} can be approximated by the expression

$$J_{nom} \cong \frac{8\pi e n_0^2 \mathscr{E}_m^2 \gamma \, \Delta \mathscr{E}}{c^2 h^3} g(\mathscr{E}_m) \tag{8.83}$$

where $\Delta \mathscr{E}$ is the gain width. In practice, a family of curves for $g(\mathscr{E})$ can be obtained at various injection levels. Figure 8.14 shows the gain spectra for an $Al_{0.08}Ga_{0.92}As$ heterostructure at room temperature derived from measured spontaneous emission spectra using the technique discussed above. Two important features of this figure are worthy of note: (1) The gain peak shifts to higher photon energies as the injection current is increased. This shift is essentially caused by the increase in ΔF with increasing I or J. The gain coefficient at the peak of the gain spectrum increases with increasing excitation level, as expected. (2) The gain width decreases with increasing excitation level. At a given excitation level, the peak value of gain decreases with increasing temperature. At a lower temperature, the tail of the Fermi–Dirac distribution becomes steeper, which results in a more efficient inversion.

Threshold current density is one of the most important considerations in the design of injection lasers. Early devices were made of homojunctions that exhibited very high J_{th} (≥ 50 kA/cm^2). As a result, excessive heating occurred in the active region and in turn, the gain degraded drastically. Two major factors contributed to these high threshold currents. The first one was the lack of carrier confinement in a homostructure laser. Diffusion of the

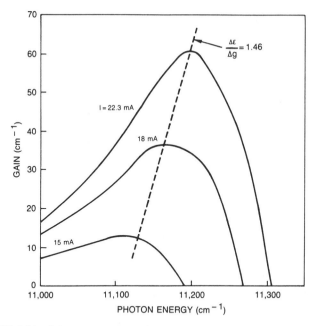

FIGURE 8.14 Gain spectra versus photon energy for an AlGaAs buried heterostructure laser converted from measured spontaneous emission spectra at currents below and at laser threshold. (After R. F. Kazarinow, C. H. Henry, and R. A. Logan, *J. Appl. Phys.*, *53*, 4631 (1982.)

injected carriers normal to the *p-n* junction plane produced a relatively thick (typically a few microns) active region, which, in turn, gave rise to high threshold current densities. The second factor was the poor optical confinement. The photons that were generated by stimulated emission were coupled into an optical field distribution that extended beyond the boundaries of the active region. As a result, only a small fraction Γ of the optical field could experience gain. The fraction $(1 - \Gamma)$ of the optical field that extended beyond the boundaries of the action region was absorbed by the material existing in the normal state. The role of heterostructures in reducing J_{th} and increasing laser efficiency will be discussed in Chapter 9.

The calculated gain coefficients for doped and undoped GaAs lasers based on a bandtailing approximation (Ref. 8.5) for the density of states to compute the transition matrix elements are shown in Figure 8.15. The calculated gain values that correspond to a relatively thick active layer of 1 micron, as a function of J_{nom} near liquid nitrogen (80°K) and room (300°K) temperatures fit Equation (8.80) quite well for *g*-values above 30 cm^{-1}. To compare *g*-values for lasing materials at a given J_{nom} value, it is important to normalize β by the active-layer thickness in microns and then multiply β by a confinement factor Γ. For undoped materials, a sharp cutoff point is

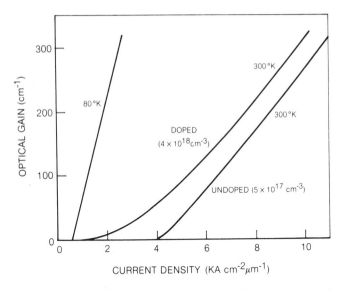

FIGURE 8.15   Calculated gain values as a function of injection current density at 80°K and 300°K. (Ref. 8.6 with permission of the Institute of Electrical and Electronics Engineers, Inc., © 1973.)

expected. As injection level increases, gain coefficient increases, and also the energy $\mathscr{E}_m$ at which the gain is maximum. This upward shift in energy or downshift in wavelength is a direct consequence of an increase in the separation energy between the quasi-Fermi levels.

For highly doped materials, cutoff is more or less gradual because bandtail states are present with significant impurity levels. In these calculations, a number of approximations must be introduced to allow for the effect of the bandtails on the density of states and on transition probability. Therefore, the results shown in Figure 8.15 should be used only as an estimate, not for precise comparisons. At low currents, higher doping increases optical gain because, the deeper the bandtail, the lower the Fermi energy required to produce inversion. However, the number of states occupied by injected electrons is much less than those in an undoped material having the usual parabolic band. At low temperatures, very little difference in gain is found between doped and undoped materials.

At high injection levels, the line width decreases considerably; however, it is compensated by a change in γ. Over a large range of currents, the product of $\gamma \Delta \mathscr{E}$ is relatively constant. Therefore, over more than a 20-dB dynamic range of gain, a linear relationship between gain and current density can be assumed.

The temperature dependence of J_{th} is shown in Figure 8.16, indicating that threshold current density increases with temperature in all types of semiconductor lasers. Because many factors are involved, no simple expres-

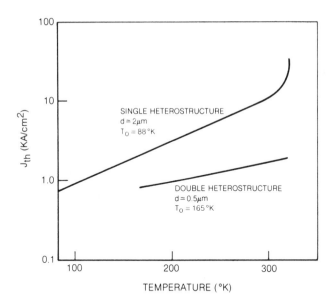

FIGURE 8.16   Temperature dependence of threshold current density for single- and double-heterostructure lasers. [After I. Hayaski, M. B. Danish, and F. K. Reinhart, *J. Appl. Phys.*, *42*, 1927 (1971).]

sion can rigorously describe the behavior over a wide temperature range. An empirical law often used to approximate the temperature effect is

$$J_{th} \propto \exp \left(\frac{T}{T_0} \right) \tag{8.84}$$

where T_0 is determined experimentally and found to be a constant over a certain operating range. As temperature increases slightly above room temperature, a superlinear increase in threshold current occurs and causes a complete cessation of laser action. For this reason, most earlier studies were done at liquid-nitrogen temperature (77°K). At 300°K the threshold current for a single-heterostructure laser is about 10,000 A/cm². It can be reduced to the 100-A/cm² level if a multiple-layer structure is employed. The details of heterostructure lasers are given in Chapter 10.

8.9 LASER OSCILLATION

As injection current is increased above J_{th}, the gain as given by Equation (8.80) becomes positive. To sustain laser oscillation, it is necessary to form an optical feedback circuit known as a laser cavity. For semiconductor lasers, the cavity is commonly formed using two parallel reflecting surfaces that are cleavage planes of the semiconductor crystal. This cavity makes semiconductor lasers different from other lasers in that optical feedback from the cleaved ends is very small, with a typical power reflection coefficient R of about 30%. The optical gain for semiconductor lasers, on the other hand, is extremely high, with a typical gain coefficient of $g \simeq 50$ cm^{-1}.

Laser threshold is defined as a condition in which loop gain is equal to the total cavity losses α_T. This condition can be established by considering a multiple reflection of a plane wave between two partially reflecting surfaces with reflectivity R_1 and R_2 of a Fabry–Perot cavity. If the gain coefficient is not saturated, the radiation intensity can grow exponentially with distance. The power at a distance z between the mirrors can be expressed by

$$P(z) = P(0) \exp[g(\mathscr{E}) - \alpha(\mathscr{E})]z$$

Making a round-trip of $2L$ inside the cavity gives

$$P(2L) = P(0)R_1R_2 \exp[2L(g - \alpha)]$$

At threshold, $P(2L) = P(0)$. This expression leads to the following condition for laser oscillation:

$$R_1R_2 \exp(g_{th} - \alpha_T)2L = 1 \qquad (8.85)$$

The most significant losses are the absorption by free carriers and the scattering by imperfections. As long as the condition of Equation (8.85) is satisfied, laser oscillation will be self-sustained. The only complication of a low-feedback laser system is that power distribution along the laser cavity length may not be uniform. Therefore, when this type of laser is analyzed, the nonuniformity condition must be taken into account. In addition, the spatial effect involving confinement of the injection current is also a very important consideration and is discussed in Chapter 9.

Equation (8.85) can be written as

$$g_{th} = \alpha_T + \frac{1}{2L} \ln \frac{1}{R_1R_2} \qquad (8.86)$$

Because the propagating mode spreads outside the active layer, g_{th} is reduced by the confinement factor Γ, as given by Equation (3.43). Substituting Equation (8.86) into Equation (8.83) provides an expression for the threshold current density:

$$J_{th}(\text{A/cm}^2) = \frac{8\pi ed \times 10^{-4} n_0^2 \mathscr{E}_m^2 \gamma \, \Delta\mathscr{E}}{\eta c^2 h^3 \Gamma} \left(\alpha_T + \frac{1}{2L} \ln \frac{1}{R_1R_2} \right) \qquad (8.87)$$

For p-type GaAs with $p_0 = 1 \times 10^{18}$ cm^{-3}, $R_1 = R_2 = 0.3$, $n_0 = 3.6$, $\alpha_T \simeq 10$ cm^{-1}, $\gamma = 9.2$, $\mathscr{E}_m = 1.4$ eV, and $\Delta\mathscr{E} = 0.06$ eV. Equation (8.87) gives a value for J_{th}/d at 300°K to be 3.5×10^3 A/cm^2-μm, assuming that $L = 400$ μm and $\eta = \Gamma = 1$. This calculated threshold value represents the theoretical limit for GaAs laser diodes, and is in agreement with measured values.

In many laser applications, it is desirable to choose an emitter with a low threshold current density so that the operating current for the device can be kept at a low level. Operating at low currents, energy dissipation and other deleterious effects that could lead to a gradual degradation of the emitter can be greatly reduced. According to the factors included in Equation (8.87), a lower threshold current density can be obtained by eliminating the leakage of injection currents and optical waves from the active layer and by reducing various losses. Several techniques have been developed to confine currents and optical waves and are discussed in Chapter 9. By improving the quality and doping level of epitaxial layers in a heterostructure and by optimizing end-face reflectivity and cavity length, a threshold current as low as 500 A/cm^2 at 300°K has been obtained.

8.10 OPTICAL MODES

The mode pattern of a semiconductor laser is determined primarily by the geometry of the laser. A typical heterostructure semiconductor laser is shown in Figure 8.17. It has a rectangular shape with multiple epitaxial layers grown on a crystalline substrate. The optical feedback is provided by the reflectivity of cleavage planes and forms a Fabry–Perot cavity. Unlike microwave resonators, optical resonators are open cavities in which only axial and small off-axis transverse modes can survive. Assuming that cavity mir-

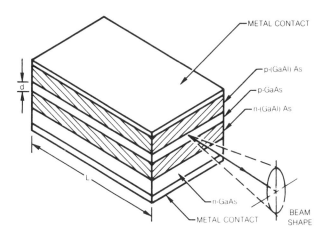

FIGURE 8.17   Structural diagram of a double-heterostructure AlGaAs laser diode showing the output radiation pattern emitted from cleaved end mirror along the junction plane.

rors are flats and separated by spacing L, the reflected and transmitted field amplitudes, A_r and A_t, from an incident plane wave having an amplitude A_i at the two surfaces, as shown in Figure 8.18, can be obtained by adding all amplitudes resulting from multiple reflections. For example

$$
\begin{aligned}
A_r &= A_1 + A_2 + A_3 + \cdots \\
&= A_i(r + tt'r'e^{i\delta} + tt'r'^3e^{2i\delta} + \cdots) \\
&= A_i\left(r + tt'r'e^{i\delta}\frac{1}{1 - r'^2e^{i\delta}}\right)
\end{aligned}
\tag{8.88}
$$

where r, r', t, and t' are reflection and transmission coefficients for the field amplitudes at appropriate boundaries. δ is the phase difference of two waves separated by a distance of one round-trip and can be written as

$$
\delta = \frac{4\pi n_0 L}{\lambda} \cos \theta
\tag{8.89}
$$

Because $\nu = c/\lambda$, the resonant frequency ν can be written as

$$
\nu = \frac{c\delta}{4\pi n_0 L \cos \theta}
\tag{8.90}
$$

Now the power reflection and transmission coefficient R and T can be defined as $R = r^2$ and $T = t^2$. For simplicity, assume that $r = r'$ and $t = t'$. Thus Equation (8.75) can be rewritten in terms of R as

$$
A_r = \frac{A_i(1 - e^{i\delta})\sqrt{R}}{1 - Re^{i\delta}}
\tag{8.91}
$$

Equation (8.91) assumed that $R + T = 1$. Similarly

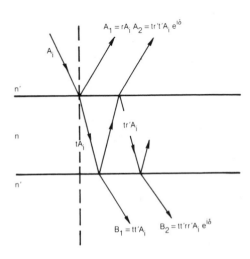

FIGURE 8.18 Multiple reflection and transmission from two planar surfaces.

$$A_t = B_1 + B_2 + B_3 + \cdots \tag{8.92}$$

$$= \frac{A_i T}{1 - Re^{i\delta}}$$

The reflected power is proportional to the product of the complex conjugate of the amplitude. Therefore

$$\frac{P_r}{P_i} = \frac{A_r^* A_r}{A_i^* A_i} = \frac{4R \sin^2(\delta/2)}{(1 - R)^2 + 4R \sin^2(\sigma/2)} \tag{8.93}$$

Similarly, for the power transmitted

$$\frac{P_t}{P_i} = \frac{(1 - R)^2}{(1 - R)^2 + 4R \sin^2(\delta/2)} \tag{8.94}$$

Equation (8.94) shows that if $\delta = 2m\pi$, the power transmitted is 100%. Substituting this value into Equation (8.90) yields resonance frequencies at which maximum transmission is

$$\nu_{max} = m \frac{c}{2n_0 L \cos \theta} \tag{8.95}$$

By letting $\theta = 0°$, the spacing between two axial modes can be obtained from Equation (8.95) as

$$\Delta \nu = \frac{c}{2n_0 L} \tag{8.96}$$

Figure 8.19 shows plots of P_t/P_i as a function of ν for various values of R. As values of R increase, the spectral width of transmission decreases. If $\Delta \nu_{1/2}$ is the full width at the half-power points on the transmission curve, it can be shown that

$$\Delta \nu_{1/2} = \frac{\Delta \nu}{F} \tag{8.97}$$

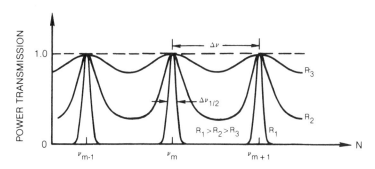

FIGURE 8.19   Transmission spectral characteristics of a Fabry–Perot interferometer.

where F is the finesse of the Fabry–Perot cavity and is given by

$$F = \frac{\pi R^{1/2}}{1 - R} \tag{8.98}$$

Equation (8.98) indicates that as R increases, finesse increases and the cavity resonances are sharply peaked, as shown in Figure 8.19. For a finesse of 100, a mirror reflectivity of 97% is required. In practice, the F value depends not only on R but also on mirror flatness as well as the angular spreading of the beam. Because of these limitations, a finesse of 100 is not easy to achieve even though $R > 97\%$.

For a typical GaAs laser whose cavity is formed by two cleaved ends, $R \simeq 30\%$. The axial mode spacing is about 2 Å for $L = 400$ μm. Figure 8.20 shows a typical emission spectrum of a room-temperature continuous-wave (CW) AlGaAs laser diode with an output that consists of many longitudinal or axial modes. In addition to the axial modes, lateral or transverse modes exist in most diode lasers. Lateral modes will be discussed in Chapter 10. An increase in the injection current above threshold corresponds to a sharp increase in the power output. Below threshold, only spontaneous radiation is emitted and the spectrum is very broad, with a spectral width of FWHM $\simeq 200$ Å. As threshold is approached, the spectrum begins to narrow by virtue of the stimulated emission through amplification. At the same time,

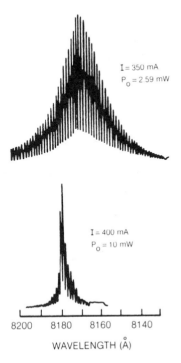

FIGURE 8.20 Output characteristics of a CW AlGaAs laser diode at different injection currents.

the emitted beam width is narrowed in the junction plane. As the current increases further, the spontaneous emission approaches a saturation value and stimulated emission dominates. The beam is then fully narrowed, as shown in Figure 8.20 by the near-field distribution. The output is usually shared among many modes, which are distributed within the line width of the gain medium. In semiconductor lasers, the gain width is homogeneously broadened primarily by inelastic collision with phonons having a line shape usually represented by a Lorentzian function.

Lasers, either with narrow stripe width (<10 μm) or utilizing distributed feedback rather than cleaved ends, can be made to operate in a single longitudinal and transverse mode over significant current ranges above threshold. The theoretical limit on the spectral broadening of a single mode can be estimated from the spontaneous emission rate R_{sp} or the photon lifetime τ_p. The line width $\delta\nu_m$ of the mode at FWHM increases with increasing photon number N_p, and can be expressed as

$$\delta\nu_m = \frac{\Delta\nu N_p}{2} \tag{8.99}$$

where $\Delta\nu$ is the spectral width of the emission line. The value of $\delta\nu_m$ is dependent on the quality factor of the laser resonator Q, and is related to the photon lifetime τ_p as

$$Q = \frac{\nu_0}{\delta\nu_m} = 2\pi\nu_0\tau_p \tag{8.100}$$

The power in this laser mode P_m is

$$P_m = \frac{N_p h\nu_0}{\tau_p} \tag{8.101}$$

By combining Equations (8.99), (8.100), and (8.101), the following results:

$$\Delta\nu = \frac{4\pi h\nu_0(\delta\nu_m)^2}{P_m} \tag{8.102}$$

In a typical GaAs laser diode, photon lifetime is on the order of 10^{-12} s; therefore,

$$\delta\nu_m \simeq \frac{10^{12}}{2\pi} = 1.6 \times 10^{11} \text{ s}^{-1}$$

At $h\nu_0 = 1.5$ eV and $P_m = 10$ mW the following is estimated:

$$\Delta\nu \simeq \frac{4\pi(2.4 \times 10^{-19})(1.6 \times 10^{11})^2}{10^{-2}}$$

$$\simeq 8 \text{ MHz}$$

which corresponds to a spectral width of 1.8×10^{-4} Å. Such a narrow width is difficult to achieve in practice, because the temperature fluctuation in this type of device introduces instability in cavity dimensions and noises in the spectrum. At room temperature, the lowest spectral width $\Delta\lambda \simeq 0.2$ Å has been achieved by operating a laser in a fundamental transverse mode. The value becomes much closer to the theoretical value by operating the laser at lower temperatures.

The physical significance of Equation (8.102) is the predictable inverse dependence of $\Delta\nu$ on laser power. A more refined expression for Equation (8.102) can be written by taking into account the effect of the spontaneous emission noise power, which is proportional to the population of the upper laser level N_2. In this case Equation (8.102) becomes

$$\Delta\nu = \frac{4\pi h\nu(\delta\nu_m)^2\mu}{P} \tag{8.103}$$

where

$$\mu = \frac{N_2}{N_2 - N_1(g_1/g_2)} \tag{8.104}$$

Because both N_1 and N_2 are power-dependent, it can be shown (Ref. 8.7) that as $P \to \infty$, a residual laser line width remains, which is power-independent. This residual width is due to the fact that as P increases, both N_1 and N_2 must increase. At sufficiently high values of P, the ratio N_2/P approaches a constant value, leading to a residual power-independent line width.

PROBLEMS

8.1. If the periodic boundary condition is imposed on the one-dimensional wave function of a system containing N atoms, show that the possible k_n values are $2\pi n/Na$, where $n = 0, 1, \ldots, N$.

8.2. Show that in the free-electron approximation, dE/dk vanishes at band edges, $k = n\pi/a$.

8.3. Using deBroglie's relation, show that the effective mass m_e of electrons in the conduction band of a semiconductor under the influence of a force field $F = m_e(dv_g/dt)$, is

$$m_e = \hbar^2/d^2E/dk^2$$

where $E = \hbar\omega$, is the energy of the electron, and $v_g = d\omega/dk$, is the group velocity associated with the electron wave packet.

8.4. Derive Equation (8.42).

8.5. Calculate the Fermi level in an n-type material when n/N_c is approximately equal to 0.1.

8.6. Calculate N_c in a p-type AlGaAs at 300°K where $m_e/m = 0.092$.

8.7. Show that the ratio of the density of states is

$$\frac{f_2(1 - f_1)}{f_1 - f_2} = \left\{ \exp \left[\frac{\mathscr{E} - (F_c - F_v)}{kT} \right] - 1 \right\}^{-1}$$

8.8. Show that the spacing of adjacent longitudinal modes in a semiconductor laser with a cavity length L is

$$\Delta\lambda = \frac{\lambda^2}{2n_0 L[1 - (\lambda/n_0) \, dn_0/d\lambda]}$$

8.9. Derive the threshold condition for laser oscillation as given by Equation (8.85).

8.10. Calculate the photon lifetime in a Fabry–Perot cavity in which a photon is lost by either absorption or transmission through the facets. Assume that a total loss coefficient for the cavity is 50 cm^{-1}.

8.11. Plot P_t/P_i as given by Equation (8.94) as a function of δ for $R = 90\%$.

REFERENCES

8.1. R. de L. Kronig and W. G. Penney, *Proc. Roy. Soc.*, *A130*, 499 (1931).

8.2. J. P. McKelvey, *Solid State and Semiconductor Physics*, Harper & Row, Publishers, Inc., New York, 1966.

8.3. G. H. B. Thompson, *Physics of Semiconductor Laser Devices*, John Wiley & Sons, Inc., New York, 1980.

8.4. C. H. Henry, R. A. Logan and F. R. Merrit, *J. Appl. Phys.*, *51*, 3042 (1980).

8.5. L. D. Landau and E. M. Lifshitz, *Statistical Physics*, Pergamon Press, Oxford (1980). Chapter V.

8.6. F. Stern, *IEEE J. Quant. Elect.*, *QE-9*, 290 (1973).

8.7. A. Yariv and K. Vahala, *IEEE J. Quantum Electron.*, *QE-19*, 889 (1983).

9

Properties and Growth of Semiconductor Heterojunctions

9.1 INTRODUCTION

This chapter makes use of the radiative recombination property of semiconductors to generate a class of light-emitting devices commonly known as light-emitting diodes (LEDs) and laser diodes (LDs) and to examine various structural properties and operating parameters of these devices. Presently, device structures of most LEDs and almost all LDs are made of heterojunctions, which are multilayered dissimilar semiconductors. The major distinction between a heterojunction and a homojunction is that two different bandgap materials are used at the junction. With properly constructed heterojunctions, it is possible to confine both injection currents and laser modes within the active layer or the recombination region.

Because two dissimilar materials are involved in such a structure, a mismatch in lattice constants can lead to a large number of nonradiative recombination centers at the junction and consequently reduce the light emission probability. Therefore, it is very important to select materials that have nearly identical lattice constants. GaAs and AlGaAs are the most widely used heterostructures and their properties have been studied extensively (Ref. 9.1) and are well understood. Another heterostructure consists of InGaAsP, and it has also received considerable attention recently because its emission spectra fall in the region of least material dispersion.

This chapter first reviews the fundamentals of a *pn* junction in a simple binary structure and discusses the effects of forward-bias on the potential barrier at the junction and on the distribution of electrons and holes in the bands. Similar considerations will be extended to a class of heterostructures made of ternary and quaternary compounds. The methods of epitaxial growth of these materials is then introduced. The relationships between the injection current and voltage and the confinement factors for currents and optical waves in these structures will be derived.

To meet the requirements of most optical fiber systems, an LED is an adequate source based on its available output power and speed of response. For long-distance and extremely high-data-rate systems, the use of LDs should definitely be considered. The trade-off for laser power and bandwidth is not just in the cost but most important, in the reliability of the devices. In general, lasers degrade much more rapidly than LEDs and the deterioration originates primarily from material imperfections. Improving the reliability of laser sources constitutes one of the active areas of research at present.

9.2 THE *pn* JUNCTION

Intrinsic semiconductors can be made into *n*-type and *p*-type materials by introducing slight deviations from stoichiometry with dopants or impurities. For an *n*-type semiconductor, the material is doped with a donor impurity, and for a *p*-type the dopant is called an acceptor. In the case of GaAs, elements such as S, Se, Te, Si, Ge, and Sn can be used to form the donor and Be, Mg, Zn, Cd, C, and also Si and Ge can be the acceptor. In heavily doped *n*-type materials, all valence-band states and to some extent conduction band states are filled up to the Fermi level F as shown in Figure 9.1(a). On the other hand, a heavily doped *p*-type semiconductor exhibits a downward shift of the Fermi energy into the valence band so that the entire conduction band and some states lying near the top of the valence band are empty, as shown in Figure 9.1(b).

When *n*-type and *p*-type semiconductors are brought into contact, the majority of carriers in each material (e.g., electrons in the *n*-type and holes in the *p*-type) diffuse into the regions at the opposite side of the junction and generate minority carriers. However, the concentration of the majority carriers is much greater than that of the minority carriers. This difference in concentration produces a depletion region in which space-charged layers are formed to sustain high electric fields. These space-charged layers arrange themselves in an electric dipole configuration with uncompensated donor ions on the *n* side and uncompensated acceptor ions on the *p* side. This charge distribution causes a potential difference ϕ at the junction, which prevents further diffusion from occurring. Beyond the depletion region, the potential remains at a constant level. The energy of an electron at the bottom of the conduction band is lower by $e\phi$ on the *n* side of the junction than that

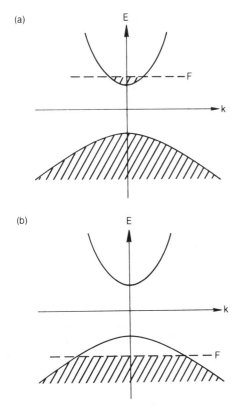

FIGURE 9.1 Electron energy diagram as a function of density of states for heavily doped (a) n-type and (b) p-type materials.

on the p side. The same is true for a hole in the valence band. The energy diagram of a pn homojunction without a forward bias is depicted in Figure 9.2. Most semiconductor devices contain at least one junction between the two types of semiconducting material. There are many ways to form a p-n junction, e.g., through the process of crystal growth, by an impurity diffusion process, or by ion implantation. The technology of junction growth is presented in Section 9.6. This section develops some useful mathematical descriptions of the p-n junction in order to gain a quantitative understanding of its properties.

At the junction, there exists a potential difference ϕ between the two types of material. As a result, a certain amount of charge transfer across the junction is expected because of the drift and diffusion of electrons and holes. The net result is that when all of the transfer processes are combined, there is no current flow under equilibrium. However, when either a forward or a reverse bias is applied to the junction, there will be a net flow of current. At equilibrium, the drift current density, $e\mu_p p(x)\mathscr{E}(x)$ must cancel exactly the diffusion current density, $eD_p \, dp/dx$, where μ_p is the mobility for holes,

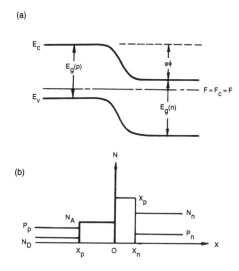

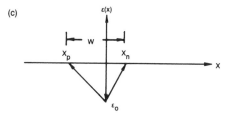

FIGURE 9.2 Abrupt *pn* homojunction without a bias: (a) energy-band diagram; (b) relative electron and hole concentrations and impurity levels in the *p* and the *n* regions; (c) electric field distributions in the *p* and the *n* regions.

$p(x)$ is the spatial distribution of holes, D_p is the diffusion coefficient, and $\mathcal{E}(x)$ is the electric field which is related to the potential differences as $-d\phi/dx$. Therefore

$$-\frac{e}{kT}\frac{d\phi}{dx} = \frac{1}{p}\frac{dp}{dx} \tag{9.1}$$

where the relation $D/\mu = kT/e$ has been used. Integrating over the entire depletion region yields

$$p_p = p_n \exp(e\phi/kT) \tag{9.2}$$

where p_p and p_n are the hole densities in the *p* and the *n* region, respectively. At equilibrium, $p_p n_p = p_n n_n = n_i^2$.

If the majority carrier concentrations are composed of N_A acceptors on the *p* side and N_D donors on the *n* side, an expression for the contact potential ϕ in terms of the concentrations of these dopants can be written as

$$\phi = \frac{kT}{e} \ln \frac{N_A}{n_i^2/N_D} = \frac{kT}{e} \ln \frac{N_A N_D}{n_i^2} \tag{9.3}$$

Because the dipole about the junction must have an equal number of charges on either side, the depletion region may extend into the p and n regions unequally, as shown in Figure 9.2(b), depending on the relative doping of the two sides. For a given cross-sectional area A, the total uncompensated charge on either side of the junction must satisfy the condition:

$$eAx_p N_A = eAx_n N_D \tag{9.4}$$

where x_p and x_n are the penetration depths of the space charges in p and n material respectively.

The electric field distribution within the depletion region must satisfy the following equation:

$$\frac{d\mathscr{E}}{dx} = \frac{e}{\epsilon}(p - n + \Delta N) \tag{9.5}$$

where

$$\Delta N = N_D - N_A \tag{9.6}$$

If the concentration of the carriers p and n can be neglected from the total space charge, Equation (9.5) can be separated into two regions as follows:

$$\frac{d\mathscr{E}}{dx} = \begin{cases} -\dfrac{e}{\epsilon} N_A & -x_p \le x \le 0 \\[2em] \dfrac{e}{\epsilon} N_D & 0 \le x \le x_n \end{cases} \tag{9.7}$$

Equation (9.7) shows that $\mathscr{E}(x)$ has two distinct slopes with a positive slope on the n side and a negative slope on the p side, as shown in Figure 9.2(c). The $\mathscr{E}_0$ value is given by the condition of Equation (9.4). The contact potential ϕ can be obtained by integrating $\mathscr{E}(x)$ over the depletion width W. Figure 9.2(c) shows a simple relationship between the contact potential to the width of the depletion region as

$$\phi = \frac{1}{2} \frac{e}{\epsilon} N_A x_n W^2 \tag{9.8}$$

Because $N_D x_n = N_A x_p$ and $W = x_n + x_p$, an expression for the depletion width W in terms of the doping concentrations and the contact potential can be obtained as follows:

$$W = \left[\frac{2\epsilon\phi}{e} \left(\frac{1}{N_A} + \frac{1}{N_D} \right) \right]^{1/2} \tag{9.9}$$

It is worth noting that the depletion width is proportional to the square root of the potential across the junction. More detailed discussions can be found in the book by Streetman (Ref. 9.2).

9.3 FORWARD-BIASED JUNCTIONS

The depletion region ($x_p < x < x_n$) in the vicinity of the junction has a much higher resistivity than any other part of the semiconductor. When an external bias voltage V is applied to the crystal, most of the voltage drop will occur across this region. A forward bias, in effect, will reduce the potential barrier height and will inject excess electrons into the p-region. The state of the system will no longer be in thermal equilibrium, or in other words, a unique Fermi level for the system is no longer defined. Two quasi-Fermi levels F_c and F_v are assigned to n- and p-regions, respectively, as shown in Figure 9.3(a). The concentrations of injection currents in both n- and p-regions contiguous to the space-changed layers are greatly perturbed from their equilibrium values, whereas the concentrations of majority carriers are not much affected. The electrons from the n-region and holes from the p-region can spill over the junction and become minority carriers on the opposite side. A condition known as population inversion can be established on the p side of the junction, as illustrated in Figure 9.3(a), by a penetration depth L of typically a few micrometers, which is characteristic of the diffusion length $L_e + L_h$ as determined by the doping gradient. Figure 9.3(b) shows a small downward step in the refractive index on the n side of the junction. This effect, commonly known as the free carrier depression, can provide a weak confinement for the optical beam. A much stronger effect can be obtained by varying the material composition in a ternary compound, and is discussed in Section 9.5.

The potential barrier at the junction is lowered by a forward bias V from the equilibrium contact potential ϕ to a smaller value $\phi - V$. As a result, the electric field also decreases with a forward bias, and consequently, the transition width W becomes narrower under a forward bias. These quantities can be calculated in a similar manner as presented in the last section simply by replacing ϕ with $\phi - V$. As shown in Figure 9.3(a), the bands are separated by $e(\phi - V)$ under a forward bias. This shifting of energy bands requires a separation of the Fermi level on either side of the junction by the same amount eV. Under a forward bias, there will be a supply of minority carriers on each side of the junction. They are composed of diffusion of majority carrier electrons on the n side, surmounting the po-

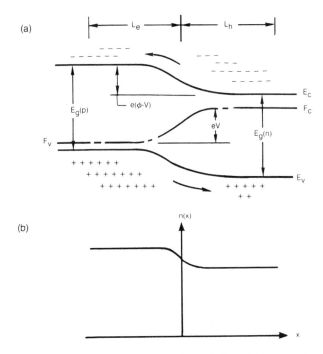

FIGURE 9.3 Abrupt *pn* junction under a forward-bias *V*: (a) energy-band diagram; (b) refractive index as a function of distance across the junction.

tential barrier to migrate to the *p* side, and holes to the *n* side. Using Equation (9.2) the minority hole concentration p' on each side of a junction can be expressed as

$$p'_p = p'_n \exp[e(\phi - V)/kT] \qquad (9.10)$$

At a low level of injection, it is reasonable to assume that the change in the majority carrier concentrations is negligible, so that $p'_p \simeq p_p$ and $n'_n \simeq n_n$. Taking the ratio of Equation (9.10) to that for the equilibrium case as described by Equation (9.2) yields

$$p'_n = p_n \exp(eV/kT) \qquad (9.11)$$

This result indicates that the probability that a carrier can diffuse under forward bias across the junction is proportional to the factor $\exp(eV/kT)$. This exponential increase of hole concentration in the *n*-type material with a forward bias is an example of minority carrier injection. The excess hole and electron concentrations Δp_p and Δn_p at the edges of the transition region x_n and x_p can be expressed by:

$$\Delta p_n = p'_n - p_n = p_n(e^{eV/kT} - 1) \qquad (9.12)$$

$$\Delta n_p = n'_p - n_p = n_p(e^{eV/kT} - 1) \qquad (9.13)$$

If the excess electron and hole distributions are assumed to be exponential functions of the distance from the junction with diffusion lengths L_e and L_h, respectively [see Figure 9.3(a)], the following expressions apply:

$$\Delta p = \Delta p_n \exp(-x/L_h) \tag{9.14}$$

and

$$\Delta n = \Delta n_p \exp(-x/L_e) \tag{9.15}$$

The current density can be calculated by taking the gradient of the distribution functions evaluated at the edges of the depletion region. Thus the current densities of holes J_h and electrons J_e at x_n and x_p are

$$(J_h)_{x_n} = -eD_h \left(\frac{\partial \Delta p}{\partial x}\right)_{x_n}$$
$$(J_e)_{x_p} = -eD_e \left(\frac{\partial \Delta n}{\partial x}\right)_{x_p} \tag{9.16}$$

where D_h and D_e are the diffusion coefficients of the holes and electrons. The total current density due to both carriers can be obtained using Equations (9.10) to (9.16) and expressed as

$$J = e \left(\frac{D_h p_n}{L_h} + \frac{D_c n_p}{L_c}\right) \left[\exp\left(\frac{eV}{kT}\right) - 1\right] \tag{9.17}$$

Equation (9.17) indicates that under a forward bias, carriers will be injected across the junction with a current density proportional to the excess carrier densities under a nonequilibrium condition. The gain coefficient and lasing threshold condition are dependent on J and will be discussed in Chapter 10.

As V increases, a breakdown value V_B will quickly be reached. At $n = 10^{18}$ cm^{-3}, the typical V_B is about 2.8 V. The junction capacitance per unit area due to voltage induced charges in the depletion region is given simply by

$$C = \frac{\epsilon}{W} \tag{9.18}$$

where W is the width of the depletion region as given by Equation (9.9) with the exception that ϕ is replaced by $\phi - V$.

The typical diffusion length in a GaAs pn junction device is on the order of several micrometers. Even this short distance is already too wide to establish a good current confinement to achieve a low current threshold laser device that operates at room temperature. Typical threshold currents of pn homojunction GaAs laser diodes are on the order of 10^3 A/cm^3 at 77°K but increase exponentially with temperature and reach values above 10^5 A/cm^2 quickly as the temperature approaches room temperature. Therefore, no homojunction laser diodes have been successfully made to operate at room temperature. However, low-cost room-temperature and CW LEDs can be made from *pn* GaAs.

Figure 9.4 shows a laser structure in the form of a parallelopiped with a planar diffused *pn* junction lying a distance of about a few micrometers below the top surface. Metal electrodes were deposited on the top and bottom surfaces [100] of the wafer, and the cleaved end-faces [110] of the crystal were used as cavity mirrors. With this simple structure, room-temperature CW operation was not possible even though a considerable amount of effort was put forth into the development of this structure by providing adequate heat sink and current confinement with the help of the strip-geometry electrode. The only mode of operation of the homostructure laser is in very short pulses with a pulse width on the order of 0.1 μs at low duty cycles of less than 0.1%.

Since the demonstration of the first working injection semiconductor laser, laser technology has undergone several stages of development before reaching its maturity as a practical source for optical fiber systems. All lasers today are made in much more complex configurations than the homostructure. However, many useful and economical LEDs are still made in this simple configuration by using zinc diffusion into *n*-type GaAs to form the *pn* junction. This simple device competes very strongly with the more sophisticated but considerably more efficient double-heterostructure AlGaAs laser diodes because of low cost and reliability. The most efficient geometry for the LED is one that allows radiation to be emitted from its surface, as shown in Figure 9.5. This geometry is commonly referred to as the Burrus type. Because *n*-type GaAs material is very conductive, it can be used as

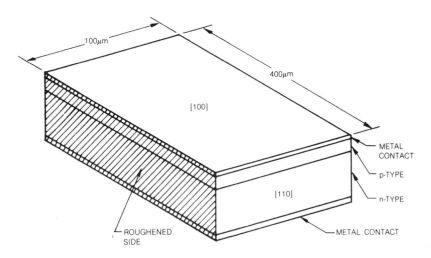

FIGURE 9.4 Schematic of a *pn* homostructure semiconductor laser.

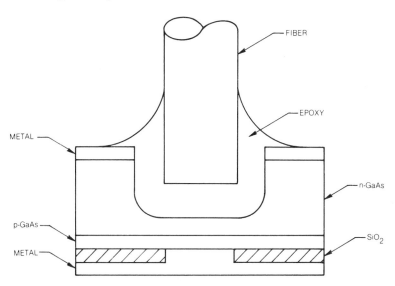

FIGURE 9.5 Schematic of a small-area Burrus-type LED-fiber coupling.

an electrode on one side of the junction without using a metallic layer, which would otherwise block the radiation emitting from the junction. It is necessary to remove as much n-type material as possible from the region above the junction by using a stop-etching technique to reduce the optical loss due to the absorption by free carriers. The emitting aperture of the Burrus-type LED is determined by the size of the opening hole in the SiO_2 layer, and can be made to match the aperture of the fiber core. Therefore, with this geometry, not only a very high radiance LED can be formed by restricting the emission to a small area but also very efficient coupling between the emitter and fiber can be obtained. The major disadvantages of the LED are (1) the wide spectral width $\Delta\lambda \simeq 200$ Å, which is more than one order of magnitude wider than that of an LD; and, (2) the low switching rate, which is governed by the spontaneous lifetime that is considerably longer than the stimulated lifetime. Both of these factors can affect the system bandwidth and impose a limit on the information-carrying capability.

9.4 SINGLE HETEROJUNCTIONS

Both injection currents and refractive indices can be controlled by varying either doping or material composition. To enhance current and optical confinement, several heterojunctions, such as pN, nP, nN, and pP in addition to the pn junction are often used in device topology. The capital letters denote large bandgap materials in forming dissimilar semiconducting junctions. One important feature of heterojunction HJ formation is the shape of the potential barrier which to a large extent is dependent on the process of expitaxial growth. In the case of very low temperature growth employing molecular beam epitaxy MBE, the transition width can be made extremely narrow ($\leq$40 Å) within which the potential barriers of the HJs are nearly abrupt, as shown in Figure 9.6(a). In other cases where the HJs are formed by diffusion, liquid phase epitaxy LPE, or metallorganic chemical vapor deposition MOCVD, the HJs are mostly all graded gap structures, as shown in Figure 9.6(b). The major factors to be considered are

1. energy bandgaps E_g of n- and p-type sides
2. transition width W
3. location of the p-n junction
4. doping concentrations N_D and N_A on n- and p-type sides
5. electron affinities χ_n and χ_p on n- and p-type sides

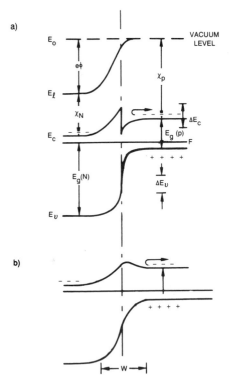

FIGURE 9.6   Energy band diagrams at thermal equilibrium of (a) an abrupt pN heterojunction, and (b) a graded pN heterojunction.

The number of variables involved and the difficulty of precisely measuring the relevant quantities make this a very difficult task to attack theoretically. The abrupt HJ's as shown in Figure 9.6(a), however, can be treated by using a simple Anderson model (Ref. 9.3). This model is restricted to an abrupt HJ and employing the depletion approximation, which assumes that there are no dipole layers or trapped charges at the heterojunction, and that the differences in electron affinities χ between two dissimilar kinds of material make up the alignment of potential energy barriers. The Anderson model predicts a discontinuity in the conduction band with a value $\Delta E_c = \Delta \chi$. This leads to a "spike-like" potential energy in the conduction band, as shown in Figure 9.6(a). This section first deals with the case of an abrupt heterojunction, e.g., $p - \text{GaAs}$ and $N - \text{AlGaAs}$ using the simple model. This model is then modified by allowing both the bandgap energy E_g and electron affinity χ to vary as a function of x within the transition region. This modification leads to a graded potential energy band profile with ϕ maxima and minima appearing in the transition region. The magnitudes of these maxima and minima depend on the material properties such as doping levels and material composition.

As shown in Figure 9.7, energy band relations for $p - \text{GaAs}$ and $N - \text{AlGaAs}$ materials before the formation of a heterojunction are given by:

$$E_g(N) = E_g(p) + \Delta E_c + \Delta E_v \tag{9.19}$$

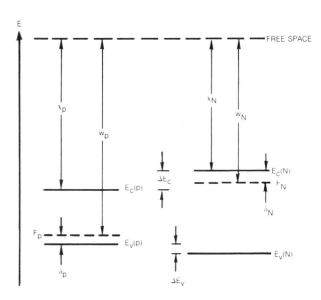

FIGURE 9.7   Energy-level diagram of p-GaAs and N-AlGaAs before forming an abrupt junction.

where

$$\Delta E_c = \chi_P - \chi_N \tag{9.20}$$

$$\Delta E_v = \chi_N + E_g(N) - \chi_P - E_g(b) \tag{9.21}$$

The bandgap energy E_g for $Al_xGa_{1-x}As$ can vary depending on the composition of aluminum fraction x. For $0 < x < 0.45$

$$E_g \text{ (eV)} = 1.414 + 1.247x \tag{9.22}$$

and for $0.45 < x < 1$, the bandgap changes from direct to indirect, and

$$E_g \text{ (eV)} = 1.424 + 1.147(x - 0.45)^2 \tag{9.23}$$

For a typical GaAs/$Al_{0.3}Ga_{0.7}As$ heterojunction, $\Delta E_g = 0.374$ eV and $\Delta E_c = 0.224$ eV.

The difference in work function W, which is the energy required to free an electron or a hole from the Fermi level to vacuum level, is

$$W_p - W_N = E_g(p) - \Delta E_c - (\delta_p + \delta_N) \tag{9.24}$$

where δ_p and δ_N are

$$\delta_p = F_p - E_v(p) \tag{9.25}$$

and

$$\delta_N = E_c(N) - F_N \tag{9.26}$$

When these two semiconductors are brought together, the electrons flow from higher to lower potential energy until an equilibrium is established. Because the two components of the heterostructure have different electron affinities and different energy gaps, the energy bands do not line up at the junction. This offset in the energy band diagram, as shown in Figure 9.6(a), plays a critical role in determining the charge rearrangement at the heterojunction. In the Anderson model, the carrier concentration in the depletion region is neglected. Therefore, the Poisson equation can be treated separately in two independent regions similar to the treatment previously for the p-n homojunction. If ϕ_p and ϕ_N are the potential barriers of the conduction band on the p-side and N-side, respectively, then

$$\frac{d^2\phi_p}{dx^2} = -\frac{e}{\epsilon_p}\Delta N_p \qquad 0 < x < x_p \tag{9.27}$$

and

$$\frac{d^2\phi_N}{dx^2} = \frac{e}{\epsilon_N}\Delta N_N \qquad -x_N < x < 0 \tag{9.28}$$

where ΔN_p is the difference in the number of acceptors and donors ($N_A - N_D)_p$ in p - GaAs, and ΔN_N is the difference in the number of donors and acceptors $(N_D - N_A)_N$ in N — AlGaAs. ϵ_p and ϵ_N are the dielectric constants: $\epsilon_p = 13.1\epsilon_0$ and $\epsilon_N = 10.1\epsilon_0$ where $\epsilon_0 = 8.85 \times 10^{-14}$ F/cm. The potentials outside the depletion region remain at constant values. At the junction, the potential ϕ_N reaches a peak value $\phi_N(0)$, and ϕ_p reaches a minimum value $\phi_p(0)$ with $\phi_N(0) - \phi_p(0) = \Delta E_c$. Integrating Equations (9.27) and (9.28) and letting the fields vanish at two edges of the depletion region yields

$$\mathcal{E}_p = -\frac{d\phi_p}{dx} = \frac{e}{\epsilon_p} \Delta N_p(x - x_p) \tag{9.29}$$

and

$$\mathcal{E}_N = -\frac{d\phi_N}{dx} = -\frac{e}{\epsilon_N} \Delta N(x + x_N) \tag{9.30}$$

Integrating Equations (9.29) and (9.30) again and using the boundary condition at $x = 0$, yields

$$\phi_p(x) = \phi_p(0) - \frac{e}{2\epsilon_p} \Delta N(x^2 - 2x_p x) \qquad \text{for } x \leq x_p \tag{9.31}$$

and

$$\phi_N(x) = \phi_N(0) + \frac{e}{2\epsilon_N} \Delta N(x^2 + 2x_N x) \qquad \text{for } x \geq -x_N \tag{9.32}$$

For $x \geq x_p$, $e\phi_p = E_c(p)$ and for $x \leq -x_N$, $e\phi_N = E_c(N)$. Therefore, the depletion depth in the p-side is

$$x_p = \sqrt{\frac{2\epsilon_p}{e^2 \Delta N_p}} [E_c(p) - e\phi_p(0)]^{1/2} \tag{9.33}$$

and in the N-side,

$$x_N = \sqrt{\frac{2\epsilon_N}{e^2 \Delta N_N}} [e\phi_N(0) - E_c(N)]^{1/2} \tag{9.34}$$

For a graded heterojunction interface, the above model must be modified. The energy bands for electrons in N-AlGaAs near a graded HJ is given by:

$$E_c(N) = E_0 - e\phi - \chi_N \tag{9.35}$$

and for the holes

$$E_v(N) = E_0 - e\phi - \chi_N - E_g(N) \tag{9.36}$$

where ϕ is the electrostatic junction potential as indicated in Figure 9.6(a).

The force on an electron in the conduction band is

$$-\nabla E_c = e\nabla\phi + \nabla\chi_N \qquad (9.37)$$

and on a hole in the valence band

$$\nabla E_v = -e\nabla\phi - \nabla(\chi_N + E_g) \qquad (9.38)$$

For simplicity, the valence band offset energy for a heterojunction is assumed to be zero. This is essentially the "common-anion" rule, which was discussed in Ref. 9.3. In this approximation, $\chi_N + E_g$ is a constant throughout the heterojunction. Consequently, the force on an electron is due to the net effect of the gradient in the electrostatic potential acting with or against the gradient in the electron affinity, or equivalently, the energy gap. The force on a hole, on the other hand, is due solely to the gradient in electrostatic potential. It is these differences in the internal forces that determine the differences in band bending, hence, the potential barriers in the heterojunctions.

With these modifications, the Poisson equation for the conduction band edge can be written as

$$\frac{\epsilon}{e^2}\frac{d^2E_c}{dx^2} = p - n + N_D - N_A - \frac{\epsilon}{e^2}\frac{d^2\chi}{dx^2} \qquad (9.39)$$

If the free electron n and hole p concentrations within the HJ transition region are neglected, Equation (9.39) reduces to

$$\frac{\epsilon}{e^2}\frac{d^2E_c}{dx^2} = N_D - N_A - \frac{\epsilon}{e^2}\frac{d^2\chi}{dx^2} \qquad (9.40)$$

This equation is similar to the homojunction depletion approximation case except for the extra term $(\epsilon/e^2)\,d^2\chi/dx^2$. This term acts as a distributed electric dipole which affects only the conduction band. The magnitude of this "effective charge" relative to the fixed charge of ionized donors and acceptors determines how the conduction band edge fluctuates within the transition region. To investigate the nature of this term, Migliorato and White (Ref. 9.4) assumed a complementary error function dependence as given by

$$\chi(x) = \text{Const} - \frac{\Delta\chi}{2}\,\text{Erfc}(x/C) \qquad (9.41)$$

where

$$\text{Erfc}(x) = \sqrt{\frac{2}{\pi}}\int_x^\infty \exp(-\xi^2/2)\,d\xi \qquad (9.42)$$

and C is a characteristic length determined by the compositional gradation within the HJ transition region. Using this function, it can be shown that

$$\frac{\epsilon}{e^2}\frac{d^2\chi}{dx^2} = -\frac{2\epsilon\,\Delta\chi}{\sqrt{\pi}e^2C^2}\left[\frac{x}{C}\exp - (x/C)^2\right] \tag{9.43}$$

The term in the square bracket is "dipole-like" with peaks of ±0.43 occurring at $x/C = \pm0.7$ as shown in Figure 9.8(b). Therefore, the maximum positive and negative values of this effective charge are given by

$$\left.\frac{\epsilon}{e^2}\frac{d^2\chi}{dx^2}\right|_{max} = \mp\frac{0.86\epsilon\,\Delta\chi}{\sqrt{\pi}e^2C^2} \tag{9.44}$$

where the $-$ sign is for $x > 0$ and the $+$ sign applies to $x < 0$. A maximum in the conduction band can occur only if

$$\frac{\epsilon}{e^2}\frac{d^2E_c}{dx^2} < 0 \tag{9.45}$$

Considering the N side of the $Al_{0.3}Ga_{0.7}As/GaAs$ junction where N_A is negligible and substituting Eq. (9.44) into (9.40), the condition for a maximum in the conduction band is

$$C < \left[\frac{0.86\epsilon\,\Delta\chi}{\sqrt{\pi}e^2N_D}\right]^{1/2} \tag{9.46}$$

Substituting the following representative values in Equation (9.46): $\epsilon_N = 8.9 \times 10^{-13}$ F/cm, $\Delta\chi = 0.24$ eV, $N_D = 3 \times 10^{17}$ cm^{-3} and $E_c - F_c = 0.046$

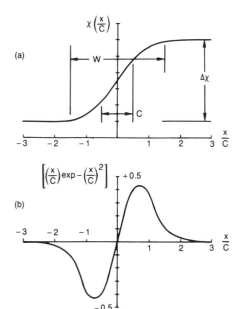

FIGURE 9.8 (a) Complimentary error function dependence of electron affinity. (b) Dipole-like effective charge distribution derives from (a).

eV, a value for $C < 0.026$ μm is obtained. This characteristic length C is related to the transition width by the equation

$$W \simeq 3C \tag{9.47}$$

Therefore, the requirement for a maximum in the conduction band on the N = side is that $W < 0.08$ μm. Similar analysis can be applied to establish a condition for a peak on the p-side of the junction. In this case, N_D is negligible and Equation (9.40) reduces to

$$\frac{\epsilon}{e^2} \frac{d^2E_c}{dx^2} = -N_A + \frac{0.86\epsilon \; \Delta\chi}{\sqrt{\pi}e^2C^2} < 0 \tag{9.48}$$

where the maximum positive value given by Equation (9.44) has been used. This leads to the following condition:

$$C > \left[\frac{0.86\epsilon \; \Delta\chi}{\sqrt{\pi}e^2N_A}\right]^{1/2} \tag{9.49}$$

Using the following representing values in Equation (9.49): $\epsilon_p = 1.16 \times 10^{18}$ F/cm, $\Delta\chi = 0.224$ eV, $N_A = 1.1 \times 10^{18}$ cm^{-3} and $F_v - E_v = 0.053$ eV, a value for $C > 0.01$ μm or $W > 0.03$ μm is obtained. These results are used to obtain qualitative energy band diagrams as shown in Figure 9.9. If $W < 0.03$ μm, a peak in the conduction band occurs on the N-side and if $W > 0.08$ μm, a peak occurs on the p-side. If $0.03 < W < 0.08$ μm, two peaks occur; one on the N-side and the other on the p-side.

The above analysis can also be applied to nP, pP and nN heterojunctions. In the cases of nN and pP, where the carrier concentration differs only very slightly from the net impurity doping level, the spatial variation of the potential curves for the conduction bands are relatively constant on either side of the junction and differ by a finite step with a height given by ΔE_c. The energy diagrams for these two isotype heterojunctions are shown in Figure 9.10. In an abrupt junction, the spike-like profile acts as a barrier to the minority carrier flow. If the potential barriers are graded over a distance comparable to the thickness of the depletion width, electrons and holes would be able to flow freely across the junction in either direction, especially under a forward bias voltage.

When a forward bias voltage is applied to a heterojunction, as shown in Figure 9.11(b), the potential barrier on the p-side is reduced by an amount equal to eV, while the N-side is not affected. The resultant energy level diagram shown in Figure 9.11(a) may be compared to the corresponding energy level diagram for the same junction in equilibrium. Under a forward bias, the electrons will be injected from the N-side into the p-side and the holes will be diffused from the p-side into the N-side. The current and voltage relationship can be derived by using the diffusion equation. In the case that

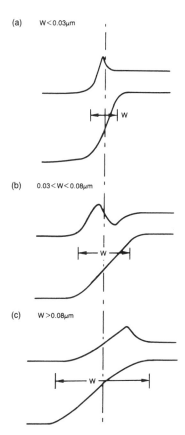

FIGURE 9.9 Graded heterojunction energy band diagrams showing variations in a conduction band energy level as the width of transition region increases.

excess electrons diffuse into the p-region, the steady-state diffusion equation is:

$$\frac{d^2n}{dx^2} - \frac{n - n_p}{L_e^2} = 0 \qquad (9.50)$$

where L_e is the diffusion length for the electrons. It is defined by

$$L_e = (D_e \tau_r)^{1/2} \qquad (9.51)$$

where D_e and τ_r are the diffusion coefficient and electron-hole recombination lifetime, respectively. The term n_p in Equation (9.50) is the excess electron concentration in the p-region at equilibrium. The solution of Equation (9.50) can be obtained by two successive integrations, as given by:

$$n(x) = C_1 \exp(x/L_e) + C_2 \exp(-x/L_e) + n_p \qquad (9.52)$$

At $x = \infty$, $n = n_p$. This implies that $C_1 = 0$. At $x = x_p$, $n = n_p \exp(e\psi/kt)$,

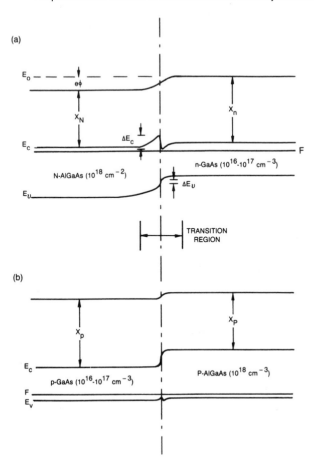

FIGURE 9.10 Energy band diagrams of two isotype heterojunctions: (a) nN heterojunction; (b) pP heterojunction.

where $\psi = \phi - V$, so that

$$C_2 = n_p[\exp(e\psi/kT) - 1]\, \exp(x_p/L_e) \tag{9.53}$$

Substituting C_2 into Equation (9.52) yields

$$n(x) = n_p[\exp(e\psi/kT) - 1]\, \exp[-(x - x_p)] + n_p \tag{9.54}$$

Equation (9.54) indicates that under a forward bias, the injected minority carries concentration decreases exponentially with increasing x from the junction ($x = 0$). At $x = x_p$, the current density can be evaluated by the equation

$$J_e = eD_e \left. \frac{dn}{dx} \right|_{x=x_p}$$

$$= \frac{eD_e n_p}{L_e} [\exp(e\psi/kT) - 1]$$ (9.55)

Similarly, diffusion of holes into the N-side is given by

$$J_h = \frac{eD_h n_N}{L_h} [\exp(e\psi/kT) - 1]$$ (9.56)

The carrier concentration in the p-side is $p \simeq N_A$ where N_A is the acceptor concentration in the narrow bandgap material. The carrier concentration in the N-side is $n \simeq N_D$, where N_D is the donor concentration in the wide bandgap material. Making use of Equation (8.28) and taking the ratio of J_e and J_h yields

$$\frac{J_e}{J_h} = \frac{D_e L_h N_D (N_c N_v)_p}{D_h L_e N_A (N_c N_v)_N} \exp(\Delta E_g/kT)$$ (9.57)

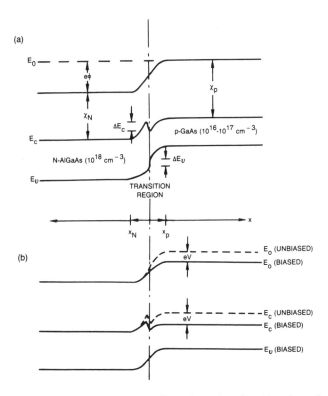

FIGURE 9.11   Energy band diagrams of a pN heterojunction: (a) at thermal equilibrium; (b) under a forward bias.

where $\Delta E_g = E_g(N) - E_g(p) = 0.37$ eV for $Al_{0.3}Ga_{0.7}As/GaAs$. At room temperature, the value of the exponential factor in Equation (9.57) is about 10^6 and is a dominating factor as compared with all others. Therefore, very high injection efficiency can be maintained in this material system.

For a $GaAs-Al_{0.3}Ga_{0.7}As$ nP heterojunction at room temperature $(297°K)$, $(N_D - N_A)_n = 1 \times 10^{18}$ cm^{-3} and $E_c - F_c = 0.042$ eV on the n side; $(N_A - N_D)_p = 2.3 \times 10^{17}$ cm^{-3} and $F_v - E_v = 0.109$ eV on the P side. For a pN heterojunction of the same composition $(x = 0.3)$ at $297°K$, $(N_A - N_D)_p = 1.1 \times 10^{18}$ cm^{-3} and $F_v - E_v = 0.053$ eV on the p side; $(N_D - N_A)_N = 3 \times 10^{17}$ cm^{-3} and $E_c - F_c = 0.046$ eV on the N side. The expressions for built-in potentials, depletion width, and injection currents under forward bias in an nP heterostructure can be obtained by following a treatment similar to the one already discussed in detail for the pN heterostructure.

The simplest single heterostructure SH laser has npP topology, which consists of three layers: n-GaAs, p-GaAs, and P-AlGaAs. It can be fabricated by the liquid-phase expitaxial LPE growth of a P-layer of AlGaAs on a heavily doped n-type GaAs substrate. Either during the growth or annealing cycle, a sufficient amount of Zn can be diffused into the n-GaAs substrate to form the pn junction. The optimum thickness of the active p-layer in this case is about 1 to 2 μm. The measured threshold current density of this laser as a function of the active layer thickness d is shown in Figure 9.12 for two different laser cavity lengths: $L = 250$ μm and $L = 400$ to 500 μm. The lowest J_{th} $(300°K)$ value is about 8.5×10^3 A/cm^2, which represents a significant improvement from the pn homostructure laser. Even then, CW operation of this laser at room temperature is very difficult. As previously mentioned, a reduction of J_{th} values for the SH lasers can be accomplished by closer optical and current confinement. By introducing optical confinement in SH lasers, the total cavity loss can be reduced from a typical value of 100 cm^{-1} to a value between 20 and 40 cm^{-1}. This reduction is simply a result of the difference in refractive index Δn at the pP heterojunction, where the Δn value is about a factor of 5 times that at the pn junction. Furthermore, the injection current is confined in the active p-layer because of the potential barrier at the heterojunction. Within this active layer, the gain profile is a slowly varying function of x. As d decreases below 2 μm, the optical confinement factor Γ decreases rapidly and consequently, J_{th} increases. Another point to be noted (Figure 9.12) is that higher electron concentrations in the n-layer yield lower threshold currents, and consequently higher emission power.

9.5 DOUBLE HETEROJUNCTIONS

Early work on semiconductor lasers dealt mostly with homostructures at liquid-nitrogen temperature $(77°K)$. The threshold current density for these

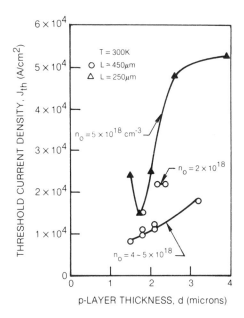

FIGURE 9.12   Variation of J_{th} at 300°K as a function of an active p-layer thickness of npP SH lasers. (From Ref. 9.1. Reprinted with permission of Academic Press, Inc., New York.)

lasers operating at room temperature was very high ($\geq 5 \times 10^4$ A/cm^2). With single heterostructures, the current threshold can be reduced by about a factor of 5 or more. Further reduction of the current threshold is possible by constructing a laser with double or multiple heterojunctions, because a further reduction of the thickness of the active layer can be activated with these structures without decreasing optical confinement. Because of strong current and optical confinement, significantly higher efficiency can be obtained from double-heterojunction (DH) lasers. There are many different types of DH structures which are used to configure semiconductor lasers. Figure 9.13(a) shows a NnP double heterojunction, and Figure 9.13(b) shows an NpP double heterojunction. In both cases, a forward bias is applied to the junction. In the NnP structure, the potential barrier between the wide bandgap P-AlGaAs and the narrow bandgap n-GaAs is reduced from ϕ to $\phi - V$, whereas in the NpP structure, the potential barrier between the wide bandgap N-AlGaAs and the narrow bandgap p-GaAs is reduced from ϕ to $\phi - V$. The potential difference between Nn and pP heterojunctions, on the other hand, is not changed very much under a forward bias. Other structures, which are made of all Al$_x$Ga$_{1-x}$As epitaxial layers with different aluminum composition x, are also used to make semiconductor laser diodes. Much discussion can be found in Chapter 10.

The most commonly used double heterostructure is the *NpP* configuration as shown in Figure 9.13(b), for which the electrons from the *N*-side can be injected into the *p*-side as a result of the reduction of the potential barrier at the *Np* junction. Most importantly is the discontinuity ΔE_c in the conduction band that provides a barrier to the electrons at the *pP* junction and thus confines the injected electrons to the *p*-GaAs active layer where the radiative recombination takes place. The thickness *d* of this active layer is usually much shorter than the diffusion length and is typically on the order of 0.1 to 0.3 μm. The valence-band discontinuity ΔE_v plus the built-in potential $e(\phi - V)$ provides a potential barrier for holes at the *Np* heterojunction, and prevents hole injection into the *N*-region. In this case, the confinement of holes is as effective as that created for the confinement of electrons by the discontinuities in the bands. Within the thin active *p*-layer, a constant level of injection carrier concentration can be assumed.

To illustrate carrier confinement, it is necessary to establish representative values of carrier concentrations in the active layer. For stripe-geometry *NpP* AlGaAs/GaAs/AlGaAs DH lasers (see Chapter 10), an injection of electron concentration of about 2×10^{18} cm^{-3} is needed to achieve lasing. The electron concentration for the direct conduction band ($x < 0.4$) is given by Equation (8.20) as the integral of the density of states $\rho_c(E - E_c)$ times the Fermi distribution function f_c as given by Equation (8.18). This product

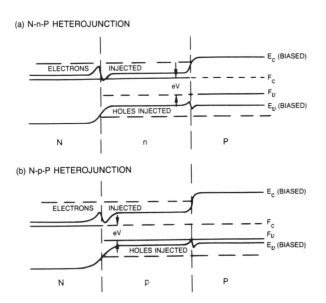

FIGURE 9.13 (a) An *NnP* double heterojunction under a forward bias. (b) An *NpP* double heterojunction under a forward bias.

$\rho_c f_c$ is the electron distribution $n(E)$ in the direct conduction band. For GaAs at $T = 297°K$, $F_c - E_c$ is 0.079 eV for $n = 2 \times 10^{18}$ cm^{-3}, and the resulting $n(E)$ is shown in Figure 9.14. For an AlAs mole fraction of 0.3, the difference between the conduction band edge of the p active layer and that of the P passive layer is $\Delta E_c = 0.318$ eV. This difference in ΔE_c is illustrated in Figure 9.6(a). $n(E)$ above ΔE_c is identified by the cross-hatched region shown in Figure 9.14. The concentration of carrier above ΔE_c denoted by n_Γ is given by

$$n_\Gamma = \frac{1}{2\pi^2} \left(\frac{2m_e}{\hbar^2}\right)^{3/2} \int_{\Delta E_c}^{\infty} [1 + \exp(E - F_c)/kT]^{-1}(E - E_c)^{1/2} \, dE \quad (9.58)$$

For this example, 1.6×10^{14} cm^{-3} electrons have energies above ΔE_c that can leak out of the active layer and contribute to the leakage current.

Next, consider the hole confinement at the Np interface at high forward bias as shown in Figure 9.11(b). The valence band discontinuity ΔE_v plus the built-in potential $e(\phi - V)$ provides a potential barrier for holes at the Np junction, and prevents hole injection into the N-region. The hole concentration can be obtained from Equation (8.23) for GaAs at $T = 297°K$. The resulting $p(E)$ is shown in Figure 9.15. In this figure, the hole barrier E_b is marked by the dashed line. $p(E)$ below the barrier is shown as cross-hatched and is given by

$$p_\Gamma = \frac{1}{2\pi^2} \left(\frac{2m_h}{\hbar^2}\right)^{3/2} \int_{-\infty}^{E_b} [1 + \exp(F_v - E)/kT]^{-1}(E_v - E)^{1/2} \, dE \quad (9.59)$$

Equation (9.59) gives a value for $p_\Gamma = 7.1 \times 10^{14}$ cm^{-3}. These holes can leak into the N passive layer and do not contribute to stimulated emission.

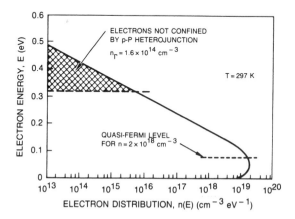

FIGURE 9.14   Electron distribution in GaAs direct conduction band to illustrate the portion of the distribution not confined by ΔE_c of 0.318 eV for $x = 0.3$. (Ref. 9.1. Reprinted with permission of Academic Press, New York, 1978).

Typical leakage currents arising from diffusion of n_Γ and p_Γ are only a few amperes per cm² which are very small values compared with the threshold currents.

The refractive index profile and light intensity distribution of an NpP DH laser are shown in Figure 9.16(b). A higher refractive index for the active p layer than that for the P- and N-cladding layers is achieved in this case by choosing a different material composition factor x. For low-doped N- or P-type $Al_xGa_{1-x}As$, the refractive index decreases linearly with increasing x from $n = 3.59$ at $x = 0$ to $n = 3.32$ at $x = 0.4$. The relation can be approximated by the equation

$$n(x) = 3.59 - 0.71x \qquad (9.60)$$

A dielectric slab waveguide with an asymmetric index profile as shown in Figure 9.16(b) can support two linearly polarized guided TE and TM wave modes, as discussed in Chapter 3. For a symmetric DH laser with $x = 0.3$, the first-order mode ($m = 1$) occurs at $d = 0.38$ μm and the second mode ($m = 2$) occurs at $d > 1.0$ μm. At larger values of x, the fundamental mode occurs at smaller d values. The variation of J_{th} with d for several x values is shown in Figure 9.17. For $d > 0.3$ μm, J_{th} is independent of x. For $d < 0.3$ μm, a considerable uncertainty exists in the measurements; nevertheless, J_{th} values decrease consistently with increasing x values. An empirical relation between J_{th} and d is found to be (Ref. 9.1)

$$J_{th} \simeq (5 \times 10^3 \text{ A/cm}^2\text{-}\mu\text{m}) \, d \qquad (9.61)$$

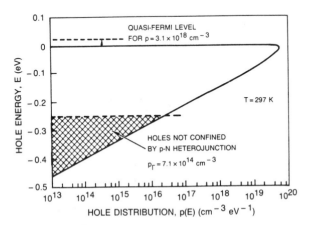

FIGURE 9.15 Hole distribution in GaAs valence band to illustrate the portion of the distribution not confined by $e(\phi - V) + \Delta E_v$ for $x = 0.3$. (Ref. 9.1. Reprinted with permission of Academic Press, New York, 1978).

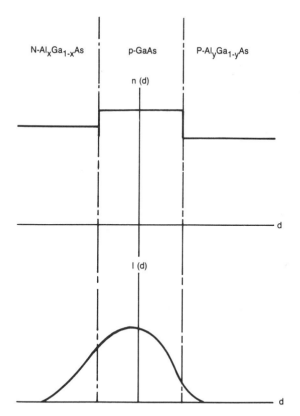

FIGURE 9.16   Refractive index and light intensity profiles of an *NpP* double heterojunction as a function of the thickness *d*.

where d is the active layer thickness in μm. For $x = 0.3$, $J_{th} \sim 1000$ A/cm^2 at $d \geq 0.15$ μm. For $x = 0.65$, $J_{th} \simeq 500$ A/cm^2 at $d \leq 0.1$ μm. The GaAs-AlGaAs DH laser is the most well developed semiconductor laser system at present and it was the first laser that was operated continuously (CW) at room temperature.

One major problem associated with DH lasers having very thin active layers is the deterioration of laser performance as a result of optical damage gradually occurring within these active regions. This problem is basically material-related; however, one way to reduce the severity of the problem is to spread light into a wider region without affecting carrier confinement. Separate confinements for light and carriers are possible but require additional epi-layers to the double heterostructure. Figure 9.18 shows the energy diagram of the bands, index profile, and light-intensity distribution for a five-layer system. This structure is known as the Large Optical Confinement

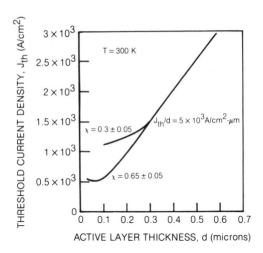

FIGURE 9.17   Variation of J_{th} at 300°K as a function of an active layer thickness for GaAs-AlGaAs DH lasers having a cavity length of 500 μm. (From Ref. 9.1. Reprinted with permission of Academic Press, Inc., New York.)

(LOC) and consists of a multilayered heterostructure grown on an N-type GaAs substrate with $y > x$. Typically, $y = 0.4$ and $x = 0.1$, which results in carrier confinement to the GaAs active region. However, the difference in the refractive index of the GaAs and $Al_xGa_{1-x}As$ layers is made sufficiently small so that the optical field overlaps both the p-GaAs and the N- and P-$Al_xGa_{1-x}As$ regions. The value of y as well as the layer thicknesses are chosen to minimize the threshold current density and to achieve a single spatial mode operation with large spot size. The active layer is typically 0.1 to 0.2 μm thick, and the $Al_xGa_{1-x}As$ guiding layer is 1 to 2 μm thick. With this type of structure a threshold current density of less than 500 A/cm² can be achieved.

9.6 MATERIAL PROPERTIES AND GROWTH OF SEMICONDUCTORS

Laser actions have been obtained from a variety of semiconductor compounds with threshold currents varying from less than 0.5 kA/cm² to greater than 100 kA/cm² at room temperature or below. This wide variation in performance is due to many factors, most of which can be related to material properties. The most efficient LD must be made of a direct bandgap material

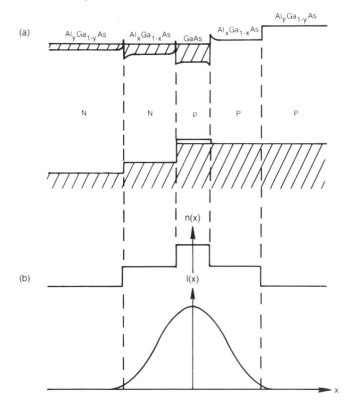

FIGURE 9.18   (a) Energy-band diagram (b) refractive index and light intensity as a function of heterostructural thickness for an LOC laser.

for its active region and have a multilayer heterostructure as discussed in Section 9.5. The major problem associated with heterostructures arises from interfacial lattice mismatch, which creates internal strain and dislocations commonly referred to as material imperfections. To increase performance and in particular, the life expectancy of an LD, material imperfections must be kept to a minimum. For this reason, lattice constants of these materials must be matched as perfectly as possible. To determine the matching compounds, it is convenient to plot the relationship between the lattice constant and the bandgap energy for a number of binary and ternary compounds. In Figure 9.19, the curves joining the binary compounds give the values for the energy gap and lattice constant of the ternary compounds. The behavior of the quaternaries can be interpolated from these ternaries. AlGaAs is the simplest system, for which a nearly perfect lattice match exists between GaAs and AlAs independent of the value of x. In contrast to this system, the lattice constant for all other ternaries varies appreciably with the com-

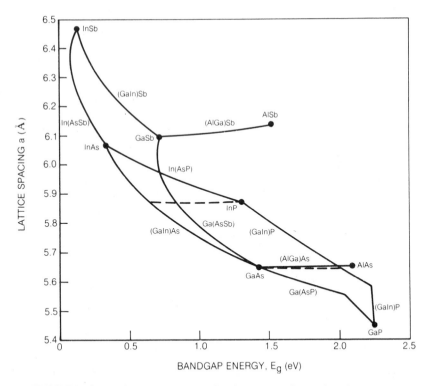

FIGURE 9.19   Lattice constant versus bandgap energy for semiconductor compounds.

position. Therefore, to make a perfect lattice match for systems other than AlGaAs, it is necessary to form a quaternary to gain the extra degree of freedom.

One interesting quaternary that has been studied extensively is the InGaAsP system emitting at wavelengths in the range of 1.1 to 1.6 μm. An epitaxial growth of this system must start with a substrate from which the lattice constant of the epitaxial quaternary layer is matched to that of the substrate. Figure 9.19 shows that two possible choices for the substrate are GaAs and InP. If GaAs is used as a substrate, the bandgap energy of the epitaxial layer can only follow the dashed line toward a higher value and terminate at the boundary curve represented by the GaInP solid solution. On the other hand, the InP substrate allows the growth of lower-bandgap quaternary material by following the dashed curve that terminates at the InGaAs solid solution. Clearly, the latter system is of more practical interest and, in principle, a quaternary compound with perfect lattice matching for the active layer can be grown to emit at the desired wavelength with minimal material dispersion.

The ternary $Al_xGa_{1-x}As$ compound changes from a direct to an indirect bandgap material as the composition factor, x, approaches 0.37, at which the bandgap energy of this compound is about 1.96 eV. The crossover points for other compounds are: 2.33 eV for AlInP, 2.25 eV for GaInP, 2.04 eV for AlInAs, and 1.97 eV for GaAsP. The composition factor x governs not only the radiative recombination rate but also the index of refraction, which decreases with increasing bandgap energy in the range of practical interest.

Three methods are commonly used for epitaxial growth of semiconductors: namely, (1) liquid-phase epitaxy (LPE), (2) chemical vapor deposition (CVD), and (3) molecular beam epitaxy (MBE). At present, LPE is the simplest and most widely used technique for producing the most efficient light-emitting devices. CVD and MBE are more accurate in reproducing layer thickness and material composition than LPE; however, they are more complicated, and in some cases many technical problems remain to be solved before superior-quality materials can be made in large quantity by these methods.

LPE can be defined, in general terms, as the growth of an oriented single crystal from a saturated or supersaturated liquid solution onto a crystalline substrate. This process is usually carried out by producing a saturated solution with appropriate composition at high temperatures and then allowing this liquid to be in contact with the substrate by some mechanical means at a lower temperature, during which epitaxial growth is instigated. To grow pure binary compound, it is necessary to provide one element from Group III and one element from Group V. The melting points for Group III elements such as Ga and In are near room-temperature; the melting points for Group V elements are at much higher temperatures. Therefore, it is only necessary to provide a sufficient amount of a Group III element to be dissolved in a Group V material at an elevated temperature that is adjusted to a growth rate. A variety of dopants can be added directly to a solution to yield the appropriate concentration. Some dopants can also be diffused into the III–V compounds in the solid phase at the growth temperature.

Ternary and quaternary semiconductors can be grown similarly to binaries, from solutions containing the appropriate constituents. To grow multilayer LPE, a common technique involves sliding a graphite boat, as shown in Figure 9.20, with multiple compartments containing different melts on top of a graphite plate with a recessed space for the substrate wafer. The boat is enclosed in a silica chamber placed within a furnace in an extremely pure hydrogen atmosphere. During growth, the melt, which is composed of accurately weighted constituent elements, must remain in the saturated condition at the growth temperature. Alternatively, a certain amount of supersaturation can be induced in each of the melts by passing them successively over the substrate at temperatures slightly (about 10 to 20°C) below the normal growth temperature to ensure nucleation of epitaxial growth of a thin

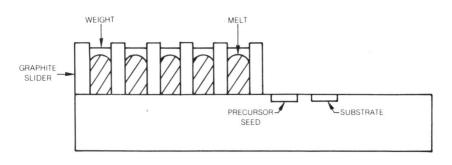

FIGURE 9.20   Sliding boat apparatus for liquid-phase epitaxy.

layer. Often a weight is applied against the upper surface of the melt to overcome the balling effect of surface tension. With some care, a very thin layer (≤ 0.1 μm) of fairly uniform thickness can be grown.

Even though excellent laser materials have been obtained using this simple method, it is, nevertheless, difficult to reliably reproduce surface quality and thickness uniformity, because very small thermal gradients and fluctuations can induce imperfections as well as strains and dislocations in epitaxial layers. Another problem associated with the LPE process is the occurrence of excess oxidation of melting compound when a solution is kept at equilibrium for a very long time. This event is particularly deleterious and can affect the life of the lasers. For these reasons, considerable efforts have been made to develop alternative techniques for more reliable epitaxial growth of large area substrates.

Vapor-phase epitaxy (VPE) and chemical vapor deposition (CVD) are also widely used techniques for growing single crystals. These methods are based on the principle of chemical reactions among various constituent elements being transported by gases over a heated substrate. Three main variants are often used for growing GaAs, AlGaAs, and InGaAsP compounds: (1) trichloride, (2) hydride, and (3) metallorganic systems. With the trichloride system single crystals of GaAs can be grown by passing arsenic trichloride gas over heated metallic gallium. The volatile gallium chloride and arsenic formed are transported in pure hydrogen gas over the substrate at a lower temperature to instigate epitaxial growth. For this system the relative fractions of Ga and As in the vapor phase are determined mainly by the temperature of the Ga solution, which must be accurately controlled. To provide better control of the concentration of constituent elements, the hydride process can be used because it bypasses the liquid phase and the major reactants are introduced directly into the gas phase. Group III elements (Ga and In) are usually introduced as monochlorides, and Group V elements are hybrids such as PH_3, AsH_3, and SbH_3. Using either LPE or CVD, both GaAs and InGaAsP have been grown successfully; however, the limitation of these systems is that they are not suitable for the growth of compounds containing aluminum. To accommodate this element, the metallorganic system has been introduced, whereby metallic compounds can be transported in a hydrogen carrier gas as metal alkyls together with the arsine gas. The metals used in this system consist of trimethyl aluminum, trimethyl gallium, diethyl zinc, and so on. Room-temperature AlGaAs/GaAs DH lasers having low threshold current have been fabricated by this technique with excellent control of thickness uniformity and concentrations.

Molecular beam epitaxy(MBE) is one of the newest methods and has gained considerable attention recently. Conceptually, it is the most direct method of material epitaxy utilizing ion beams directly from heated sources in an ultrahigh vacuum. Very accurate control over alloy composition and the deposition rate can be obtained with this technique. Low-threshold and single-transverse-mode DH lasers have been produced by this technique, and the performance has been brought up to the standards made by other methods. However, MBE has problems associated with it that are of a fundamental nature and involve kinetic effects such as surface migration and adsorption under nonequilibrium growth conditions.

By heating a substrate to a high temperature above 600°C, good single-crystal material with the desired composition can be grown with correct stoichiometry by controlling the relative arrival rates of the constituents from different ion beams. MBE is considered to be the most precisely controlled deposition process, which offers not only precise dimensional controls in the growth direction but can also achieve three-dimensional structures. To obtain low-threshold current density for DH lasers prepared by MBE, the growth conditions and substrate preparation must be handled with care. Of particular importance is the elimination of residual gases such as water vapor, CO, and O_2 in the growth chamber. To achieve this balance, an interlock chamber for sample exchange is used so that the growth chamber is always kept under ultrahigh vacuum. The threshold current density lowers significantly with increasing substrate temperature, T_s. The optimum T_s is around 650°C. Although MBE appears very attractive for present-day material research, the practicality of this technique for device production remains to be seen.

PROBLEMS

9.1. Calculate the depletion width of a *pn* junction under no bias voltage. *Hint:* Use the charge neutrality condition $N_D x_n = N_A x_p$.

9.2. Explain physically why the depletion width of an abrupt *pn* junction is proportional to the square root of the potential drop $\phi - V$.

9.3. Construct an energy-band diagram for an *nP* heterojunction at equilibrium by showing the spike and notch and the step in the band edges.

9.4. Calculate the optimum waveguide thickness for the fundamental mode in a symmetric DH laser with $\Delta n = 0.01$ and $\lambda = 0.9$ μm.

REFERENCES

9.1. H. C. Casey, Jr., and M. B. Parrish, *Heterostructure Lasers*, Part A: *Fundamental Principles*, and Part B: *Materials and Operating Characteristics*, Academic Press, Inc., New York (1978).

9.2. B. G. Streetman, *Solid State Electronic Devices*, 2nd Edition, Prentice Hall, Inc. (1980).

9.3. R. L. Anderson, *Solid-State Electron.*, 5, 341 (1962).

9.4. P. Migliorato and A. M. White, *Solid State Electron.*, 26, 65 (1983).

10

Semiconductor Lasers

10.1 INTRODUCTION

The sources for transmitters of optical fiber communication systems consist primarily of LEDs and semiconductor injection lasers. These devices emit in the wavelength range of 0.75 to 1.6 microns. In the lower portion of the spectra (0.75 to 0.95 microns) AlGaAs heterojunction devices are the most widely used sources. In longer wavelength regions (1 to 1.6 microns), InGaAsP devices are now available for system applications. This chapter introduces a class of laser diodes having a variety of features governing their operation and performance. It first discusses the most common stripe-geometry semiconductor lasers, which can be operated in either gain-guided or index-guided configurations. Various techniques are introduced to achieve low-current threshold, high-power and single spatial mode operation for these lasers. In addition, several specialized laser structures including the distributed feedback DFB, cleaved coupled cavity C^3, phase-locked array PLA, and quantum-well QW semiconductor lasers are described in great detail. The most desirable features of a semiconductor laser are those offering highest output power and lowest current threshold.

For most optical fiber systems, an LED may be an adequate source based on its available output power and speed of response. For long-distance and extremely high-data-rate systems, laser diodes (LDs) are often employed. The trade-off for laser power and bandwidth is not only in cost but most importantly, in the reliability of the devices. In general, lasers degrade much more rapidly than LEDs and the deterioration originates primarily from material imperfections. Improving the quality and reliability of laser sources constitutes one of the active areas of research at present.

10.2 STRIPE-GEOMETRY LASERS

A variety of semiconductor lasers are available with either symmetric or asymmetric multiple-heterojunction topology to provide separate optical and carrier confinements in the transverse direction. For lateral confinement in the junction plane, either a stripe or the buried geometry shown in Figure 10.1 is commonly used. The simplest configuration is that shown in Figure 10.1(a), where the optical gain region is confined by a stripe electrode with a width of typically 5 to 30 μm. In this case, the P-AlGaAs layer of the heterostructure with a relatively low doping concentration of $5 \times 10^{17} \, \text{cm}^{-3}$ has been used to reduce the current spreading in a thickness direction. The top GaAs layer is necessary to provide a good electrical contact with the electrode, which is formed by first opening a narrow stripe in the SiO_2 layer and then diffusing a shallowed depth of Zn through this window before metallic deposition. Alternative methods, which have also been used to provide the stripe-geometry, consist of either etching the top GaAs layer to form a shallow mesa, or ion implantation to reduce the carrier mobility in the region outside the electrode.

A more sophisticated structure is made by forming a buried p-GaAs active layer in a stripe configuration completely surrounded by AlGaAs, as shown in Figure 10.1(b). Although the fabrication of this device is compli-

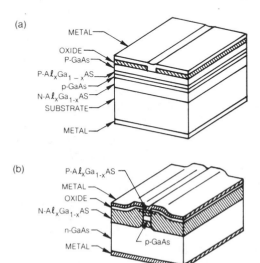

FIGURE 10.1 Topology of semiconductor heterojunction lasers: (a) stripe-geometry DH laser; (b) buried heterostructure BH laser.

cated, it basically involves preferential etching and LPE growth around the mesa. Such a structure can provide very stable output power in a single transverse mode when the width of the active layer in the junction plane is made as narrow as 2 to 3 μm. Table 10.1 summarizes performance characteristics of a representative class of stripe-geometry AlGaAs double-heterostructure lasers. In general, the performance of these lasers depends mainly on the stripe width, length, and amount of current spreading. For a stripe width greater than 10 μm, the laser output usually consists of many transverse modes, in some cases a nonuniform distribution, and kinks that are usually unstable can be developed in the output. By reducing the width down below 5 μm, these nonlinearities can be avoided; however, threshold currents become poorly defined. Stripe lasers with a narrow width differ in another respect from those having wider stripes. When a fast-rise current pulse is applied to wider-stripe heterostructure lasers, the laser output usually exhibits a transient oscillation, commonly known as relaxation oscillation. This kind of pulsation can be suppressed by using very narrow stripe-geometry.

Although the stripe-geometry laser has many attractive features, its threshold current density increases rapidly when the stripe width is reduced below 10 μm. Also, the astigmatic effect, which will be discussed later, increases rapidly with decreasing stripe width. These effects can be explained by the spatial variations of carrier concentration and gain along the junction plane. This situation can be analyzed by using a model in which electrons diffuse away from the active region along the junction plane defined by the stripe. This diffusion model can provide an expression for the carrier concentration profile. The calculated profile can be compared to the measurements of the spatial variation of spontaneous emission along the junction plane. Excellent agreement has been found to exist between the computed carrier concentration profile and the spontaneous emission measurements (Ref. 10.1). Figure 10.2 shows the spatial variation of current density, carrier concentration, refraction index, and gain coefficient of a stripe-geometry laser. The current density is assumed to be constant over the stripe width, as depicted in Figure 10.2(b). Furthermore, no variation is assumed to take place along the length of the electrode in the z direction.

The leakage current density outside the electrode can be expressed by

TABLE 10.1   Typical Output Characteristics of Some Semiconductor Lasers

Laser Type	Stripe Width (μm)	I_{th} (mA)	Peak Power (mW)	Output
Planar	10	100–125	10–50	Multimode
Ion implanted	12	100–150	5–10	Multimode
Zn diffused	5	30–70	5–10	Single mode
Buried	2	5–50	1–5	Single mode

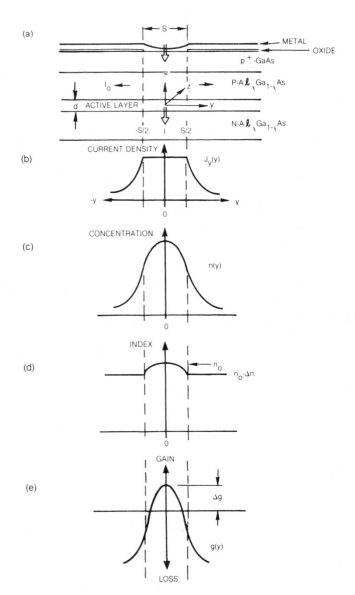

FIGURE 10.2 Characteristics of a stripe-geometry DH laser: (a) laser configuration; (b) lateral current density distribution; (c) carrier concentration profile; (d) index profile; (e) gain and loss profile.

a simple exponential decay function as

$$J(y) = J_0 \exp \left[- \frac{|y| - S/2}{l_0} \right] \qquad \text{for } |y| > \frac{S}{2} \qquad (10.1)$$

where J_0 is the current density inside the electrode, S the stripe electrode width, and l_0 a characteristic length as defined by

$$l_0 = \frac{2L}{\beta r I_0} \qquad (10.2)$$

The quantities in Equation (10.2) are the junction parameter β, the resistivity r, the electrode length L, and the leakage current I_0 at the edge of the electrode, respectively. Because the thickness of the active layer d is usually very small in comparison with the stripe width S and the diffusion length L_e, a one-dimensional diffusion equation along the y axis is sufficient to describe the lateral carrier concentration profile. This equation is

$$\frac{d^2 n}{dy^2} = \frac{n}{L_e^2} - R \qquad (10.3)$$

where R is the spontaneous recombination rate divided by the diffusion coefficient D as given by

$$R = \frac{J_0}{edD} \qquad (10.4)$$

In Equation (10.4), J_0 is the uniform current density inside the electrode ($-S/2 < y < S/2$), and e and d are the electronic charge and active-layer thickness, respectively. Assuming that the leakage current I_0 is very small compared to the injection current I, a solution for the electron density profile for $|y| < S/2$ is given by

$$n(y) = RL_e^2 + [n(0) - RL_e^2] \cosh \frac{y}{L_e} \qquad (10.5)$$

where $n(0)$ is the electron concentration at $y = 0$.

For $|y| > S/2$, electrons diffuse with a different diffusion constant D' and diffusion length L_e' into a profile that is characterized by the exponential decay function:

$$n(y) = n(S/2) \exp \left(- \frac{y}{L_e'} \right) \qquad (10.6)$$

Using the conditions that $n(y)$ and dn/dy must be continuous at $|y| = S/2$, explicit expressions for $n(0)$ and $n(S/2)$ can be derived in terms of diffusion constants and diffusion lengths, which are measurable quantities, obtainable from the spontaneous emission profile. The results are

$$n(0) = RL_e^2 \left[1 - \left(\cosh \frac{S}{2L_e} + \zeta \sinh \frac{S}{2L_e} \right)^{-1} \right] \qquad (10.7)$$

where

$$\zeta = \frac{DL_e'}{D'L_e} \qquad (10.8)$$

Substituting Equation (10.7) into (10.5) yields the following carrier concentration inside the electrode ($| y | < S/2$):

$$n(y) = RL_e^2 \left[1 - \frac{\cosh(y/L_e)}{\cosh(S/2L_e) + \zeta \sinh(S/2L_e)} \right] \qquad (10.9)$$

Evaluating Equation (10.9) at $| y | = S/2$ gives

$$n(S/2) = RL_e^2 \zeta \sinh \frac{S}{2L_e'} \exp\left(\frac{S}{2L_e} \right) \left(\cosh \frac{S}{2L_e} + \zeta \sinh \frac{S}{2L_e} \right)^{-1} \qquad (10.10)$$

From the measurement of spontaneous emission along the lateral junction plane, the diffusion lengths L_e and L_e' can be obtained assuming that the local spontaneous emission intensity is a linear function of local carrier concentration. This assumption is found to be valid for cases both below and above threshold. A quantity that is useful for determining the diffusion length from local measurements of spontaneous emission is the ratio $n(S/2)/n(0)$. Equations (10.7) and (10.10) provide

$$\frac{n(S/2)}{n(0)} = \zeta \sinh \frac{S}{2L_e} \left[\cosh \frac{S}{2L_e} + \zeta \sinh \frac{S}{2L_e} - 1 \right]^{-1} \qquad (10.11)$$

The local gain must obey the same functional relationship for the electron concentration. With this assumption the gain profile can be related to the electron concentration profile by a simple transformation:

$$g(y) = an(y) - b \qquad \text{for } | y | < \frac{S}{2} \qquad (10.12)$$

where $n(y)$ is given by Equation (10.9) and a and b are parameters dependent on injection currents. Equation (10.12) provides

$$\Delta g = g(0) - g(S/2) = a \exp\left(-\frac{S}{2L_e} \right) \left(\cosh \frac{S}{2L_e} - 1 \right) \qquad (10.13)$$

At a given current, the calculated gain width is in agreement with the measured beam width obtained by an extrapolation from the far-field beam divergence.

The optical field inside a diode laser cavity is characterized by three-dimensional spatial variations of the index and gain distributions that determine the longitudinal and transverse optical modes of the cavity. They are solutions of the wave equations given by Equations (2.16) and (2.17) subject to appropriate boundary conditions. According to device geometry, it is convenient to separate the field dependence in the forms given by Equation (2.18) where β is the propagation constant along the laser longitudinal axis and is limited to a set of discrete frequencies associated with the Fabry–Perot or longitudinal modes. The wave equation representing the transverse modes is then reduced to a two-dimensional equation in the form

$$\nabla^2 E + [k_0^2 \epsilon(x, y) - \beta^2]E = 0 \qquad (10.14)$$

In the case of a planar waveguide, e.g., $\partial/\partial y = 0$, the field reduces to two independent sets of TE and TM modes as discussed in Chapter 3. In practice, two groups of current-injected semiconductor lasers have been developed for lateral confinement of spatial modes. The first group is called gain-guided lasers, in which injected carriers are laterally confined to a narrow stripe by means of a narrow electrode contact as shown in Figure 10.1(a). The lateral optical confinement is primarily dictated by the gain distribution along the y-axis, which is parallel to the active layer with a profile that closely follows the carrier distribution. The second group of stripe-geometry diode lasers is called index-guided. In index-guided structures, as shown in Figure 10.1(b), the photons and carriers are laterally confined by using two-dimensional heterostructure configurations, which will be discussed in Section 10.5.

10.3 GAIN-GUIDED STRIPE-GEOMETRY LASERS

Different methods can be used to achieve lateral confinement of injected carriers. Figure 10.3 shows the schematic cross-sections for three representative DH gain-guided laser structures. Figure 10.3(a) shows an oxide stripe laser structure with a narrow stripe opening (<20 μm) in an insulating oxide film deposited on top of the contact layer for current confinement. The layers located above the active region are made sufficiently thin so that for all practical purposes they do not alter the gain distribution. Figure 10.3(b) shows a proton-implanted laser structure for which proton-implantation is used to create highly resistive regions on both sides of the conductive stripe. This method makes it possible to adjust the degree of current confinement by varying the implantation depth. Figure 10.3(c) shows a V-groove laser structure, in which a narrow current path (3–5 μm wide) is formed by diffusing Zn, a P-dopant, into the structure through an etched V-groove.

In gain-guided stripe-geometry lasers, the unpumped regions on both sides of the stripe are very lossy with α values in the range of 200–400 cm^{-1}. This attribute leads to the lateral gain profile shown in Figure 10.2(e). The

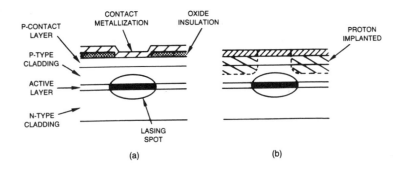

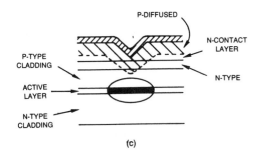

FIGURE 10.3   Schematics of three representative DH gain-guided heterojunction laser structures: (a) An oxide stripe laser structure. (b) A proton-implanted laser structure. (c) A V-groove laser structure.

photons generated in the region with gain values below the threshold are absorbed by the medium and those that propagate along the stripe contribute to the maximum optical power. For stripe widths greater than the carrier diffusion length (>10 µm), a dip in the carrier density distribution near the center of the stripe often occurs. Such a change in the spatial gain profile often results in a shift in the gain-guided laser beam and a corresponding "kink" in the light versus current as illustrated in Figure 10.4. These undesirable characteristics of wide-stripe gain-guided lasers can be avoided by using very narrow ($\sim$3 to 5 µm) stripe contacts. With narrow stripes, carrier diffusion in the active layer region can smooth out the dip that otherwise develops in wider stripe geometry. The problem associated with very narrow stripes is the rapid increase in threshold current density with decreasing stripe width. Another problem is the increase in astigmatism in the output of very narrow stripe gain-guided lasers as discussed later in this section.

To treat the lateral optical modes in a gain-guided laser, Equation (10.14) is solved by introducing medium gain into the expression for $\epsilon(x, y)$, which is a function of not only the dielectric constant but also the gain (loss) coefficient g of the medium. Therefore, it is necessary to introduce a complex

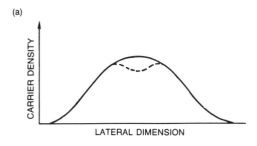

(a)

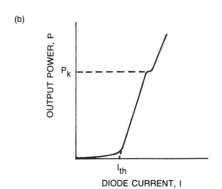

(b)

FIGURE 10.4 (a) Carrier density versus lateral dimension of a stripe-geometry laser. A dip in the profile represents the hole-burning phenomenon. (b) The output power of a stripe-geometry laser versus the injection current.

refractive index, e.g., $n = \text{Re}(n) + i \, \text{Im}(n)$, where $\text{Im}(n)$ is related to the gain or loss of the medium. The E-field in Equation (10.14) is denoted by $E_n(x, y)$ for the nth mode, which is governed not only by the abrupt change in material compositions along the x-direction, but also by the gain medium along the y-direction. To a first approximation, a solution of the following form is assumed

$$E_n(x, y) = E_n(x)E_n(y) \exp(-i\beta z) \qquad (10.15)$$

Because $E_n(y)$ is a slowly varying function, it is reasonable to assume that $E_n(x)$ is not significantly affected by its confinement along the y-direction, therefore, $E_n(x)$ satisfies the one-dimensional wave equation

$$\frac{d^2 E_n(x)}{dx^2} + (k_0^2 n_x^2 - \beta^2 - \beta_x^2)E_n(x) = 0 \qquad (10.16)$$

where β_x^2 is a constant resulting from the separation of variables and

$$n_x^2 = \begin{cases} n_1^2 & x < 0 \\ n_{2r}^2 & 0 < x < d \\ n_3^2 & x > d \end{cases}$$

The wave equation for $E_n(y)$ is decomposed as follows:

$$\frac{d^2E_n(y)}{dy^2} + (k_0^2 n_y^2 + \beta_x^2)E_n(y) = 0 \tag{10.17}$$

where

$$n_y^2 = \begin{cases} 0 & x < 0 \\ -n_{2i}^2(y) + i2n_{2r}n_{2i}(y) & 0 < x < d \\ 0 & x > d \end{cases}$$

Equation (10.16) is identical to that for TE modes in one-dimensional, three-layer waveguides, whose solutions were given in Chapter 3. The only difference in this two-dimensional problem is that the propagation constant β can only be found after β_x values have been determined by solving Equation (10.17). Because n_y^2 in Equation (10.17) depends on x and y, it cannot be solved directly; however, by integrating Equation (10.17) for $x = -\infty$ to ∞ after multiplying $|E_n(x)|^2$, an equation containing only y results. This procedure gives

$$\frac{d^2E_n(y)}{dy^2} + [\beta_x^2 + i2n_{2r}\Gamma k_0^2 n_{2i}(y)]E_n(y) = 0 \tag{10.18}$$

where Γ is the confinement factor along the thickness direction as defined by

$$\Gamma = \int_0^d |E_n(x)|^2\, dx \Big/ \int_{-\infty}^{\infty} |E_n(x)|^2\, dx \tag{10.19}$$

To solve Equation (10.18), $n_{2i}(y)$ must be related to the medium gain function. As discussed in Section 10.2, the gain profile is related to the carrier concentration $n(y)$ which is given by Equation (10.9). For simplicity, a parabolic gain profile is often used. In this case, $n_{2i}(y)$ can be written as $ay^2 - b$, and n_{2r} is a constant that is equal to n_2. With this simplification, the solution of Equation (10.18) is expressed in terms of the familiar Gaussian–Hermite functions. If the transverse guided-wave mode m has an electric field distribution, $E_m(y)$, along the active region, then the modal gain is defined by:

$$G_m = \int_{-\infty}^{\infty} g(y)E_m^2(y)\, dy \Big/ \int_{-\infty}^{\infty} E_m^2(y)\, dy \tag{10.20}$$

where $g(y) = an(y) - b$. Note that a is a constant that determines the rate at which the gain decreases, and $n(y)$ is the carrier concentration, which is given by Equation (10.9). For convenience, the Gaussian–Hermite modal distribution functions will be replaced by simple trigonometric functions in the forms described by Equations (3.34) and (3.36). For the fundamental mode, Equation (10.20) becomes

$$G_0 = \frac{1}{S} \int_{-s}^{s} g(y) \cos^2 k_0 y \, dy \qquad (10.21)$$

where $k_0 = \pi/2S$. Substituting Equation (10.9) into Equation (10.21) yields

$$G_0 = aRL_e^2 \left[1 - \frac{2L_e/S}{(S/2\pi L_e)^2 + 1} \frac{\tanh(S/2L_e)}{1 + \zeta \tanh(S/2L_e)} \right] - b \qquad (10.22)$$

For high-order modes ($m \neq 0$), the modal gain is given by

$$G_m = aRL_e^2 \left[1 - \frac{2(m + 1)^2 L_e/C}{(S/2L_e)^2 + (m + 1)^2} \frac{\tanh(S/2L_e)}{1 + \zeta \tanh S/2L_e} \right] - b \qquad (10.23)$$

where ζ is defined by Equation (10.8).

For a lightly doped p-layer at 10^{17} cm^{-3} with $S/L_e \simeq 2$, the parameters a and b are found (Ref. 10.1) to be $(1.08 \pm 0.06) \times 10^{-16}$ cm^2 and 146 cm^{-1}, respectively. At threshold, the current density in this case is 4.45×10^3 A/cm^2. The gain profile at threshold is

$$g_{th}(y) = 146 \left[1 - 0.72 \cosh \frac{y}{L_e} \right] \text{cm}^{-1} \qquad y < \frac{S}{2}$$

$$g_{th}(y) = 322 \exp \left(-\frac{y}{L_e'} \right) - 146 \text{ cm}^{-1} \qquad y > \frac{S}{2} \qquad (10.24)$$

A plot of g_{th} is shown in Figure 10.5.

It is interesting to see how the spatial distribution of modal gain can affect current threshold. Equation (10.22) shows that the threshold modal gain G_{th} corresponds to a threshold recombination rate R_{th}. If the ratios of

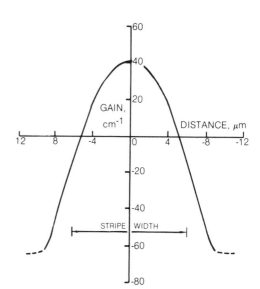

FIGURE 10.5 Threshold gain as a function of distance for a lightly doped stripe-geometry AlGaAs DH laser. (From Ref. 10.1 Reprinted with permission of J. Appl. Phys.)

the threshold current density for a stripe-geometry laser having a stripe width S are related to a laser with an infinite electrode width, the following results:

$$\frac{R_{th}(S)}{R_{th}(\infty)} = \left[1 - \frac{2L_e/S}{(S/2\pi L_e)^2 + 1} \frac{\tanh(S/2L_e)}{1 + \zeta \tanh(S/2L_e)} \right]^{-1} \quad (10.25)$$

Figure 10.6 shows a plot of the injection rate required for lasing in the fundamental mode as a function of the stripe width normalized to diffusion length for two cases: $\zeta = 1$ and $\zeta = 0.25$. It is clear that in both cases the threshold increases rapidly with decreasing stripe width. The increase in threshold is due to two major factors: (1) a loss of electrons is due to out diffusion, and (2) a reduction in mode coupling leads to a nonuniform gain profile.

Another interesting observation from the output of a stripe-geometry laser is the nonplanar wavefront, which is cylindrical concave in the direction of laser propagation. This shape is completely different from the planar wavefront for the case of a planar waveguide of an infinite extent. This behavior can be attributed to the spatial variation of gain medium, as shown

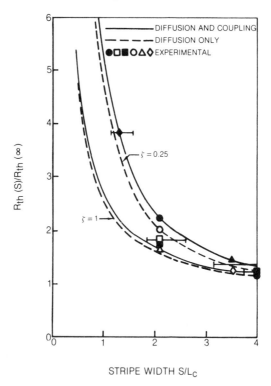

FIGURE 10.6 Threshold recombination rate as a function of normalized stripe-electrode width for a stripe-geometry laser normalized to that of an infinitely wide electrode. (From Ref. 10.1 Reprinted with permission of J. Appl. Phys.)

in Figure 10.2(e), and is commonly referred to as gain-guiding, a phenomenon in which field distribution is dependent more on the gain of the medium than on the usual refractive index. The wavefronts for stripe-geometry lasers are illustrated in Figure 10.7. The phase fronts for $E_n(x)$ as shown in Figure 10.7(a) are determined by a constant Δn across the junctions. However, the field $E_n(y)$ is influenced by gain-guiding and has a cylindrical phase front as shown in Figure 10.7(b), resulting from a virtual beam waist occurring behind the facet. This output is therefore astigmatic because for the field confined perpendicular to the junction plane, the beam waist is at the facet, and is located behind the facet for the field confined along the junction plane. Measurements (Ref. 10.2) of the far-field beam divergence and beam width confirm the fact that optical confinement along the junction plane is caused by gain profile alone. Below threshold, the measured intensity profile of spontaneous emission corresponds well with the carrier distribution $n(y)$, as given by Equations (10.9) and (10.6).

The strength of astigmatism can be expressed in terms of a K factor as defined by

$$K = \frac{[\int |E(y)|^2 \, dy]^2}{|\int E^2(y) \, dy|^2} \tag{10.26}$$

where $E(y)$ denotes the complex modal field distribution in the junction

(a) SIDE VIEW

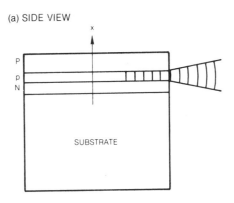

(b) TOP VIEW

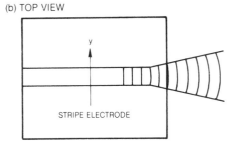

FIGURE 10.7 Schematic diagrams showing astigmatic wavefronts from (a) a side view and (b) a top view of a stripe-geometry laser.

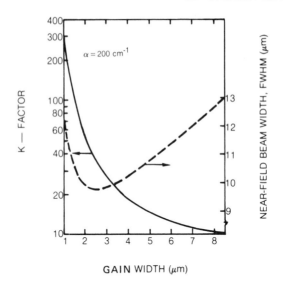

GAIN WIDTH (μm)

FIGURE 10.8 Variation of astigmatic K factor and beam width as a function of gain width. (From Ref. 10.4 Reprinted with permission of IEEE.)

plane. If a wave is index-guided, $E(y)$ is real; therefore, $E^2(y) = |E(y)|^2$. In this case it is clear that $K = 1$. K exceeds unity when $E(y)$ is complex, as for the case of gain-guided lasers. The astigmatism is shown (Ref. 10.3) to have a great influence on the spontaneous emission factor, which is defined as a ratio of the rate of spontaneous emission into one oscillating mode to the total emission rate R_{sp}. It has been shown (Ref. 10.3) that by narrowing the stripe width, the spontaneous emission factor increases much faster than that expected from the corresponding decrease of the active layer volume. As a result, narrow stripe-geometry lasers exhibit a much broader spectrum than index-guiding lasers with comparable active volumes.

To extend this concept further, Streifer, Scifres, and Burnham (Ref. 10.4) utilized a more accurate representation of the lateral mode distribution to calculate the K factor. Without going through the details, the numerical results for K are plotted in Figure 10.8 as a function of the lateral gain width W_g, where W_g is determined by a combined effect of stripe width, current spreading, and lateral charge diffusion. Roughly speaking, $W_g \simeq 2S$, where S is the stripe width. These results were obtained by assuming a band-to-band absorption coefficient $\alpha = 200 \text{ cm}^{-1}$. Also shown in Figure 10.8 is the near-field beam width as a function of the gain width.

10.4 POWER SPECTRUM OF DH LASERS

The spectral characteristics of a 4-μm stripe-geometry DH laser are shown in Figure 10.9. Because of a narrow electrode width, laser oscillation is limited to only a single transverse TE_0 mode. However, many longitudinal

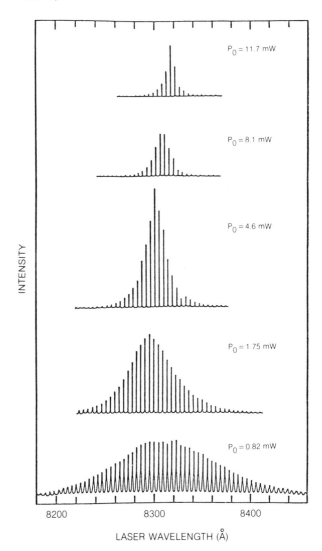

FIGURE 10.9 Spectral characteristics of a 4-μm stripe-geometry DH laser at several injection current levels. (From Ref. 10.4 Reprinted with permission of IEEE.)

modes still exist with a spectral separation $\Delta v \simeq 2$ to 5 Å, as indicated by Equation (8.96). At low power ($P = 0.82$ mW), the spectral envelope λ, is approximately equal to the width of the homogeneously broadened spontaneous emission line λ_h, which can be characterized by a Lorentzian profile. Within this width, many longitudinal modes oscillate simultaneously. As power increases, the spectral width decreases and its shape is no longer symmetric, with a upshift toward the longer wavelengths.

To explain some of these behaviors, Streifer, Scifres, and Burnham (Ref. 10.4) have presented a model based on spontaneous emission coupling into longitudinal modes. The results of their analysis are in excellent agreement with experimental data for both gain-guided and index-guided DH lasers. The theory assumes that homogeneously broadened spontaneous emission is coupled to all longitudinal modes by a different amount depending on the emission line shape. Below threshold, the emission envelope is very broad and is equivalent to that of an LED. Above threshold, the stimulated emission causes injected carriers to recombine, hence the gain begins to saturate and the spectral envelope narrows significantly. Using a self-consistent approach, requiring that the field must reproduce itself in a round-trip within the cavity, a simple result has been obtained from a rather complicated analysis for the dependence of the full width λ_s of the longitudinal spectral envelope at half-maximum power (FWHM) on the output power P_T. It is (Ref. 10.4)

$$\lambda_s = \lambda_h \left[\left(\frac{P_T^2}{4P_i^2} + 1 \right)^{1/2} - \frac{P_T}{2P_i} \right] \tag{10.27}$$

where λ_h is the homogeneous spontaneous line width, P_T is the total power of the fundamental mode transmitted through facet 1, and P_i is the internal circulating power of the mode just below threshhold. They are related by the expression

$$P_T = 2P_i(1 - R_1)\sqrt{R_2}\ \frac{(1 + R_1)\sqrt{R_2} + (1 + R_2)\sqrt{R_1} - 4\sqrt{R_1R_2}}{(1 + R_2^2)R_2 + (1 + R_2^2)R_1 - 4R_1R_2} \tag{10.28}$$

where

$$P_i = \frac{1 - R_1}{R_1}\ \frac{\sqrt{R_1} + \sqrt{R_2}}{2\sqrt{R_2}}\ \frac{1 - \sqrt{R_1R_2}}{\ln(1/R_1R_2)}\ \frac{hc\lambda_0 K}{4\pi n^2 A} \tag{10.29}$$

R_1 and R_2 are the facet power reflectivities, λ_0 is the free-space wavelength, and

$$A = \frac{ed}{\Gamma}\frac{dg}{dJ} \tag{10.30}$$

In Equation (10.30), dg/dJ is the slope of the gain versus current density below threshold and is approximately a constant for most semiconductor lasers. For example: If $\Gamma = 0.3$, $d = 0.1$ μm, and $dg/dJ \simeq 100$ cm^{-1}/kA/cm^2, then $A \simeq 0.6 \times 10^{-24}$ cm^2-s.

At low output levels (i.e., $P_T < P_i$), Equation (10.27) indicates that

$$\lambda_s \simeq \lambda_h \left(1 - \frac{1}{2}\frac{P_T}{P_i} \right) \tag{10.31}$$

Above threshold λ_s can be written in another form:

$$\lambda_s \simeq \frac{\lambda_h P_i}{P_T} \qquad (10.32)$$

Equation (10.32) indicates that in the lasing regime, the spectral envelope varies inversely with the output power. Because the expression above is independent of lasing parameters, it is applicable to all semiconductor lasers regardless of the guiding mechanism.

Figure 10.10 shows the variations of λ_s as a function of P_T for a number of K values. The measured λ_s values for 4- and 8-μm stripe lasers agree quite well with the calculated values, $K = 30$ and 20, respectively. At higher power levels, the spectra are no longer symmetric, and longitudinal modes tend to lase on the short-wavelength side of the dominant mode. Such behavior is probably caused by nonlinear effects, which will not be treated

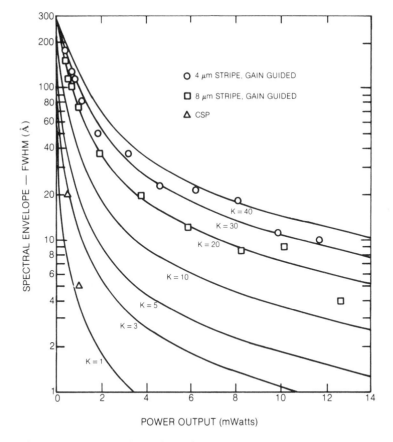

FIGURE 10.10 Theoretical and experimental results of spectral envelope width (FWHM) versus output power. (Ref. 10.4 Reprinted with permission of IEEE.)

here. Also shown in Figure 10.10 are measurements of an index-guided channeled substrate planar CSP laser. Because $K = 1$ for this device, these measurements are in agreement with the theory. A slight phase-front distortion in the index-guided lasers could result in larger K values and broaden the spectrum.

10.5 INDEX-GUIDED STRIPE-GEOMETRY LASERS

The laterally guided wave modes in index-guided lasers can be obtained by solving the wave equation (10.14) with the effective-index method as discussed in Chapter 3. This method is especially appropriate for highly asymmetric index profiles such as those in double-heterostructure semiconductor lasers whose layered dielectric structures are made of mostly discrete steps in refractive index with dimensions on the order of one wavelength in the thickness (x) direction, but a slowly varying function for the refractive index in the transverse or lateral (y) direction. In this case, a solution of Equation (10.14) is assumed of the form

$$E(x, y) = F(x, y)G(y) \tag{10.33}$$

Substituting Equation (10.33) into (10.14) yields

$$G \frac{\partial^2 F}{\partial x^2} + F \frac{\partial^2 G}{\partial y^2} + [k_0^2 \epsilon(x, y) - \beta^2]FG = 0 \tag{10.34}$$

where the terms involving derivatives of F with respect to y have been neglected. To make the analysis more general, the dielectric constant of a three-layer system is written as

$$\epsilon_i(0, y) = \epsilon_i(0) + \epsilon_i(y) \qquad (i = 1, 2, 3) \tag{10.35}$$

Substituting Equation (10.35) into (10.34), multiplying by F^*, and integrating over x from $-\infty$ to ∞ gives

$$\frac{\partial^2 G}{\partial y^2} - \beta^2 G + \sum_i [k_0^2 \Gamma_i(y)\epsilon_i(y) - \Gamma_i(y)h^2(y)]G = 0 \tag{10.36}$$

where

$$h^2(y) = -\frac{1}{F} \frac{\partial^2 F}{\partial x^2} \tag{10.37}$$

and

$$\Gamma_i(y) = \int_{i\text{th layer}} |F(x, y)|^2 \, dx \tag{10.38}$$

Γ_i is the overlap parameter. For a normalized $F(x, y)$, $\sum \Gamma_i = 1$. If an effective dielectric constant, ϵ_{eff}, is defined as

$$\epsilon_{\text{eff}} = \sum_i [\Gamma_i(y)\epsilon_i(y) - \Gamma_i(y)h^2(y)/k_0^2] \qquad (10.39)$$

Equation (10.36) becomes

$$\frac{\partial^2 G}{\partial y^2} + [k_0^2\epsilon_{\text{eff}}(y) - \beta^2]G = 0 \qquad (10.40)$$

In principle, ϵ_{eff} can be obtained by solving

$$\frac{\partial^2 F}{\partial x^2} + [k_0^2\epsilon(x, y) - k_0^2\epsilon_{\text{eff}}]F = 0 \qquad (10.41)$$

for each value of y. By taking advantage of the slowly varying $\epsilon(x, y)$ function with y, it is possible to obtain ϵ_{eff} by using a perturbation theory (Ref. 10.5) as follows:

$$\epsilon(x, y) = n^2(x, y) = n_0^2(x) + \Delta n(x, y) \qquad (10.42)$$

where $n_0(x)$ is the refractive index of a slab waveguide with values given by

$$n_0(x) = \begin{cases} n_1 & x < 0 \\ n_2 & 0 < x < t \\ n_3 & x > t \end{cases} \qquad (10.43)$$

The unperturbed field $F_0(x) = F(x, 0)$ satisfies the equation

$$\frac{\partial^2 F}{\partial x^2} + [k_0^2 n_0^2(x) - \beta_{x(0)}^2]F = 0 \qquad (10.44)$$

where the zero-order propagation constant $\beta_{x(0)}$ is directly related to the effective index of the perturbed waveguide at $y = 0$ as given by

$$\beta_{x(0)} = k_0 n_{\text{eff}}(0) \qquad (10.45)$$

The first-order correction to $\beta_{x(0)}$ resulting from the y variation of $\Delta n(x, y)$ can be expressed by

$$\beta_x^2(y) = k_0^2 n_{\text{eff}}^2(y) \simeq k_0^2[n_{\text{eff}}^2(0) + \sum_i \Gamma_i \Delta n^2] \qquad (10.46)$$

For example, consider a laser structure that employs thickness variations as means to achieve optical confinement, e.g., $\Delta t = \Delta t(y)$. Using known solutions for a three-layer, unperturbed waveguide gives

$$\beta_x^2(y) = 2k_0^2[n_2^2 - n_{\text{eff}}^2(0)]\frac{\Delta t}{t_{\text{eff}}(0)} \qquad (10.47)$$

where $n_{\text{eff}}(0)$ is the modal effective index for the unperturbed waveguide and

$$t_{\text{eff}}(0) = t + \frac{1}{k_0\sqrt{n_{\text{eff}}^2(0) - n_1^2}} + \frac{1}{k_0\sqrt{n_{\text{eff}}^2(0) - n_3^2}} \tag{10.48}$$

The second and third terms in Equation (10.48) are the reciprocals of the decay constants γ_1 and γ_3 of the evanescent waves extending into media n_1 and n_3, respectively. The result of the first-order perturbation calculation indicates that the variation in n_{eff}^2 is proportional to the variation in the guiding-layer thickness and that a tapering of the guiding-layer thickness results in a decrease in n_{eff} and leads to lateral optical confinement.

More specifically, consider a crescent-shaped waveguide (Figure 10.18) with $n_1 = n_3$. The active region tapers down to zero thickness at $y = \pm W/2$ where W is the width of the waveguide. If a parabolic variation in thickness is assumed, e.g., $t(y) = t(0)[1 - 4y^2/W^2]$, then

$$n_{\text{eff}}^2(y) = n_{\text{eff}}^2(0) - py^2 \tag{10.49}$$

where

$$p = 8[n_2^2 - n_{\text{eff}}^2(0)] \frac{t(0)}{W^2 t_{\text{eff}}(0)} \tag{10.50}$$

If Equation (10.49) is substituted into (10.40), the wave equation that has an analytic solution can be expressed in terms of Gaussian Hermite polynomials G_{nm}, when n is the lateral mode index for each m value (m is the unperturbed slab mode order). The lateral field can be expressed by

$$G_{nm}(y) = H_n(\sqrt{k_0}\, p^{1/4} y) \exp(-k_0 p^{1/2} y^2/2) \tag{10.51}$$

and the propagation constants β_{nm} are

$$\beta_{nm}^2 = \beta_{x(0),m}^2 - (2n + 1)k_0\sqrt{p_m} \tag{10.52}$$

For the fundamental lateral mode ($n = 0$), Equation (10.51) shows that the field assumes a Gaussian profile with a near-field spot size ω_0 (full width at half-power) as given by

$$\omega_s = [4 \ln 2/k_0\sqrt{p}]^{1/2} \tag{10.53}$$

For a crescent-shaped GaAs/AlGaAs laser with $n_1 = 3.4$, $n_2 = 3.6$, $\lambda = 0.88\ \mu\text{m}$, $t(0) = 0.21\ \mu\text{m}$, and $W = 8\ \mu\text{m}$, Equation (10.52) yields a ω_0 value of 1.41 μm.

The far-field patterns of a stripe-geometry DH laser are shown in Figure 10.11. As the width of the stripe S is increased beyond 10 μm, higher-order transverse modes supersede the lower-order modes even if the injection current is low. At high currents, the laser output usually consists of multi-modes. The field distribution of these transverse modes can be generated

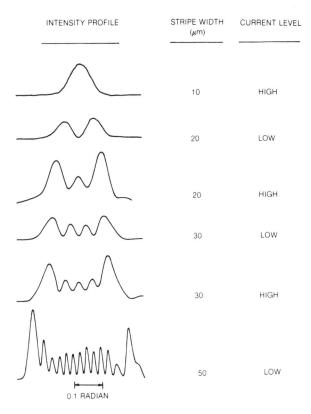

INTENSITY PROFILE	STRIPE WIDTH (μm)	CURRENT LEVEL
	10	HIGH
	20	LOW
	20	HIGH
	30	LOW
	30	HIGH
	50	LOW

0.1 RADIAN

FIGURE 10.11   Far-field radiation patterns of a stripe-geometry DH laser.

using the following expression:

$$H_n(y) = (-1)^n \exp(y^2) \frac{d^n}{dy^n} \exp(-y^2) \qquad (10.54)$$

where H_n is the Hermite polynomial of order n. The first three Hermite polynomials are $H_0(y) = 1$, $H_1(y) = 2y$, and $H_2(y) = 2y^2 - 2$. To calculate the far-field intensity distribution, the square of the fields must be multiplied by the gain profile. These mode patterns are governed to a large extent by the properties of the laser resonator. The resonator can be designed to limit the number of possible modes. In the thickness direction, it is normally done by reducing the active layer thickness to a size equivalent to only one-half period of the wave. In the lateral direction along the junction plane, the width of the electrode is important, because the oblique angles permit excitation of higher-order modes. To achieve single transverse mode operation, either the width of the electrode for gain-guided lasers or the width of the constricted channel for index-guided lasers must be kept below 10 microns.

10.6 HIGH-POWER AND SINGLE MODE SEMICONDUCTOR LASERS

The demand for long-haul, high-data-rate communication, high speed optical recording, printing, data bus distribution systems, local area networks, optical pump sources for other solid state lasers, and many other industrial applications has brought the high-power III-V compound semiconductor lasers successfully into the marketplace. This section summarizes some of the important aspects of high-power semiconductor lasers. The effects of spatial hole-burning, catastrophic damage, and techniques for minimizing these effects are also discussed. A number of single mode and high-power laser structures, including the use of dielectric coating and non-absorbing mirrors are presented as well.

It was recognized fairly early in laser development that the simple stripe-geometry configuration could not provide high-power output. The simple stripe-geometry has many problems associated with beam steering, dynamic instabilities, and beam astigmatism. These anomolous effects are related to the physical nature of optical waveguiding in a stripe-geometry laser and were already discussed. As injection current is increased in a stripe-geometry laser with a stripe width in the range of 10 to 20 microns, the output often contains a series of instabilities in the form of nonlinear pulsations in the output spectrum as shown in Figure 10.4((b). These pulsations cause excess noise and mode shifts in the power spectrum. A closer examination of the carrier concentration on the gain profile indicates that the strong stimulated emission near the center of the stripe begins to produce a localized depletion of the gain profile as shown in Figure 10.4(a) and spatial-hole burning occurs. This dip in the gain profile has several important implications: (1) The dip ΔG creates a region of increased refractive index due to free carrier and dispersion effects. (2) The dip can produce an unstable guided mode that tends to move toward the high gain region. (3) If the gain dip becomes sufficiently large, the mode can switch to the next-higher-order mode.

The instabilities of the output power spectrum can be explained by the spectral broadening mechanism. As the power level of the laser is increased, spatial hole-burning leads to a nonuniform gain profile. This phenomenon implies that a nonuniform quasi-Fermi level separation exists and gives rise to a nonuniform spectral gain profile and a spread of the peak gain with respect to wavelength. The magnitude of the spectral broadening $\Delta\lambda$ is proportional to ΔG and can be simply expressed as

$$\Delta\lambda = k\lambda \frac{\Delta G}{G_{max}} \tag{10.55}$$

where the measured k values are 5×10^{-3} for GaAs and 4×10^{-2} for GaInAsP, and λ is the wavelength that corresponds to the maximum gain

G_{max}. For a 300 micron GaAs laser diode, a 10% gain dip can produce a spectral broadening of 4 Å, which is sufficient to support two longitudinal FP modes. Equation (10.55) provides useful design criterion for high-power lasers operating in a single longitudinal mode. High-power lasers must have a high G_{max} value and a low ΔG value.

To minimize the spatial hole-burning effect, one method has been introduced whereby a built-in index difference Δn_B is provided in the plane of the stripe so that the index of refraction has a minimum at the stripe's center and increases toward the edges, which is commonly known as a negative index waveguide. This built-in difference must be sufficiently large to overcome the effects of gain for all driving levels. Two criteria for preserving single mode operation and mode stability in such a guide are given by

$$\frac{\Delta n_B}{n} > \frac{\Delta G}{G_{max}} \qquad (10.56)$$

and

$$W < \frac{\lambda}{(8n\,\Delta n_B)^{1/2}} \qquad (10.57)$$

where n is the index of the active layer, G_{max} is the maximum gain, and W is the width of the guide. To achieve stable, single mode operation, many approaches have been attempted and they all exhibited some degree of success. One approach used a negative index waveguide as already mentioned with a Δn_B value between 10^{-2} to 5×10^{-3}. Other approaches to reduce the effect of spatial hole-burning are to increase either the rate of diffusion or the losses near the edge of the stripe. The former can be accomplished by simply decreasing stripe width and the latter can be accomplished by forming three-dimensional waveguide configurations. One such configuration is the channel substrate planar CSP laser that have high absorption losses at the edges of the stripe.

An alternative but very effective way to obtain single mode operation is to restrict the confinement factor Γ to a very small value by using a very thin active layer with a thickness of 500 to 1000 Å. However, it is difficult to control thickness uniformity with the most common LPE growth technique. With the recent advances in metalorganic chemical vapor deposition (MOCVD) and in molecular beam epitaxy (MBE), thin uniform active layers in large areas can be produced with good control and thus could revolutionize laser diode production.

Figure 10.12 shows a number of index-guided, single mode CW lasers. In addition to the buried-heterostructure BH laser, which was introduced previously, Figure 10.12(a) shows a transverse junction stripe TJS laser, and Figure 10.12(b) shows a plane-convex waveguide PCW laser. Figure 10.12(c) shows a channeled substrate planar CSP laser, and Figure 10.12(d) shows a constricted double-heterostructure CDH laser. The output of these lasers

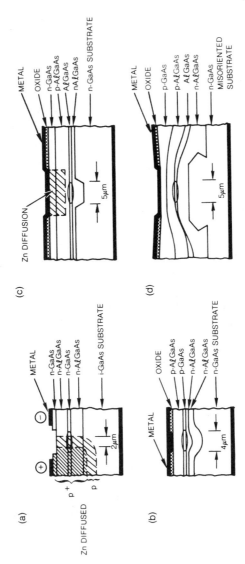

FIGURE 10.12 Single-mode DH lasers: (a) transverse junction stripe TJS lasers; (b) planar convex waveguide PCW lasers; (c) channeled substrate planar CSP lasers; (d) constricted double-heterostructure CDH lasers. (From Ref. 10.6. Reprinted with permission of North-Holland Publishing Company, Amsterdam.)

260

is typically in the range of 1 to 7 mW. With the exception of the TJS laser, all the lasers noted above are fabricated by LPE growth of heterojunctions over a channeled or ridged substrate with a width varying from 1 to 5 μm. The lateral variation in thickness is equivalent to a lateral variation in refractive index that provides optical confinement along the junction plane.

For the TJS laser, lateral waveguiding is obtained by two consecutive Zn diffusions that produce a variation in the index between p and p^+ material; therefore, both the cathode and the anode can be placed on the top surface. The CDH laser is grown on a "double-dovetail" channel configuration. By placing a 10-μm-wide electrode on the top layer, a constricted active region is formed that resembles a leaky waveguide. Figure 10.13 shows the power spectrum of a CDH laser. At a low injection level, the output consists of a number of longitudinal modes. As the injection current increases above 100 mA, a single longitudinal mode at a power of ~10 mW is obtained with its wavelength shifting toward the longer region. This shift is attributed to Joule heating in the device junction.

The CW power-current (P–I) characteristics of a CDH laser operated at various temperatures between 20 and 70°C are shown in Figure 10.14. At all temperatures the curves are linear and "kinkless." As temperature is increased from 20°C to 70°C, the threshold current increases by only 22%. Higher outputs can be obtained from these lasers by changing the laser structures to the LOC configurations shown in Figure 10.15. This alteration involves the growth of additional cladding layers of intermediate index. LOC CDH lasers have been operated successfully to yield single-longitudinal-mode CW output at 50 mW per facet.

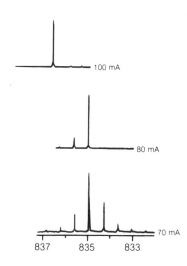

FIGURE 10.13 Power spectrum of a CDH laser. (From Ref. 10.6. Reprinted with permission of North-Holland Publishing Company, Amsterdam.)

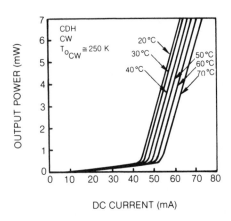

FIGURE 10.14 CW power output as a function of injection current for a CDH laser operated at temperatures varied from 20 to 70°C. (From Ref. 10.6. Reprinted with permission of North-Holland Publishing Company, Amsterdam.)

The ultimate limit in output power for GaAlAs/GaAs semiconductor lasers is the catastrophic damage of the laser facets. This phenomenon occurs when the laser intensity in the active layer is raised to a critical level beyond which a localized melting of the facets occurs. For GaAlAs/GaAs lasers the critical intensity at an uncoated facet is about 2 to 3 $\times$ 10^6 W/cm². It is somewhat higher for GaInAsP/InP lasers. One of the most common methods to reduce these catastrophic damages is to use a dielectric coating on the laser facets. Coatings such as Al_2O_3, Si_3N_4, and SiO_2 at a thickness of $\lambda/2n_d$ where n_d is the index of these dielectric materials have been used and shown to yield beneficial effects. Al_2O_3 appears to be the most useful coating on GaAs facets due to its close thermal expansion match and its high thermal conductivity. An improvement factor of 2 to 3 in power damage threshold

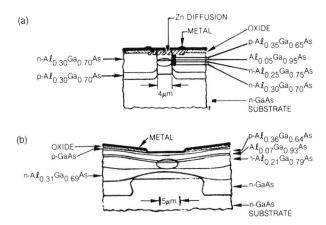

FIGURE 10.15 Single-mode and high-power large-optical-cavity DH lasers: (a) LOC BH laser topology; (b) LOC CDH laser topology. (From Ref. 10.6. Reprinted with permission of North-Holland Publishing Company, Amsterdam.)

with the Al_2O_3 coating is believed to be due to a reduction in surface recombination velocity at the interface.

Another method for increasing catastrophic intensity is to reduce the absorption near the laser facets with a nonabsorbing mirror NAM. Nonabsorbing mirrors can be made by selective diffusion of Zn along the length of the stripe except near the facets. This diffusion creates a bandgap difference between the facets and the gain medium and permits a 3- to 4-fold

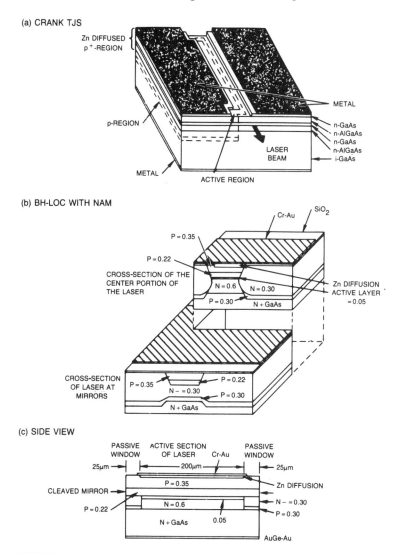

FIGURE 10.16 Schematics for various NAM structures. (After L. Figueroa—Ch. 2, Fig. 2.26 in *Handbook of Solid State Lasers,* Marcel Dekker, Inc., NY (1988).)

increase in the CW facet damage threshold. A commercial laser using this structure is the CRANK TJS laser as shown in Figure 10.16(a). Figure 10.16(b) shows a NAM BH-LOC laser structure. This device is fabricated by first epitaxially growing a BH-LOC structure, then etching channels in a direction perpendicular to the buried mesa. A passive waveguide is regrown in the trenches by liquid-phase epitaxy and a NAM device is finally obtained by cleaving the structure in the passive NAM region. Although other means can be used to produce NAM devices, they all involve complex fabrication procedures, and therefore, tend to have low yields. All available results seem to indicate that all lasers with a NAM structure show an increase in catastrophic damage threshold and all lasers without a NAM structure show a decrease in damage threshold as the device degrades. Table 10.2 summarizes various high power GaAlAs/GaAs lasers with maximum output power greater than 100 mW. For single spatial mode and stable operation, these lasers can provide output power in the 50 to 100 mW range.

TABLE 10.2 Summary of Mode Stabilized High Power GaAlAs/GaAs Laser Characteristics

Manuf.	Type	Processing	Max Power (mW)	Single Spat. (mW)	I_{th} (mA)	Slope Eff (mW/mA)	Subst.
Hitachi[1]	CSP	1 step LPE	100	40	75	0.5	n
MATS.[2]	TRS	1 step LPE	115	80	90	0.43	n
MATS.[3]	BTRS	2 step LPE AR/HR	200	100	50	0.8	p
RCA[4]	CC-CDH	1 step LPE HR	165	50	50	0.77	n
RCA[5]	CPS	1 step LPE AR/HR	190	70	50	—	n
SHARP[6]	VSIS	2 step LPE	100	50	50	0.74	p
SHARP[7]	BVSIS	2 step LPE AR/HR	100	70	50	0.80	p
TRW[8]	ICSP	2 step MOCVD AR/HR	175*	150*	75	0.86	n

* 50% duty cycle

1. K. Aiki, et al., J. Appl. Phys., *30* 649 (1977).
2. M. Wada, et al., J. Appl. Phys. Lett., *42* 853 (1983).
3. K. Hamada, et al., IEEE J. Quant. Elect., *21* 623 (1985).
4. D. Botez, et al., Elect. Lett., *19* 882 (1983).
5. H. Kumabe, et al., J. Appl. Phys. *21* (1982).
6. T. Hayakawa, et al., J. Appl. Phys. *53* 7224 (1983).
7. Y. Yamamato, et al., J. Appl. Phys. Lett., *46* 319 (1985).
8. J. Yang, et al., Elect. Lett., *21* 751 (1985).

10.7 LONG-WAVELENGTH SOURCES

The interest in obtaining low material dispersion has led to rather significant development in recent years of the InGaAsP system, which emits at long wavelengths ranging from 1.1 to 1.6 μm. This section examines several low-threshold InGaAsP/InP lasers and their performance characteristics. With conventional LPE methods, the growth of an InGaAsP active layer encounters a serious problem due to the fact that the active layer can easily be melted back into the Indium solution during the subsequent growth of the cladding layers. To prevent meltback, subsequent growth must occur at relatively low temperatures. Several anti-meltback methods have been introduced: for example, the use of an anti-meltback layer between the active and InP cladding layers. Such a structure is shown in Figure 10.17(a). The InGaAsP layer is grown on an Sn-doped InP substrate oriented in the (100) plane with a carrier concentration of 2×10^{18} cm^{-3}. The growth temperature and rate must be controlled very carefully to prevent the loss of the active layer. The lowest threshold current density of approximately 1.2 kA/cm^2 was obtained (Ref. 10.7) from an active-layer thickness of 0.2 μm. A similar

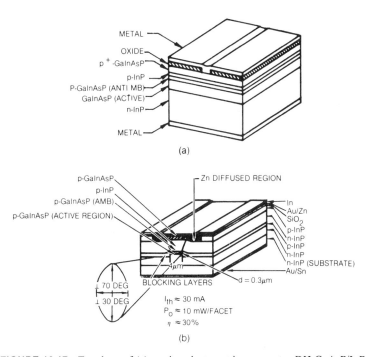

FIGURE 10.17 Topology of (a) semiconductor stripe-geometry DH GaAsP/InP laser, and (b) buried DH InGaAsP/InP laser. (From Ref. 10.7. Reprinted with permission of IEEE, © 1981.)

technique (Ref. 10.7) has been used to fabricate InGaAsP/InP buried heterostructure lasers, as shown in Figure 10.17(b). These lasers can be operated continuously at room temperature with single-transverse-mode output up to 10 mW, at a differential quantum efficiency as high as 43%.

Operating in the 1.3-μm region, a crescent-shaped active layer, as shown in Figure 10.18, has been fabricated by using a two-step LPE technique (Ref. 10.7). Low threshold current and single-transverse-mode operation have also been obtained from this laser. As shown in Figure 10.18(a), a crescent-shaped InGaAsP active region is embedded in InP by LPE growth on a dovetail channeled substrate. A double-current confinement scheme is incorporated with two reverse-biased *pn* junctions at both sides of the active layer. The reverse-biased InP junction serves as a barrier against the spreading of current, even when it is very close to the active region. The active region has a parabolic cross-section and is completely surrounded by InP.

The fabrication procedure of the structure described above is shown in Figure 10.18(b). A planar InGaAsP heterostructure is first grown on a (100) *n*-InP substrate by LPE. Subsequently, a dovetail-shaped channel is etched along the (011) direction into the upper two layers, with a depth of approximately 2 μm. Then four additional layers, as shown in Figure 10.18(b), are grown successively on the channeled substrate. The crescent region has a refractive index equivalent to a cladded parabolic index wave-

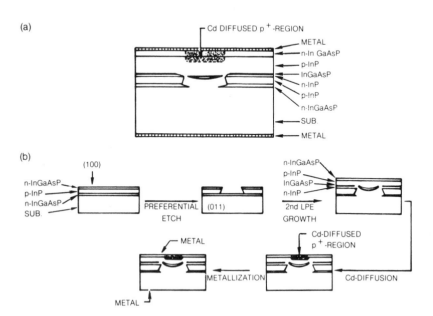

FIGURE 10.18   (a) Topology of a buried crescent InGaAsP/InP laser; (b) fabrication procedure of a BC laser. (From Ref. 10.8. Reprinted with permission of IEEE, © 1981.)

guide. With an active width of 2 μm and an active thickness of 0.1 μm, only the fundamental transverse mode is allowed, while all higher-order modes are beyond cutoff. Figure 10.19(a) depicts the $L–I$ curve and Figure 10.19(b) shows the output spectral characteristics of a crescent-shaped InGaAsP laser. The I_{th} of this laser is about 20 mA and the output increases with increasing currents and exhibits no "kinks." At high currents, the $L–I$ curve saturates due to the temperature effect. The maximum CW power obtainable from this laser without causing any catastrophic damage is about 25 mW per facet. At low injection levels, several longitudinal modes exist; however, as the current level increases, the output power begins to concentrate in one mode as the current reaches the $1.1I_{th}$ level. With a further increase in the injection current, the output spectrum shifts toward the longer wavelength, again due to the temperature effect; however, the output remains in a single longitudinal mode for currents up to twice the threshold value.

Since the work in the early eighties (Refs. 10.7 and 10.8), much progress has been made in producing high-power GaInAsP/GaAs lasers operating at 1.3 microns. A summary of various high-power, long wavelength diode lasers is given in Table 10.3. Using a buried-crescent geometry, researchers from OKI (Ref. 10.9) have achieved 200 mW output in a single spatial mode. The life-test results showed that the mean time to 0.75 P_{max} for these lasers operating at T = 20°C and λ = 1.3 microns was about 7×10^4 hours. It appears that 1.3 micron lasers are more reliable than GaAlAs/GaAs lasers for high-power (>50 mW) applications.

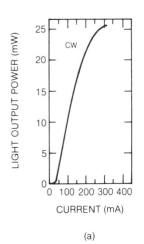 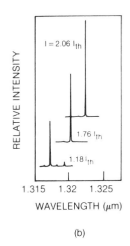

(a)   (b)

FIGURE 10.19   (a) $L–I$ curve for a BC InGaAsP/InP laser; (b) output spectral characteristics of a BC InGaAsP/InP laser. (From Ref. 10.8. Reprinted with permission of IEEE, © 1981.)

TABLE 10.3   Summary of Mode Stabilized High-Power GaInAsP/InP Laser Characteristics

Manuf.	Type	Processing	Max Power (mW)	Single Mode Power (mW)	Spot Size (μm^2)	I_{th} (mA)
Mitsubishi	PBC[1]	2 step LPE (P-subst)	140	70	0.9×2.2	10–30
NEC	DC-PBH[2]	2 step LPE	140	140	1.1×2.5	10–30
OKI	PBC[3]	2 step LPE (P subst)	200	200		10–30
TRW/EORC	DC-PBH[4]	2 step LPE	110	70	1×2.5	10–30

1. Y. Kawahara, et al., Proc. OFC, p. 12 (1985).
2. M. Yamaguchi, et al., Proc. CLEO, p. 180 (1985).
3. Ref. 10.9.
4. E. Rezek, et al., Proc. Topic Meet. on Semicond. Lasers, Albuquerque, NM (1987).

10.8 DISTRIBUTED FEEDBACK LASERS

Distributed feedback semiconductor laser structure is the one utilizing a periodic corrugation spatially distributed along the length of a gain medium to produce a feedback mechanism for laser oscillation. Figure 10.20 illustrates the structural configuration of a typical distributed feedback (DFB) laser. The fabrication of a desired corrugating structure commonly known as phase grating on a semiconducting layer between heterojunctions usually involves a very high-resolution material-processing technology.

Even though the physical model of this laser has been established for a long time, it has become commercially available only very recently. Kogelnik and Shank (Ref. 10.10), analyzed this laser system using a coupled-wave theory. A simple physical model was established by assuming a coupling between two counter-running waves caused by an interaction known as backward Bragg scattering. Therefore, this laser is sometimes referred to as a distributed Bragg reflection (DBR) laser. In this model, the field

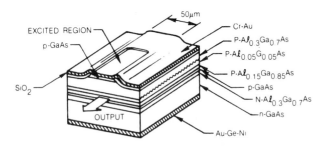

FIGURE 10.20   Topology of an AlGaAs DH DFB laser.

amplitude of each of the two oppositely traveling waves is superimposed with a backward Bragg scattered component of the other wave of a particular order that must satisfy the Bragg condition. Because the medium provides optical gain along its path length, such an interaction is capable of generating sufficient feedback to sustain laser oscillation. The major advantage of this laser structure is the high degree of spectral selectivity by the phase grating.

For simplicity, this analysis deals only with a linear system in which any nonlinear effect such as gain saturation is ignored. Therefore, this model is valid only near threshold. For a medium having gain or loss, the wave equation as given by Equation (2.16) must be modified to include the effects of source or sink. This modification is made by adding a term involving the electrical conductivity, $\mu\sigma(\partial \mathbf{E}/\partial t)$ into the wave equation for a passive medium. In this case the wave equation takes the form

$$\nabla^2 \mathbf{E} = \mu\sigma \frac{\partial \mathbf{E}}{\partial t} + \mu\epsilon \frac{\partial^2 \mathbf{E}}{\partial t^2} \tag{10.58}$$

For a plane wave polarized in the x-direction and its spatial variations occurring only in the z-direction, $\partial/\partial x = \partial/\partial y = 0$. $\mathbf{E}$ can be replaced by $\mathbf{E}_x$. Equation (10.58) reduces to

$$\frac{\partial^2 E_x}{\partial z^2} = \mu\sigma \frac{\partial E_x}{\partial t} + \mu\epsilon \frac{\partial^2 E_x}{\partial t^2} \tag{10.59}$$

Substituting the time-dependent term, $e^{-i\omega t}$, into Equation (10.59) yields a scalar wave equation of the form

$$\frac{\partial^2 E_x}{\partial z^2} + k^2 E_x = 0 \tag{10.60}$$

where

$$k^2 = i\omega\mu(\sigma - i\omega\epsilon) \tag{10.61}$$

It is customary to express the propagation constant k in terms of a complex refractive index $\mathcal{N}$ as

$$k^2 = \mathcal{N}^2 k_0^2 \tag{10.62}$$

where $\mathcal{N}$ has a real part n, and an imaginary part n_i. Now, assume that the real part has a sinusoidal variation with z as

$$n_r(z) = n + n_m \cos K_z \tag{10.63}$$

where $K = 2\pi/\Lambda$, with Λ being the period of the phase grating, n_m the maximum modulation index, and n the real index of the material. The imaginary part is related to the field gain coefficient, g_F, where g_F must be distinguished from the usual power gain coefficient g (i.e., $g = 2g_F$) by the expression

$$n_i = -g_F k_0 \tag{10.64}$$

Substituting Equations (10.63) and (10.64) into (10.62) provides

$$k^2 = k_0^2[(n + n_m \cos Kz)^2 - (k_0 g_F)^2 + 2(g_F K_0)i(n + n_m \cos Kz)] \quad (10.65)$$

For all practical purposes, assume that the gain is small over a distance of the order of a wavelength, λ_0, and also that the perturbation of refractive index n_m is very small compared with n. In other words,

$$g_F \ll \frac{2\pi n}{\lambda_0} = \beta \qquad n_m \ll n \quad (10.66)$$

With these assumptions, Equation (10.65) reduces to

$$k^2 = \beta^2 + 2i\beta g_F + 2k_0\beta n_m \cos Kz \quad (10.67)$$

Substituting Equation (10.67) into (10.60) gives

$$\frac{\partial^2 E_x}{\partial z^2} + (\beta^2 + 2i\beta g_F)E_x = -2k_0\beta \cos(2\beta_B z)E_x \quad (10.68)$$

where

$$\beta_B = \frac{2\pi n}{\lambda_B} \quad (10.69)$$

is a number that corresponds to half Bragg wavelengths in the medium. If the periodicity of the phase grating Λ is $\lambda_B/2n$, so that the Bragg condition is satisfied, only two opposite waves are synchronized in phase, while all other diffraction orders can be neglected in the coupled-wave model. These two counter-running waves are denoted by $\mathscr{E}_+(z) \exp(i\beta_B z)$ and $\mathscr{E}_-(z) \exp(-i\beta_B z)$. The total electric field is the sum of these two waves as given by

$$E_x(z) = \mathscr{E}_+(z) \exp(i\beta_B z) + \mathscr{E}_-(z) \exp(-i\beta_B z) \quad (10.70)$$

where $\mathscr{E}_+$ and $\mathscr{E}_-$ are complex amplitudes. To simplify the analysis further, assume that the transfer of power between these two waves occurs slowly and therefore the second derivatives $\partial^2 \mathscr{E}_-/\partial z^2$, $\partial^2 \mathscr{E}_+/\partial z^2$ can be neglected. In this approximation, upon substituting Equation (10.70) into (10.68), the following pair of coupled-wave equations is obtained:

$$-\frac{\partial \mathscr{E}}{\partial z} + (g_F - i\delta) = iK_c \mathscr{E}_+$$
$$\frac{\partial \mathscr{E}_+}{\partial z} + (g_F - i\delta) = iK_c \mathscr{E}_- \quad (10.71)$$

where the parameter δ is a normalized frequency defined by

$$\delta \equiv \frac{\beta^2 - \beta_B^2}{2\beta_B} \simeq \beta - \beta_B = \frac{n}{c}(\omega - \omega_B) \quad (10.72)$$

and the parameter K_c is the coupling coefficient, defined by

$$K_c = \frac{\pi n_m}{\lambda_0} \qquad (10.73)$$

The coupled-wave equations (10.71) describe wave propagation in the DFB structure in the presence of gain and a periodic perturbation in the refractive index. As illustrated by Figure 10.21, laser oscillation builds up from zero amplitudes with waves reflected at the device boundaries. The boundary conditions for the wave amplitudes are

$$\mathscr{E}_- \left(\frac{L}{2}\right) = \mathscr{E}_+ \left(-\frac{L}{2}\right) = 0 \qquad (10.74)$$

To solve the coupled-wave equations, assume a general solution of the form

$$\begin{aligned}\mathscr{E}_- &= C_1 e^{\gamma z} + C_2 e^{-\gamma z} \\ \mathscr{E}_+ &= d_1 e^{\gamma z} + d_2 e^{-\gamma z}\end{aligned} \qquad (10.75)$$

The symmetric and antisymmetric requirements [i.e., $\mathscr{E}_-(z) = \pm \mathscr{E}_+(-z)$] lead to the following relations:

$$C_1 = \pm d_2, \qquad C_2 = \pm d_1 \qquad (10.76)$$

Furthermore, the boundary conditions require that

$$\frac{C_1}{C_2} = \frac{d_2}{d_1} = -e^{\gamma L} \qquad (10.77)$$

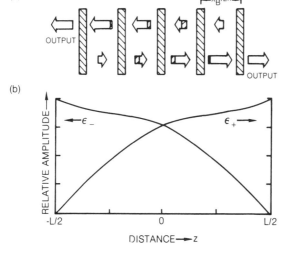

FIGURE 10.21   (a) Illustration of laser oscillation in a periodic structure; (b) field amplitudes of a left-traveling wave $\mathscr{E}_-$ and a right-traveling wave $\mathscr{E}_+$ versus distance. (From Ref. 10.10 Reprinted with permission of J. Appl. Phys.)

Using these relations, Equation 10.75 becomes

$$\mathscr{E}_- = \sinh \gamma \left(z + \frac{L}{2} \right)$$

$$\mathscr{E}_+ = \pm \sinh \left(z - \frac{L}{2} \right)$$

(10.78)

To determine the allowed eigenvalues corresponding to the longitudinal modes in the DFB structure, substitute the expressions of Equation (10.78) into the coupled-wave equations (10.71) to obtain a pair of transcendental equations

$$-\gamma \sinh \frac{\gamma L}{2} + (g_F - i\delta) \cosh \frac{\gamma L}{2} = \pm iK_c \cosh \frac{\gamma L}{2}$$

$$-\gamma \cosh \frac{\gamma L}{2} + (g_F - i\delta) \sinh \frac{\gamma L}{2} = \mp iK_c \sinh \frac{\gamma L}{2}$$

(10.79)

Performing the sum and difference of the equations above yields

$$\gamma + (g_F - i\delta) = \pm iK_c e^{\gamma L}$$

$$\gamma - (g_F - i\delta) = \mp iK_c e^{-\gamma L}$$

(10.80)

Multiplying these two equations gives a dispersion relation for the complex propagation constant γ:

$$\gamma^2 = K_c^2 + (g_F - i\delta)^2$$

(10.81)

Adding these two equations gives an eigenvalue equation for γ:

$$K_c = \pm \frac{i\gamma}{\sinh \gamma L}$$

(10.82)

Equation (10.82) indicates that in general, values of γ are complex and each corresponds to a different branch of the complex hyperbolic sine functions. Furthermore, each eigenvalue γ associates with a threshold gain constant and a resonant frequency δ. The values for g_F and δ can be evaluated from the following relation, which is obtained by taking the difference of Equation (10.80):

$$g_F - i\delta = \gamma \coth \gamma L$$

(10.83)

Because of the dispersive nature of the DFB structure, "stop bands" of frequencies exist in which mode propagation is forbidden. Equations (10.70) and (10.75) and the dispersion relation as given by Equation (10.81) show that four waves exist in the structure. Each traveling wave $\exp(i\beta_B \pm \gamma)z$ has two γ values, which are complex quantities, determined by the dispersion relation in terms of K_c, g_F, and δ. The eigenvalues γ can be obtained only by solving the complex transcendental equation (10.82) numerically, with the help of a computer. It is possible, however, to obtain approximations in the limits of high and low threshold gain. In the high-gain limit, where $g_F \gg K_c$, the complex propagation constant γ as given by Equation (10.81) becomes

$$\gamma \simeq g_F - i\delta \qquad (10.84)$$

and the eigenvalue equation (10.80) can be written as

$$2(g_F - i\delta) \simeq \pm iK_c \exp(g_F - i\delta)L \qquad (10.85)$$

Multiplying the complex conjugate of Equation (10.85) yields

$$\frac{(\exp 2g_F L)K_c^2}{4(g_F^2 + \delta^2)} = 1 \qquad (10.86)$$

Equation (10.86) is equivalent to the oscillation condition for semiconductor lasers with end mirrors as given by Equation (8.85), where $g - \alpha$ corresponds to $2g_F$ and the product of mirror reflectivities corresponds to $K_c^2/4(g_F^2 + \delta^2)$. Equating the phase of Equation (10.85) gives

$$\left(m + \frac{1}{2}\right)\pi + \tan^{-1}\frac{\delta}{g_F} = \delta L \qquad (10.87)$$

where $m = 0, \pm 1, \ldots$ Near the Bragg frequency, Equation (10.73) gives $K_c = \pi n_m/\lambda_0$. Substituting these K_c values into Equations (10.86) and (10.87) and letting $\delta = 0$ yields

$$\left(\frac{\pi n_m}{\lambda_0}\right)^2 \exp(2g_F L) = (2g_F)^2 \qquad (10.88)$$

and

$$\delta = \frac{(m + \frac{1}{2})\pi}{L} \qquad (10.89)$$

Equation (10.89) provides the expression for the resonance frequencies

$$\nu = \nu_B \pm \frac{c}{2nL}\left(m + \frac{1}{2}\right) \qquad (10.90)$$

Equation (10.90) shows that the resonances are spaced approximately $c/2nL$ apart, which is the same as in a usual two-mirror laser cavity of length L. It should be noted that no resonance is found at ν_B, the Bragg frequency, and the width of the stop band is $2K_c$. Within the stop band, waves are evanescent. The threshold gain for a given resonance frequency can be calculated from Equation (10.89). To verify the spectral selectivity of a DFB laser, Equation (10.86) shows that for a wavelength that deviates from the Bragg wavelength by $\delta = g_F$, K_c^2 must be doubled to keep the threshold gain the same. This strong spectral selectivity is a direct consequence of the dispersive property of the DFB laser structure.

To obtain approximate formulas for the low-gain limiting case, where $g_F \ll K_c$, take only the real part of Equations (10.82) and (10.83) and expand in a power series near $g_F = 0$ to get

$$\delta \simeq K_c \qquad (10.91)$$

for the first resonance that occurs again just outside the stop band. The threshold condition in this case is

$$g_F L \simeq \left(\frac{\lambda}{n_m L}\right)^2 \qquad (10.92)$$

Figure 10.22 shows the numerical results for the resonance spectrum and threshold gain of the periodic structure obtained by Kogelnik and Shank (Ref. 10.10). The dashed lines indicate the basic frequency spacing $\Delta\nu = c/2nL$. In terms of the Bragg wavelength,

$$\Delta\lambda = \frac{\lambda_B^2}{2nL} \qquad (10.93)$$

This result is exactly the one that emerges in the high-gain limit. Figure 10.22 indicates that the mode spectrum is symmetric with respect to the Bragg frequency ν_B and no resonance occurs at ν_B. Within the stop band, all oscillations cease. The stop band increases with increasing K_c, and eventually becomes comparable to $2nL$. At this point it starts to push the resonances away from ν_B. The threshold gain as shown in Figure 10.22 also increases with the frequency spacing from ν_B. As a result, the DFB laser structure provides the highest spectral selectivity.

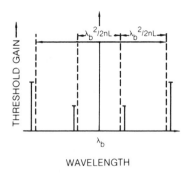

FIGURE 10.22 Calculated mode spectrum and threshold gain of a DFB laser. (From Ref. 10.10 Reprinted with permission of J. Appl. Phys.)

10.9 CLEAVED COUPLED-CAVITY SEMICONDUCTOR LASERS

Another interesting development in semiconductor laser technology is an electronically tunable and single frequency laser source, commonly called the cleaved coupled-cavity (C^3) laser. This laser consists of two optically interacting cavities and can be batch manufactured by cutting a planar heterojunction device along its cleavage plane and realigning the two parts. Each part is electronically controlled and independent of the other. Because of the spectral purity and extremely broad frequency tunability of its output, this laser has the potential to increase data rate and transmission length, or both, significantly. Using this laser, it has been shown (Ref. 10.11) that single frequency operation can be maintained under 2 Gb/s direct modulation with error rates of less than 10^{-10}. It has also been demonstrated that a C^3 laser can transmit information at 420 Mb/s through a 119 km unrepeated length of fiber with a BER $< 10^{-9}$. In this case the C^3 laser was made to operate at 1.55 μm with a chromatic dispersion of 2.08×10^{-3} ps/km. Because the frequency of this laser can be tuned, it is possible to design a system involving multichannel optical frequency shift-keying and a new approach to optical data switching and routing.

Figure 10.23 is a schematic of a C^3 laser. The basic laser material can be either a gain-guided or index-guided heterojunction. Two standard Fabry–

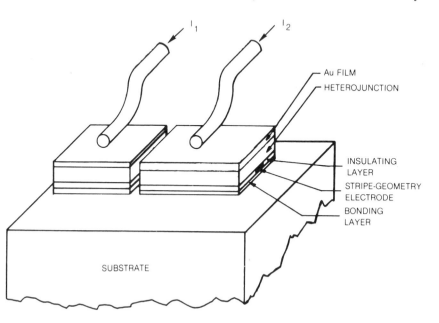

FIGURE 10.23   A schematic diagram of a cleaved coupled-cavity laser.

Perot (FP) cavities with two completely separate stripe-geometry electrodes are involved. These two laser diodes have slightly different lengths, typically about 125 μm, and are strongly coupled optically with each other through a separation of less than 5 μm. All reflecting facets are formed by cleaving along crystallographic planes so that they are perfectly parallel. The electrodes that are buried underneath the laser material must be perfectly aligned with respect to each other, and must be electrically isolated from each other. To achieve the above, a heterojunction epilayer is cleaved at two ends to form a standard FP laser cavity of approximately 250 μm length. It is then covered with a thick (~5 μm) electroplated Au layer. The device is recleaved near the middle to form two separate FP diodes. Because of the thick Au layer, these two diodes remain hinged together and are bonded with indium upside down on a Cu heat sink. Two stripe-geometry electrodes must be deposited on the epilayer before bonding. The separation between the two diodes is only a few microns, nevertheless it can be varied by stretching the thick Au film. This fabrication technique can be extended to the manufacture of an array of coupled-cavity laser diodes.

The C^3 laser can be operated in either a tuning or an untuning mode. In either case, the first diode is always operated above threshold. The second diode can either be above or below threshold. In a tuning mode, the laser wavelength can be controlled by varying the injection current level, which is always kept below threshold. A change in injection currents causes a change in carrier density, which affects the refractive index of the medium and causes a shift in the cavity modes in the second diode. Typically, a frequency excursion of 150 Å and a tuning rate of 10 Å/mA can be achieved. Let n_1 and n_2 be the effective refractive indices in the first and the second diodes, respectively. Then the mode spacings for the two diodes in terms of wavelength can be expressed as

$$\Delta\lambda_1 = \frac{\lambda_0^2}{2n_1 L_1}$$

and

$$\Delta\lambda_2 = \frac{\lambda_0^2}{2n_2 L_2}$$

Figure 10.24(a) shows the allowed Fabry–Perot modes for the two independent diodes. Because these two diodes are strongly coupled, the modes from each cavity that coincide spectrally will interfere constructively to form the resultant modes of the coupled cavity, while all others interfere destructively and are suppressed. The spectral spacing $\Delta\lambda_C$ of the coupled-cavity modes, therefore, is significantly larger than those of the two individual diodes and can be approximated by

$$\Delta\lambda_C = \frac{\Delta\lambda_1 \Delta\lambda_2}{|\Delta\lambda_1 - \Delta\lambda_2|} = \frac{\lambda_0^2}{2|n_1 L_1 - n_2 L_2|} \tag{10.94}$$

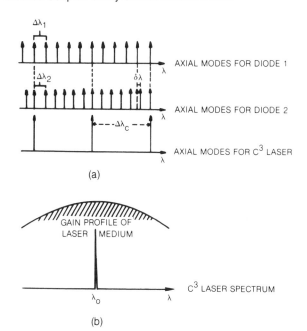

FIGURE 10.24   Spectral characteristics of cleaved coupled-cavity lasers. (a) Allowed Fabry–Perot modes for the first and the second diodes, and for the coupled-cavity. (b) The output spectrum of the C^3 laser.

As a result, the spectrum of the coupled-cavity laser becomes extremely pure and the only one that occurs near the peak of the gain, as shown in Figure 10.24(b), can survive. As the injection current in Diode 2 varies by a small amount equivalent to a shift in wavelength by $\delta\lambda$, the wavelength of the C^3 laser will be shifted by exactly one FP mode spacing ($\Delta\lambda_1 \simeq \Delta\lambda_2$), which is about 15Å, depending on the cavity length. This type of stepwise tuning will continue over a very large spectral width of the gain profile ($\sim$150Å). The ability to tune over such a wide range of wavelengths naturally leads to system applications involving frequency shift-keying of wavelength division multiplexing.

The C^3 laser also has the ability to perform a set of basic logic operations, including AND, OR, INVERT, and EXCLUSIVE OR. In these operations, both diodes are operated above threshold with a pulsed current superimposed with a dc bias. Therefore, the laser output spectra can be in either λ_1, λ_2, or $\lambda_1 + \lambda_2$ states. By a superposition of injection current pulses I_1 and I_2 with a varying pulse width and time delay, it is possible to achieve all of the above operations. For example, the overlap of two current pulses produces an increase of output power in the overlapping period. One simple method for obtaining AND or OR logic functions is to limit the levels in the detection channel that are sensitive to all wavelengths. Another method is analyze the output spectra using a diffraction grating with an array of detectors. This method provides the flexibility of obtaining a variety of logic functions by virtue of the fact that both the power and wavelength are monitored independently. A detailed analysis of GaAlAs C^3 lasers incorporating gain dependence on pumping and wavelength of both laser segments as well as the effects of spontaneous emission in the lasing diode section and the known dependence of the refractive index on carrier density can be found in the paper written by Streifer, et al. (Ref. 10.12).

10.10 SEMICONDUCTOR LASER ARRAYS

The most practical method for increasing the output power of a semiconductor laser involves the use of a monolithic array of phase-locked laser diodes. The first reported laser array (Ref. 10.13) consisted of five closely coupled proton bombarded lasers. Much progress has been made on monolithic laser arrays since the first reported work on the phase-locked operation. The original coupling scheme involved branched waveguides, but it is now believed that the evanescent field coupling among an array of index-guided lasers is a better approach. Recent effort has been focused on achieving high CW output power while controlling the far-field distribution corresponding to the fundamental array mode with all elements locked in-phase. To excite the fundamental array mode, it is necessary to introduce either optical gain in regions between the elements of an array or to vary the width of the array elements.

To describe the operation of a laser array, Butler, Ackley and Ettenberg (Ref. 10.14) analyzed a system of N equally spaced and weakly-coupled lasers in which the elements are assumed to be identical. If the separation between elements is s, then $\psi^1(x, s_1, z) = \psi^n(x, s_n, z)$. Each element satisfies the wave equation

$$\nabla^2\psi^m + k_0^2 K^m(x, y)\psi^m = 0 \tag{10.95}$$

where K^m is the dielectric function that describes the mth element. The wavefunction is assumed to be of the form

$$\psi^m(x, y, z) = U^m(x, y)V^m(y) \exp(-\gamma_m z) \tag{10.96}$$

where U^m, which is a function of both x and y, describes the transverse ($\perp$ to the junction) profile, V^m represents the lateral mode profile, and γ_m is a complex propagation constant. The wavefunction $V^m(y)$ satisfies

$$\frac{d^2}{dy^2} V^m + [\gamma_m^2 - \gamma_{0m}^2(y) + k_0^2\Gamma^m(y)K^m(y)]V^m = 0 \tag{10.97}$$

where Γ^m is the optical confinement factor of the field. The quantity γ_{0m} can be expressed as

$$\gamma_{0m} = \alpha(y) + ik_0 n(y) \tag{10.98}$$

where α and n are the absorption coefficient and effective index of refraction. The quantity K^m is the perturbation of the active layer dielectric constant due to injected carriers and also contains a real and an imaginary part of the dielectric constant, which are related to the gain of the medium. An equation similar to Equation (10.97) has been treated in Section 10.3. Because adjacent elements are linked by the optical fields through the lateral dielectric function, the nature of their coupling is dependent on both the gain and refractive index variation. A specification of a lateral gain profile allows the calculation of the array mode discrimination via the coupled mode analysis. The following analysis shows that the propagation constant of the array mode will split into a series of complex numbers that represent the propagation constants of individual array modes for phase-locked operation. The imaginary part of the splitting affects the laser frequency while the real part characterizes the array modal gain.

The array mode is written as a superposition of the modes of $2N + 1$ individual elements as

$$\phi(x, y, z) = \sum_{m=-N}^{N} A^m(z)\psi^m(x, y, z) \tag{10.99}$$

where $A^m(z)$ is a complex coefficient that determines the amplitude and phase of each of the individual elements in the allowed array modes. Substituting Equation (10.99) into the usual wave equation and applying the weak coupling condition, the following eigen-equation for the amplitude coefficient $A^n(z)$ is found to be (Ref. 10.14)

$$\sum_m k_0^2 \frac{C_{nm}}{2\gamma_n} A^m(z) = \frac{\partial A^n}{\partial z} \tag{10.100}$$

where the complex coupling coefficient for identical elements in the array is given by

$$C_{nm} = \frac{\iint (K - K^m)(U^m U^n V^m V^n) \, dx \, dy}{\iint (U^m U^n V^m V^n) \, dx \, dy} \tag{10.101}$$

The array modes can be obtained by solving the set of simultaneous equations given by Equation (10.100). Because A^m is a weakly varying function of z under weak coupling, Equation (10.100) can be simplified by letting

$$A^m(z) = A_0^m \exp(-\delta_m z) \tag{10.102}$$

Note that if $\delta_m = 0$, no coupling occurs. If δ_m is element-dependent, it would be impossible to establish the resonant condition simltaneously for each laser element, hence, an array mode would not be allowed. Therefore, the subscript m is eliminated from δ. Substituting Equation (10.102) into Equation (10.101) yields

$$\sum_m C_{nm}A_0^m + \frac{2\gamma_n\delta}{k_0^2} A_0^n = 0 \tag{10.103}$$

With the assumption that in an array of elements only the two nearest neighbors are coupled, the above eigenvalue equation can be expressed in a simple matrix form as

$$[C][A] = \mu[A] \tag{10.104}$$

where μ is the eigenvalue and

$$[C] = \begin{bmatrix} 0 & C_{12} & 0 & \cdots & 0 \\ C_{21} & 0 & C_{23} & \cdots & 0 \\ 0 & C_{32} & 0 & \cdots & 0 \\ \cdot & \cdot & \cdot & \cdots & \cdot \\ 0 & \cdot & \cdot & \cdots & 0 \end{bmatrix} \tag{10.105}$$

and

$$[A] = \begin{bmatrix} A_0^1 \\ A_0^2 \\ \vdots \\ A_0^N \end{bmatrix} \tag{10.106}$$

Because $[C]$ is bidiagonal, a generalized second-order difference equation for the eigenvalue problem can be written as

$$C_{n,n-1}A_0^{n-1} - C_{n,n+1}A_0^{n+1} = \mu A_0^n \tag{10.107}$$

Equation (10.107) can be greatly simplified if the array geometry is specified.

For example, when elements are uniformly spaced, all elements of the $[C]$ matrix are identical, e.g., there is equal coupling to each nearest neighbor. In this case, Equation (10.107) becomes

$$CA_0^{n-1} - \mu A_0^n + CA_0^{n+1} = 0 \tag{10.108}$$

where C is the coupling coefficient. Now, letting

$$A_0^n = B \exp(in\varphi) \tag{10.109}$$

where B are φ are unknown constants, the resulting characteristic equation becomes

$$\exp(-i\varphi) - \mu/C + \exp(i\varphi) = 0 \tag{10.110}$$

which yields a solution for the eigenvalues

$$\mu = 2C \cos \varphi \tag{10.111}$$

Because Equation (10.108) is a second-order difference equation, its second solution is of the form

$$A_0^n = B \exp(-in\varphi) \tag{10.112}$$

The linear combination of Equations (10.109) and (10.111) is

$$A_0^n = B_1 \cos n\varphi + B_2 \sin n\varphi \tag{10.113}$$

Because only the A_0^n values lying in the range $1 < n < N$ are applicable to this discussion, $A_0^0 = A_0^{n+1} = 0$. Substituting this condition into Equation (10.112) provides

$$(N + 1)\varphi = p\pi \qquad p = 1, 2, \ldots, N \tag{10.114}$$

where the integer p specifies the array modes. The amplitude of the nth element of an array operating in array mode p becomes

$$A_{0,p}^n = B_2 \sin \left(\frac{n\pi p}{N + 1} \right) \tag{10.115}$$

The corresponding eigenvalue is

$$\mu_p = 2C \cos \left(\frac{p\pi}{N + 1} \right) \tag{10.116}$$

The constant B_2 can be calculated by using the normalization conditions. Figure 10.25 is a plot of the normalized amplitude distributions for each of the allowed array modes for a 10-element array. It is evident that the $p = 1$ and $p = 10$ modes correspond to what are commonly known as the in-phase (0° phase shift between elements) and the out-phase (π phase-shift) amplitudes, which can be obtained using simple diffraction theory. The amplitude of the individual elements of an array operating in the pth array mode will lie on an envelope function $\sin[p\pi y/(N + 1)s]$ where s is the interspacing between elements.

The splitting of the complex propagation constant can be calculated from the eigenvalues μ_p as given by Equation (10.116), which yields the value $\delta = -k_0^2\mu_p/2\gamma_p$, representing the splitting of the propagation constant from that of the single element value γ. Hence, the array modes have propagation constants given by

$$\gamma_p = \gamma \left[1 - \frac{Ck_0^2}{\gamma^2} \cos \left(\frac{p\pi}{N + 1} \right) \right] \tag{10.117}$$

If $\gamma = \alpha + i\beta$ and $C = C_r + iC_i$ where α is related to the modal gain

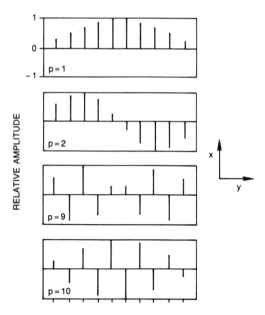

FIGURE 10.25 The normalized amplitude distribution for a 10-element array. (Ref. 10.14. Reprinted with permission of *Appl. Phys. Lett.*)

coefficient and C_r and C_i are the real and imaginary parts of the coupling coefficient, respectively, Equation (10.117) becomes in the case that $\beta \gg \alpha$,

$$\gamma_p = \left[\alpha - \frac{k_0 C_i}{n_{\text{eff}}} \cos\left(\frac{p\pi}{N+1}\right) \right] + i \left[\beta + \frac{k_0 C_r}{n_{\text{eff}}} \cos\left(\frac{p\pi}{N+1}\right) \right] \quad (10.118)$$

where $n_{\text{eff}} = \beta/k_0$ is the mode effective index. The wavelength and gain coefficient of the pth mode can readily be obtained from Equation (10.118) as

$$\frac{\lambda_p - \lambda_0}{\lambda_p} = - \frac{C_r}{n_{\text{eff}}^2} \cos\left(\frac{p\pi}{N+1}\right) \quad (10.119)$$

and

$$G_p = G_0 + \frac{2k_0 C_i}{n_{\text{eff}}} \cos\left(\frac{p\pi}{N+1}\right) \quad (10.120)$$

Note that for $C_r > 0$, the wavelength of the $p = 1$ mode is smaller than that of the single element laser, while for $p = N$, the array mode wavelength increases relative to the single element. For $C_i > 0$, the fundamental $p = 1$ mode is favored, while for $C_i < 0$ the highest-order mode $p = N$ is the most preferred mode to lase.

With the above results, the eigen-function of the pth array mode can be written as

$$\psi_p(xyz) = \sum_{m=1}^{n} \sin(m\theta_p) U^m(x, y) V^m(y) e^{-(\gamma + \delta\gamma_p)z} \qquad (10.121)$$

where

$$\theta_p = p\pi/(N + 1) \qquad (10.122)$$

and

$$\delta\gamma_p = -\frac{k_0^2 C}{\gamma} \cos m\theta_p \qquad (10.123)$$

The envelope function in Equation (10.121) that results from the eigenvalue equations provides the necessary weighing process for the contribution from the individual element field functions. Note that in the absence of coupling, the array mode does not exist because the amplitude distribution function is a component of an eigenvector that is a solution of a matrix equation in terms of the coupling coefficient.

The far-field distribution can be calculated by summing the field distributions of the array including the phase variation across the array. The total intensity in the far field resulting from N identical radiators separated by D, $2D$, $3D$, . . . is given by (Ref. 10.15)

$$I(\theta) \propto \cos^2 \theta \mid E(\theta) \mid^2 \left[\frac{1 - \cos(Nk_0 D \tan \theta)}{1 - \cos(k_0 D \tan \theta)} \right]$$

where $E(\theta)$ is the far field of each radiator and is proportional to the Fourier transform of ψ_p of the pth array mode. The features of the far-fields corresponding to the array modes can be demonstrated by using the closed-form solution for the relative far-field distribution for the pth array mode with no interelement phase shift as given by (Ref. 10.14)

$$I_p(u) \propto \left[\frac{\sin N(u + \theta_p)/2}{\sin(u + \theta_p)/2} - (-1)^p \frac{\sin N(u - \theta_p)/2}{\sin(u - \theta_p)/2} \right]^2$$

$$= \frac{\sin^2 \left[\dfrac{N + 1}{2} (u + \theta_p) \right]}{\left[\sin^2 \left(\dfrac{u}{2} \right) - \sin^2 \left(\dfrac{\theta_p}{2} \right) \right]^2} \qquad (10.124)$$

where

$$u = k_0 s \sin \phi \qquad (10.125)$$

and ϕ is the angle with respect to the normal of the facet. Figure 10.26 shows

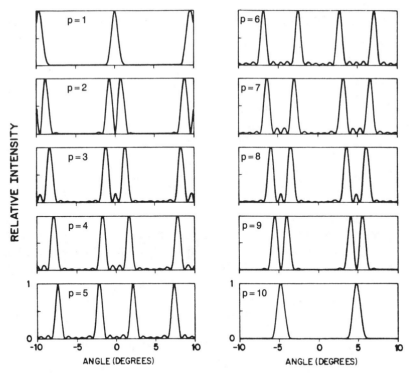

FIGURE 10.26 Plots of the calculated far-field distributions for the various array modes. The array has 10 elements located on 5 μm centers. (Ref. 10.14. Reprinted with permission of *Appl. Phys. Lett.*)

the lateral far-fields of the ten allowed modes of a ten-element array with the elements located on 5 micron centers. This array is fairly typical of an index-guided laser structure. Figure 10.27 compares the coupled mode results with those predicted by the simple diffraction theory. It shows the far-field distribution (solid curve) of the lowest-order array mode ($p = 1$), and also shows the simple diffraction pattern resulting from 10 slits on 5 micron centers. It is clear that some spatial broadening exists in the array mode distribution accompanied by a reduction in the sidelobes. This broadening is a result of a narrowing of the effective aperture by the amplitude function $A_p^m(z)$, which also leads to a reduction of the sidelobes. While the simple diffraction theory can predict the $p = 1$ and $p = 10$ far-fields fairly accurately, it cannot predict the far-field for other modes. The main lobe of mode $p = 1$ has a 35% wider beam width than the simple diffraction theory lobe. The first null of the $p = 1$ mode occurs at an angle ~1.5 λ/a, where a is the array aperture. This angle is about 50% larger than that obtained by simple diffraction theory. The ratio of the beam full width at the first nulls to the beam full width at half power is approximately 2.5 and stays at the same

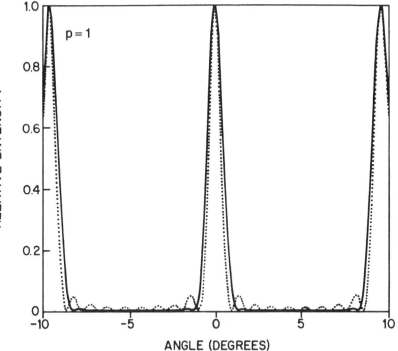

FIGURE 10.27 Comparison between the lowest-order-mode, far-field distributions obtained by coupled-mode analysis (solid line) and by the simple diffraction theory (dashed line). Ten elements on 5 μm centers. (Ref. 10.14. Reprinted with permission of *Appl. Phys. Lett.*).

value as N decreases to a value as low as 10. Because in coupled mode theory the field goes to zero at the aperture edges, the sidelobes are relatively small, -23 dB below the main lobe peak for the first sidelobe.

If a phase shift $\Delta\varphi$ exists between adjacent array elements across the entire array, the expression for the far-field can be obtained from Equation (10.124) simply by adding $\Delta\varphi$ to u. Thus,

$$I_p(u) \propto \frac{\sin^2\left[\dfrac{N+1}{2}(u + \theta_p + \Delta\varphi)\right]}{\left[\sin^2\left(\dfrac{u + \Delta\varphi}{2}\right) - \sin^2\left(\dfrac{\theta_p}{2}\right)\right]^2} \qquad (10.126)$$

If the spacing between array elements is a variable, the coupling coefficients C_{nm} are not equal, and are determined by the overlap integrals of the individual element wavefunctions at various element spacings. This method has been used to control the coupling between array elements (Ref. 10.16).

An important advantage of a variable-spacing array structure is that the fundamental array mode can be readily matched to a uniform pumping profile across the entire array aperture. This method allows for the optimization of the device at high-power operation. More recent results (Ref. 10.17) show that laser arrays employing Y-junctions to coherently couple an emitter in-phase can provide stable pulsed output power up to 400 mW with 10 elements and can center the far-field pattern normal to the facet. Such Y-junction structure provides strong discrimination against out-of-phase field components. The in-phase locking of the Y-junctions operates equally well in both buried heterostructure (BH) and inverted channel substrate planar (CSP) waveguides. The light from out-of-phase modes that destructively interferes in the Y-junction, is either absorbed in the lateral waveguide cladding material or propagates outside the waveguide resulting in radiative loss. The power in the satellite diffraction lobes can be reduced by flaring the waveguides at the facet, thereby increasing the near-field fill factor to greater than 80%.

In practice, a number of problems associated with the excitation and stability of array modes exist. Various designs have been proposed for achieving high-power phase arrays that operate predominantly in the fundamental array mode, which has only a single lobed far-field pattern. However, even the best reported phase arrays show a limited range of stable operation, especially when the number of elements becomes large. The instability of a phase-locked diode array is attributed to array-mode competition, which is more severe than that of stripe-geometry diode lasers because the modal gains of the array modes are densely spaced. One convenient technique that can be used to enhance the gain of the $p = 1$ mode relative to $p = N$ modes is to introduce optical gain in the region between elements. With gain added between elements, a significant increase in the overlap integral for nearest neighbor interactions is expected. This result arises from the exponentially growing evanescent fields as opposed to evanescent decaying fields for cases without gain. For these cases, the coupled-mode formulation as presented above will very likely break down and a more exact formulation will be required. Other techniques such as fabricating arrays with multiple contacts to control the gain, variable channel width, chirped array structures, and the use of index-guided structures have all been explored. All of these methods have shown some degree of success, nevertheless, further research efforts are needed to optimize diode laser phased arrays.

10.11 QUANTUM-WELL LASERS

Quantum-well lasers (Ref. 10.18) with either single or multiple ultrathin (<500 Å) active layers L_y have shown superior characteristics such as ultralow threshold current, less temperature dependence, and narrow gain spectrum, compared to conventional DH lasers. Quantum size effects not typical of the bulk material occur when the thickness L_y of a semiconductor layer is reduced to the order of a carrier de Broglie wavelength ($\lambda = h/p \sim L_y$). The energy spectrum of carriers in such a thin layer is typically determined using the Hamiltonian for a single particle in a one-dimensional potential well. With this approximation, the wavefunction can be separated into a y-component normal to the layer and another component expressed in terms of the usual Bloch function $U_B(x, z)$ in the plane of the layer. The wavefunction of an electron confined in a well of the coordinate system shown in Figure 10.28 can be written as

$$\psi_{nB} = \phi_{nB}(y)U_B(x, z) \exp i(k_x x + k_z z) \tag{10.127}$$

where the subscript nB denotes the subband in the quantum well, for which n denotes the number of the quantized level and B can be either the conduction c or the valence v band. $U(r)$ is the usual Bloch function of the bulk crystal. $\phi_n(y)$ is the envelope function that varies very slowly when compared to $U_B(r)$, and is a solution of Schrödinger's equation involving a square

(a)

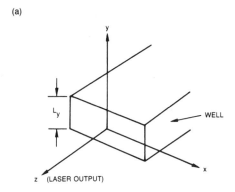

(b)

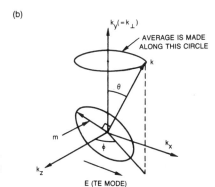

FIGURE 10.28   (a) The coordinate system used in the analysis. (b) The relation between directions of wavevector $\hat{k}$ and the dipole moment $\hat{m}$, which is rotating in the plane perpendicular to $\hat{k}$.

potential well $V(y)$ of the form

$$-\frac{\hbar^2}{2m_c^*}\frac{\partial^2}{\partial y^2}\phi_{cn} + V(y)\phi_{cn} = E_{cn}\phi_{cn} \qquad (10.128)$$

where ϕ_{cn} is the envelope function for the conduction band, m_c^* is the effective mass of the electron, which is distinguished between those inside m_{c1}^* and outside m_{c2}^* of the well, and E_{cn} is the quantized energy level to be determined. The origin of the energy is located at the bottom of the conduction band well. For a simple square well, as shown in Figure 10.29, the solution of Equation (10.128) for the particle inside the well, $y \le L_y/2$, is

$$\phi_{cn} = A \begin{Bmatrix} \cos \\ \sin \end{Bmatrix} \sqrt{m_{c1}^* E_{cn}} \,|\, y \,|/\hbar \qquad \begin{matrix} n \text{ is even} \\ n \text{ is odd} \end{matrix} \qquad (10.129)$$

and for the particle outside the well, $y \ge L_y/2$, is

$$\phi_{cn} = B \exp(\sqrt{2m_{c2}^*(\Delta E_c - E_{cn})} \,|\, y \,|/\hbar) \qquad (10.130)$$

Applying the continuity conditions for ϕ_{cn} at the heterojunction, e.g., ϕ_{cn} and ϕ'_{cn} must be continuous, the following eigenvalue equations are obtained:

$$\left[\frac{m_{c2}^*}{m_{c1}^*}(E_c - E_{cn})/E_{cn}\right]^{1/2} = \begin{Bmatrix} \tan \\ -\cot \end{Bmatrix} L_y\sqrt{2m_{c1}^* E_{cn}}/2\hbar \qquad \begin{matrix} n \text{ is even} \\ n \text{ is odd} \end{matrix} \qquad (10.131)$$

If the barrier height is sufficiently large,

$$E_{cn} = (n\pi\hbar)^2/2m_{c1}^* L_y^2 \qquad (10.132)$$

where $n = 0, 1, \ldots$. Because electrons are free in the direction parallel to the heterojunction, the total energy of an electron E is a sum of the free-particle energy and the bound energy by the well as given by

$$E = \hbar^2 k_c^2/2m_{c1}^* + E_{cn} \qquad (10.133)$$

In the case of multiple quantum wells, barrier thickness also plays an

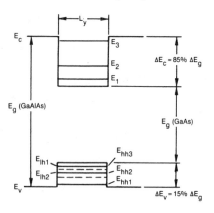

FIGURE 10.29 Square potential well of an AlGaAs/GaAs/AlGaAs quantum-well heterostructure, in which a series of discrete energy levels exists. E_n are the bound states for the electrons in the conduction band, E_{khn} and E_{lhn} are the states for the heavy holes and the light hole, respectively. (Ref. 10.17. Reprinted with permission of IEEE.)

important role. If the barrier thickness is thin or its barrier height is small, coupling between adjacent wells occurs. In this case, the degeneracy of individual well-quantized energy levels is removed and each single well level splits into N different energy levels. To simplify this discussion, the following description is restricted to the weak coupling limit of QW structures in which the energy broadening due to coupling between wells is small compared to the energy broadening due to intraband relaxation time τ_r.

The probability of optical transition between the subbands of the conduction and valence bands in a quantum well is proportional to the absolute square of the matrix element of the momentum or dipole moment $\mathbf{m}$, which is defined by

$$\mathbf{m}_{nl} = \langle \psi_{cn} \mid e\mathbf{r} \mid \psi_{vl} \rangle \tag{10.134}$$

Because the focus here is on the parallel component of wave vector $\mathbf{k}$ and position vector $\mathbf{r}$ to the interface of the heterojunction, e.g., within the x-z plane (see Figure 10.28) Equation (10.134) becomes

$$\mathbf{m}_{nl} = \left(\int_{-\infty}^{\infty} \phi_{cn}^* \phi_{vl} \, dy \right) \delta(k_{c\parallel}, k_{v\parallel}) \mathbf{R} \tag{10.135}$$

where

$$\mathbf{R} = \int_{\text{unit all}} U_c^* e\mathbf{r} U_v \, dr \tag{10.136}$$

Because the potential wells in the conduction and valence bands have the same symmetry, the overlapping integral in Equation (10.135) is, to a good approximation, equal to δ_{nl},

$$\delta_{nl} = \begin{cases} 1 & n = l \\ 0 & n \neq l \end{cases} \tag{10.137}$$

The R-integral is the momentum matrix element between an s-like and a p-like Bloch state, and has been calculated (Ref. 10.19) using the $\mathbf{k}.\mathbf{p}$ method. The $\mathbf{k}.\mathbf{p}$ method in the envelope function approximation takes into account not only the orientation properties of the Bloch function at the zone center ($k = 0$), but also the band mixing that occurs in accordance with the strict k-selection rules. In other words, the calculations of the eigenfunctions and the eigenvalue (confinement energies) are done using the Hamiltonian in terms of the corresponding $\mathbf{k}.\mathbf{p}$ matrix, which has been formulated using an eight band model. This model takes into account not only the electrons, light holes, and heavy holes, but also the spin-orbit split-off interactions and the coupling to all other bands. The calculated E versus k dispersion results indicate that $E(k)$ curves for the valence bands are nonparabolic due to the effects of band mixing. These results make significant changes in both the minimum confinement energies and the density of state functions obtained by conventional approach. The calculated optical matrix elements are also quite different from those calculated with simple models using parabolic bands and bulk material effective mass values. The results for m^2 between subbands of a 34 Å thick GaAs/Ga$_{0.75}$Al$_{0.25}$As quantum well for $c1 \rightarrow v1$, and $c1 \rightarrow v2$ transitions are plotted in Figure 10.30. The dotted curves give corresponding matrix elements calculated for the same quantum well by ignoring the band mixing effects within the well. The $k_\parallel$ dependence of the matrix elements is solely due in this case to the angular dependence of the k projections on the respective axes as illustrated in Figure 10.28(b). Note that the calculations are made for the components of the wave vectors $k_\parallel$ and the position vector $r_\parallel$, both of which are components parallel to the

(a)

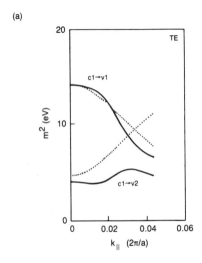

(b)

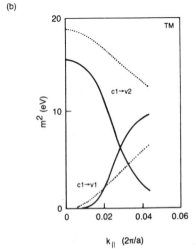

FIGURE 10.30 Squared momentum matrix elements for the transitions between the subbands of the 34 Å thick QW. $v1$ and $v2$ correspond to the heavy hole and light hole subbands, respectively. (Ref. 10.18. Reprinted with permission of IEEE.)

interface of the heterojunction. Also note that the values for m^2 given in Figure 10.30 are not averaged over the angle ϕ. The difference between the results obtained with and without band mixing is due to the fact that the energy bands in the k-space are no longer parabolic when the effect of subband mixing is accounted for.

The subbands in a quantum well affect the density of states that can lead to lower current threshold, as well as other improvements. Using Equations (8.49) and (8.56), and the k-selection rule, the optical gain $g(\mathscr{E})$ resulting from subband to subband transition can be expressed as

$$g(\mathscr{E}) = \int A(E) \sum_{j=h,l} \sum_n \frac{\rho_r^j(E)}{L_y} [f_c(\mathscr{E}_{cn}) - f_v(\mathscr{E}_{vn})] \frac{\hbar/\tau_r}{(\mathscr{E} - E)^2 + (\hbar/\tau_r)^2} dE$$

$$(10.138)$$

where $\mathscr{E} = h\nu$ the proton energy, j designates either the light holes (l) or heavy holes (h), ρ_r is the reduced density of states defined by

$$\rho_r = [\rho_{cn}^{-1} + \rho_{vn}^{-1}]^{-1} \qquad (10.139)$$

and $A(E)$ is the coefficient related to the square of the dipole matrix element m^2 as given by

$$A(E) = \frac{\pi e^2 h}{m_0^2 c n_0 E_g} m^2 \qquad (10.140)$$

where m_0 is the mass of electrons, n_0 is the refractive index of GaAs, and e is the electronic charge. E_g is the bandgap energy and c is the velocity of light. In Equation (10.138) $\mathscr{E}_{cn}$ and $\mathscr{E}_{vn}$ are given by

$$\mathscr{E}_{cn} = \frac{m_c^* E_{vn}^j + m_v^* \mathscr{E} + m_v^* E_{cn}}{m_c^* + m_v^{*j}}$$

and $\qquad (10.141)$

$$\mathscr{E}_{vn} = \frac{m_c^r E_{vn}^j - m_c^* \mathscr{E} + m_v^{*j} E_{cn}}{m_c^* + m_v^{*j}}$$

where E_{cn} and E_{vn} are the energy levels of the nth subband of the electrons and holes. Using the relationship between the injected current density J and the optical gain (see Equation (8.83)), Arakawa and Yariv (Ref. 10.20) have calculated the maximum gain as a function of the injected current density J, as shown in Figure 10.31, for various number N of quantum wells. In these calculations, the QW thickness L_y and τ_r are assumed to be 100 Å and 0.2 ps. The energy bandgap discontinuity E_c and E_v of the conduction band and valence band at the interface of GaAs and $Al_{0.3}Ga_{0.7}As$ are $0.8\Delta E$ and $0.2 \Delta E$, respectively, where ΔE is the total band discontinuity and is assumed to be 333 meV. These calculated results indicate that the maximum gain of a multiple quantum well with an N period at high injection current is N times larger than that available with $N = 1$. The enhancement in gain is a result of the increase in gain saturation for MQW devices. Figure 10.31 shows that if cavity loss is very low, a single quantum well ($N = 1$) has the lowest current threshold. As the loss increases, more gain is needed, therefore, MQW is the most desirable structure. For $\alpha \geq 50$ cm^{-1}, the lowest threshold case is $N = 5$.

Another important parameter is the thickness of the quantum well. Figure 10.32 shows threshold current density as a function of QW thickness. In this calculation, the number of QWs with each QW thickness is optimized so that the threshold current is minimum. This figure shows that for thinner QW lasers ($L_y = 50$ to 100 Å), the threshold currents are very low ($J_{th} \leq 200$ A/cm^2 for $\alpha = 30$ cm^{-1}). The optimum N of thinner QWs is larger than that of thicker QWs because the gain saturation is more enhanced for thinner L_y. At $\alpha = 50$ cm^{-1} $N_{opt} = 4$ for $L_y = 50$ Å, while $N_{opt} = 2$ for $L_y = 100$ Å and $N_{opt} = 1$ for $L_y = 200$ Å.

Finally, this chapter will examine QW laser linewidth and show that a substantial reduction of the linewidth can be obtained with an optimum quantum-well structure. The expression for power-dependent linewidth of semiconductor lasers is given by Equation (8.90). This power dependence is due to the spontaneous emission that modulates the laser field both in intensity and frequency. Therefore, laser linewidth is proportional to the ratio of spon-

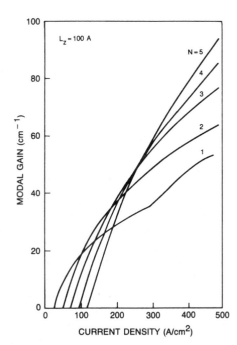

FIGURE 10.31 The modal gain as a function of the injected current density with various number N of quantum wells. In this case, the QW thickness L_y is assumed to be 100 Å. (Ref. 10.19. Reprinted with permission of IEEE.)

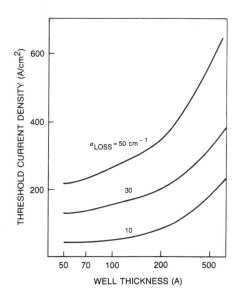

FIGURE 10.32 The threshold current density as a function of the quantum-well thickness L_y with various loss α_{loss}. The number N of quantum wells is chosen so that the threshold current is minimized. (Ref. 10.19. Reprinted with permission of IEEE.)

taneous emission and stimulated emission rates R_{sp}/R_{st}. At g_{max}, when $\mathscr{E} = \mathscr{E}_L$, this ratio can be approximated by

$$\frac{R_{sp}}{R_{st}} = [1 - \exp(\mathscr{E}_L - \Delta F)]^{-1} \qquad (10.142)$$

where ΔF is the difference between the quasi-Fermi levels of the electrons and holes. Equation (10.142) shows that as ΔF approaches $\mathscr{E}_L$, this ratio increases substantially and approaches infinity as the current density reaches J_{th} at which transparency occurs. For quantum-well lasers, this ratio depends on QW parameters such as L_y and N. The spectral broadening due to the coupling between the amplitude fluctuation and FM noise for quantum-well lasers has also been calculated by Arakawa and Yariv (Ref. 10.20). Figure 10.33 shows the calculated laser spectral width as a function of the number of QWs for various α values. Results indicate that Δv increases monotonically with increasing N. This behavior is due to the fact that as ΔF increases, both the ratio of spontaneous to stimulated emission rates and spectral broadening decrease with the decreasing number of QWs. Therefore, a single quantum-well laser provides the narrowest linewidth. In this calculation, QW thickness and τ_r are assumed to be 100 Å and 0.2 ps. The minimum Δv appears to occur at $L_y = 80$ Å. Generally speaking, both the threshold current and laser linewidth can be significantly reduced by using quantum-well structures compared to the conventional double-heterostructure.

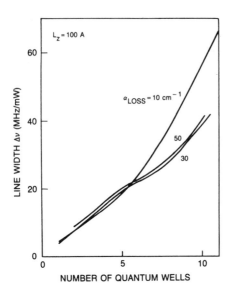

FIGURE 10.33 The spectral linewidth $\Delta \nu$ as a function of the number of quantum wells with various loss α_{loss}. (Ref. 10.19. Reprinted with permission of IEEE.)

PROBLEMS

10.1. Calculate the center frequency of a light source emitting at $\lambda = 0.9$ microns.

10.2. Derive the carrier distribution function as given by Equation (10.5) under the stripe electrode.

10.3. If $n(S/2)/n(0) = 0.5$, compute L_c and L_c'.

10.4. For a stripe-geometry laser emitting at $\lambda_0 = 0.83$ microns, having a gain width of about 4 microns, an active layer thickness of 0.1 micron, and assuming that $A = 10^{-24}$ cm^2-s, calculate the internal circulating power and output power if $R_1 = R_2 = 0.3$.

10.5. For an asymmetric planar index-guided laser, the wavefunctions for the TE modes are given by Equations (3.34) to (3.37). Using the boundary conditions requiring that E_y and H_z be continuous, show that the eigenvalue equation is

$$\tan k_1 d = \frac{2 k_1 \gamma_2}{k_1^2 - \gamma_2^2}$$

10.6. Using the appropriate approximation (Ref. K. Chen and S. Wang, IEEE J. Quant. Electron., QE-19, 1355 (1983)), show that the effective index of the fundamental TE$_0$ mode of an asymmetric index-guided laser is given by

$$n_{eff} = n_1^2 - \frac{2}{k_0^2 d^2} \ln \left[1 + \frac{k_0^2 d^2}{2} (n_1^2 - n_2^2) \right]$$

10.7. Derive an oscillation condition for a DFB laser.

10.8. Show that the increase in threshold gain Δg_F for a DFB laser when the resonance wavelength λ is exceeded over one spectral bandwidth $\Delta \lambda / \lambda$ is $g_F = 2\pi n / \lambda_B$.

10.9. Assume N identical emitters aligned along a horizontal axis separated by D, $2D$, $3D$, ... The far-field diffraction pattern obtained by using simple diffraction theory is given by

$$I(\theta) \sim \frac{1 - \cos(Nk_0D \tan \theta)}{1 - \cos(k_0D \tan \theta)}$$

where $k_0 = 2\pi/\lambda$ and θ is the observation angle. Plot $I(\theta)$ as a function of $k_0D \tan \theta$ for $N = 10$.

REFERENCES

10.1. B. W. Hakki, *J. Appl. Phys.*, **44**, 5021 (1973).

10.2. D. D. Cook and F. R. Nash, *J. Appl. Phys.*, **46**, 1660 (1975); T. L. Paoli, *IEEE J. Quantum Electron.*, *QE-13*, 662 (1977).

10.3. K. Petermann, *IEEE J. Quantum Electron.*, *QE-15*, 566 (1979).

10.4. W. Streifer, D. R. Scifres, and R. D. Burnham, *Appl. Phys. Lett.*, **40**, 305 (1982); *IEEE J. Quant. Electron.*, *QE-17*, 736 (1981); *Electron. Lett.*, *17*, 933 (1981).

10.5. E. Kapon, A. Hardy, and A. Zussman, *IEEE J. Quantum Elect.*, *QE-19*, 1618 (1983).

10.6. D. Botez, *J. Opt. Commun.*, *1*, 2 (1980).

10.7. S. Arai, M. Asada, T. Tanbunek, Y. Suematsu, Y. Itaya, and K. Kishino, *IEEE J. Quantum Electron.*, *QE-17*, 640 (1981).

10.8. E. Oomura, T. Murotani, H. Higuchi, H. Namizaki, and W. Susaki, *IEEE J. Quantum Electron.*, *QE-17*, 646 (1981).

10.9. M. Kawahara, S. Oshiga, A. Matoba, Y. Kawai, and Y. Tamura, Proc. Opt. Fiber Conf., Paper ME-1 (1987).

10.10. H. Kogelnik and C. V. Shank, *J. Appl. Phys.*, **43**, 2327 (1972).

10.11. W. T. Tsang, N. A. Olsson, and R. A. Logan, *IEEE J. Quantum Electron.*, *QE-19*, 1621 (1983).

10.12. W. Streifer, D. Yevick, T. L. Paoli, and R. D. Burnham, *IEEE J. Quant. Elect.*, *QE-20*, 754 (1984).

10.13. D. R. Scifres, R. D. Burnham, and W. Streifer, *Appl. Phys. Lett.*, *33*, 1015 (1978).

10.14. J. K. Butler, D. E. Ackley, and M. Ettenberg, *IEEE J. Quant. Elect.*, *QE-21*, 458 (1985); J. K. Butler, D. E. Ackley and D. Botez, *Appl. Phys. Lett.*, *44*, 293 (1984).

10.15. M. Born and E. Wolf, *Principles of Optics, 4th Ed.* Pergamon, England, 1970. p. 41.

10.16. D. E. Ackley, J. K. Butler, and M. Ettenberg, *IEEE J. Quant. Elect.*, *QE-22*, 2204 (1986).

10.17. D. F. Welch, W. Streifer, P. S. Cross, and D. R. Scifres, *IEEE J. Quant. Elect.*, *QE-23*, 752 (1987).

10.18. N. Holonyak, Jr., R. M. Kolbas, R. D. Dupuis, and P. D. Dapkus, *IEEE J. Quant. Elect.*, *QE-16*, 170 (1980).

10.19. S. Colak, R. Eppenga, and M. Schuurmans, *IEEE J. Quant. Elect.*, *QE-23*, 960 (1987).

10.20. Y. Arakawa and A. Yariv, *IEEE J. Quant. Elect.*, *QE-21*, 1666 (1985).

11

Optical Transmitters

11.1 INTRODUCTION

The transmitter is one of the major components in an optical fiber system. It usually consists of a semiconductor light source, a driving circuit or an optical modulator, and an electronic interface with an input terminal. Its main function is to convert an input electrical waveform into an identical optical waveform at a high rate. Exact conversion may not be achievable because of a device's inadequate frequency response. Besides the limitation of the transmitter, fiber dispersion and receiver noise can also cause a degradation in transmitted waveform and consequently, reduce the integrity of the input. This chapter discusses the limiting factors responsible for the frequency response of a transmitter and introduces electronic circuits commonly used to modulate the LED and LD.

A variety of signal coding schemes are used in communication systems today. The performance of various optical fiber components dictates the choice of input waveform and modulation scheme. Only a few simple digital and analog signal codes are considered. The modulation can be obtained either by direct switching of the light source or by using an external modulator whose refractive index can be changed in the presence of an applied electric field or acoustic wave. Because the primary concern here is to transmit information, some attention is given to the information theory of optical communication. However, this chapter introduces only the theoretical aspects of the transmitter, leaving the receiver portion to Chapter 13. In particular, various factors such as signal errors and transmitter noise, which have a profound effect on the fidelity of a communication system are discussed.

11.2 FREQUENCY RESPONSE

The simplest way to modulate a semiconductor light emitting device is to modulate the diode directly with a time varying current pulse. In most applications, the modulated envelope of the output laser power is of prime interest but it is important to recognize that output frequency also varies in response to the signal component of the drive current. This frequency deviation manifests itself as an undesirable frequency chirp that introduces a dispersion penalty in high-data-rate direct detection systems. This section considers the basic limitations on the frequency response of laser output due to both the laser's intrinsic parameters and the parasitics of the device.

The effect of intrinsic parameters on a laser's frequency response can be determined from the following rate equations

$$\frac{dN}{dt} = \frac{I}{eV} - \frac{N}{\tau} - G(N - N_{th})(1 - \epsilon\rho)\rho \tag{11.1}$$

and

$$\frac{d\rho}{dt} = \Gamma G(N - N_{th})(1 - \epsilon\rho)\rho - \frac{\rho}{\tau_p} + \Gamma\beta\frac{N}{\tau} \tag{11.2}$$

where N, ρ, and I are electron density, photon density, and injection current, respectively. N_{th} is the inversion electron density at threshold. V is the active volume, Γ is the confinement factor, G is the gain coefficient, ϵ is a small number (with units of volume) that specifies the gain compression characteristics of the active region, β is the fraction of the spontaneous emission coupled into the laser mode, τ is the spontaneous recombination lifetime, and τ_p is the photon lifetime, given by

$$\tau_p = \frac{1}{v_g}\left(\alpha + \frac{1}{L}\ln\frac{1}{R}\right)^{-1} \tag{11.3}$$

where v_g is the group velocity of the light, α represents the distributed losses, L is the cavity length, and R is the mirror reflectivity.

The term $G(N - N_{th})$ gives the net rate per unit volume of induced transition. The term $\Gamma\beta N/\tau$, which is the contribution of spontaneous emission to the photon density, is usually a very small fraction of the total laser power. The approximation that $\epsilon = 0$, and $\beta = 0$ allows the determination of the frequency response by separating N, ρ, and I into two parts: the steady-state and a frequency-dependent variable as follows:

$$x = x_0 + x_1 \exp i\omega t \tag{11.4}$$

where x represents N, ρ, or I and x_0 represents N_0, ρ_0, or I_0, which are the DC components and can be determined from the steady-state part of Equations (11.1) and (11.2). x_1 represents N_1, ρ_1, and I_1, which are the magnitudes of the AC components. If $dN/dt = d\rho/dt = 0$ the following emerges:

$$I_0/eV - N_0/\tau - G(N_0 - N_{th})\rho_0 = 0 \tag{11.5}$$
$$G(N_0 - N_{th})\rho_0\Gamma - \rho_0/\tau_p = 0$$

Substituting Equations (11.4) and (11.5) into (11.1) and (11.2) yields

$$-i\omega N_1 = -I_1/eV + (1/\tau + G\rho_0)N_1 + \rho_1/\Gamma\tau_p \tag{11.6}$$
$$i\omega\rho_1 = G\rho_0\Gamma N_1$$

Eliminating N_1 from Equation (11.6) gives

$$\rho_1(\omega) = -\frac{I_1 G\rho_0\Gamma/eV}{\omega^2 - i\omega(1/\tau + G\rho_0) - G\rho_0/\tau_p} \tag{11.7}$$

Equation (11.7) shows that the frequency-dependent photon density peaks at the resonance or the relaxation frequency ω_r, which can be determined by minimizing the magnitude of the denominator of Equation (11.7). Thus,

$$\omega_r = [G\rho_0/\tau_p - \tfrac{1}{2}(1/\tau + G\rho_0)^2]^{1/2} \tag{11.8}$$

For a typical semiconductor laser with $L = 300$ μm, $\tau_p = 10^{-12}$ sec, $\tau = 4 \times 10^{-9}$ sec, and $G\rho_0 = 10^9$ sec. Therefore, to a good approximation, Equation (11.8) can be written as follows, assuming the second term is neglected:

$$\omega_r = (G\rho_0/\tau_p)^{1/2} \tag{11.9}$$

Equation (11.9) suggests three ways to increase ω_r: (1) increase the optical gain coefficients, (2) increase the photon density, or (3) decrease the photon lifetime. The gain coefficient G can be increased by a factor of five by cooling the laser from room temperature to liquid nitrogen temperature (77°K). Biasing the laser at higher currents would increase the photon density in the active region, which would simultaneously increase the optical output power density $P/A(\text{W/cm}^2)$ according to

$$P/A = \tfrac{1}{2}\rho_0\hbar\omega v_g \ln(1/R) \tag{11.10}$$

However, an upper limit on the maximum permissible photon density is set by the catastrophic facet damage that occurs at the 10^9 W/cm^2 level. Photon lifetime can be reduced by reducing the length of the cavity (see Equation (11.3)). For example, a typical semiconductor laser with a cavity length of

300 μm operating at an output optical power density of 0.8 MW/cm² possesses a bandwidth of 5.5 GHz. The corresponding pump current density is 3 kA/cm². For a shorter laser with a cavity length of 100 μm and operating at a pump current density of 6 kA/cm², the achievable bandwidth is 8 GHz. Recently, bandwidths of 16 GHz have been achieved at room temperature and bandwidths of 22 GHz have been achieved at lower temperatures. The overall response is dependent on a combination of intrinsic parameters of the laser and the parasitics of the packaged device.

Parasitics vary widely among different laser structures. The parasitics of interest are those that divert high frequency components of the drive current away from the intrinsic device. In practice, they take the form of a resistance in series with the intrinsic device combined with a shunt capacitance. A simple circuit model of structural parasitics is shown in Figure 11.1. These structural parasitics can be combined with package parasitics that are caused by the bondwire inductance L_p, small loss resistance R_p, and the capacitance C_p associated with the contact pad. The capacitance C_s is the effective shunt parasitic capacitance and R_s is the resistance in series with the intrinsic laser. The main sources of the shunt capacitance are: (1) the space-charge capacitance C_L of the p-n junction, with a typical value of 100 pF; (2) the metal-insulator-semiconductor (MIC) capacitance C_n with a typical value of 10 pF for a 500×250 μm clip area and a 0.2 μm-thick insulator layer; and, (3) the resistive p-layer capacitance C_J associated with the forward-biased region adjacent to the active layer, with values as large as 1000 pF, depending on device gometry, the magnitude of the DC leakage current, and the resistance R_J of the p-layer.

The signal generator I_s has a source resistance R_{IN} that has a typical value between 50 to 100 Ω. The circuit is terminated with a laser. A convenient method for determining the values of the parasitics in Figure 11.1 is to fit the equivalent circuit model with measured electrical scattering parameters over a range of frequencies. The transfer function of the model can then be calculated and the influence of the parasitics evaluated. In the following. the main features of various parasitics are summarized.

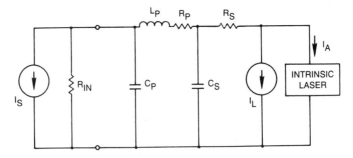

FIGURE 11.1　Circuit model of a semiconductor laser chip's parasitics. (Ref. 11.1. Reprinted with permission of IEEE.)

1. The stand-off capacitance C_p has little effect on response for low R_{IN} (<50 Ω) but can resonate with L_p to give a weakly enhanced response for large R_{IN} (>100 Ω)

2. If the bonding wire inductance L_p is small (<0.2 nH), it has little effect on response up to 20 GHz. Low inductances of this order can be achieved using short (<0.5 mm) wire and/or tape or mesh to replace the usual thin wire. If L_p > 1 nH, L_p can cause significant roll-off in the response above 6 GHz with R_{IN} = 50 Ω. If R_{IN} < 50 Ω, this inductive roll-off occurs at lower frequencies

3. For small L_p, the dominant elements affecting the high frequency parasitic roll-off are the $R_c C_s$ parasitics in the chip. It is essential that these parasitics be minimized if high-speed operation is to be achieved.

A number of low-parasitic devices have been reported. A summary of these devices can be found in Reference (11.1). One class of such device structures uses a semi-insulating SI substrate in place of the conducting substrate to form a buried heterostructure BH short cavity laser. With the BH or SI AlGaAs laser structure, the effective capacitance C_s and series resistance R_s can be reduced to as low as 3 pF and 2.5 Ω, respectively. The corresponding −3 dB roll-off frequency due to the parasitics is approximately 20 GHz. Low parasitics can also be achieved on conducting substrates by forming constricted InGaAsP mesa laser structures as shown in Figure 11.2. A number of features are found in this laser structure that can provide very low parasitics:

1. The quaternary material on each side of the active region is removed

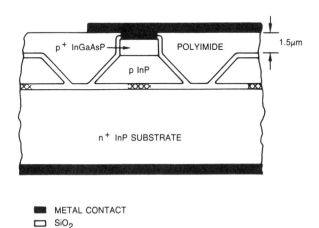

- ■ METAL CONTACT
- ☐ SiO_2
- ⊠ InGaAsP ACTIVE REGION

FIGURE 11.2   Cross-section of a constricted mesa InGaAsP laser. (Ref. 11.1. Reprinted with permission of IEEE.)

by etching. The gap under the "mushroom" strip is filled with SiO_2 to block leakage current. The capacitance between the mushroom strip and the n^+ InP substrate is very small (~0.2 pF).

2. Removal of the quarternary material beside the active layer eliminates any forward-biased junction adjacent to the active layer. Consequently, the capacitance C_J is zero for this laser.

3. The bonding pad capacitance is made small by minimizing the pad area and using a relatively thick (~1.5 μm) polyimide layer under the contact.

4. The p^+ InGaAsP cap layer helps to minimize parasitic contact resistance. The parasitic $R_s C_s$ product for this constricted mesa device corresponds to a -3 dB roll-off frequency of more than 24 GHz.

11.3 HIGH-SPEED DIRECT MODULATION OF SEMICONDUCTOR LASERS

This section considers the transmitter response performance characteristics of high-speed direct modulation of semiconductor lasers, assuming that electrical parasitics are negligible. It focuses on two types of direct modulation schemes: (1) small signal Intensity Modulation (IM) and Frequency Modulation (FM), and (2) large signal dynamics involving switching transients and chirping.

11.3.1 Intensity Modulation

The transfer function for intensity modulation is defined as

$$M(i\omega) = \frac{P(i\omega)}{I(i\omega)} \tag{11.11}$$

where $P(i\omega)$ is the small signal output power and $I(i\omega)$ is the injection current waveform. Equations (11.1) and (11.2) can be solved for the transfer function by using a first-order perturbation method (Ref. 11.2). The result for $M(i\omega)$ normalizing to the DC response $M(0)$ is given by

$$\frac{M(i\omega)}{M(0)} = \frac{B\omega_r^2}{(i\omega)^2 + i\omega\left[\dfrac{\beta'}{\rho_0} + \dfrac{1}{\tau} + \rho_0\left(G + \dfrac{\epsilon}{\tau_p}\right)\right] + \dfrac{\beta'}{\tau\rho_0} + \dfrac{\beta + \epsilon\rho_0}{\tau\tau_p} + B\omega_r^2} \tag{11.12}$$

where ω_r is given by Equation (11.9) and $\beta' = \beta\Gamma I_{th}/eV$, $B = 1 - \epsilon\rho_0$. The DC response is $M(0) \simeq \eta h\nu/2e$. Figure 11.3 illustrates the general form of the transfer function versus frequency. Above the resonance peak ω_p, the

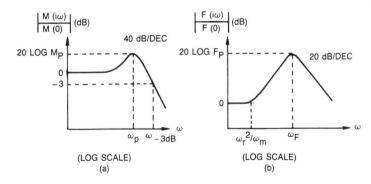

FIGURE 11.3 Small signal transfer functions (a) IM, and (b) FM responses. (Ref. 11.2. Reprinted with permission of IEEE.)

magnitude of the transfer function approaches asymptotically a slope of -40 dB/decade. The phase of $M(i\omega)/M(0)$ is zero at low frequencies, but undergoes a π radian shift near the resonance frequency.

Several important properties govern the frequency response:

1. Damping of the resonance is controlled by the coefficient of the $i\omega$ term in the denominator of Equation (11.12). If this damping coefficient is small, the height of the resonance peak is large and the resonance frequency ω_p is close to ω_r. If the damping coefficient is large, the height of the peak is reduced and $\omega_p \neq \omega_r$.

2. For low values of ρ_0 (low output power), the spontaneous emission term is large and dominates.

3. At large values of ρ_0, the spontaneous emission term becomes small and the gain compression damping term (proportional to ϵ) dominates. The $1/\tau$ and $G\rho_0$ terms in the damping coefficient are small and can be neglected.

The relative significance of the spontaneous emission and gain compression terms is shown in Figure 11.4, which shows the calculated and measured resonance peak height M_p for an InGaAsP ridge laser as a function of bias current. In this case, I_{th} was measured to be 45 mA and the parameters β, Γ, and V were calculated from known device dimensions and properties. The coefficient, $\epsilon = 6.7 \times 10^{-23}$ m^3, was obtained by fitting the circuit model to the measurements. Near threshold, the M_p value is low because the contribution comes mainly from spontaneous emission terms in the damping coefficient. As bias current increases, M_p reaches a peak value and then decreases with further increasing of the bias current. This reduction in M_p at high current is due to the increasing gain compression term in the damping coefficient. Also shown in Figure 11.4 is the zero gain compression

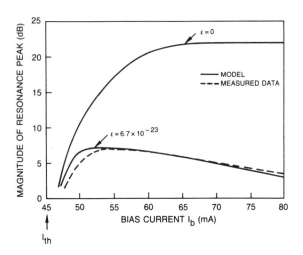

FIGURE 11.4 IM resonance peak height M_p for a ridge laser versus the bias current I_b. (Ref. 11.1. Reprinted with permission of IEEE.)

case ($\epsilon = 0$), which reaches a maximum M_p value of 22 dB at high currents. This value is much greater than that observed experimentally, indicating that the product $\epsilon\rho_0$, even though only a few percent at maximum bias current, has a significant effect on the dynamic response of the laser. This effect is often referred to as hole-burning in the gain medium. With the approximations $\beta' = 0$ and $B = 1$, and by neglecting other small terms, the transfer function reduces to a simple form as given by

$$\frac{M(i\omega)}{M(0)} = \frac{1}{\left(\dfrac{i\omega}{\omega_r}\right)^2 + \left(\dfrac{i\omega}{\omega_m}\right) + 1} \tag{11.13}$$

where $\omega_m = G/\epsilon$, which is a bias-independent parameter and is related to the damping time constant τ_d as $\tau_d = \omega_m/\omega_r^2$. Note that ω_m is a useful normalization parameter and serves as a measure of the maximum achievable bandwidth as shown below. Equation (11.13) can be used to calculate the resonance frequency ω_p, the -3 dB frequency $\omega_{-3\,\text{dB}}$, and the resonance peak M_p. They are

$$\left(\frac{\omega_p}{\omega_m}\right)^2 = \left(\frac{\omega_r}{\omega_m}\right)^2 - \frac{1}{2}\left(\frac{\omega_r}{\omega_m}\right)^4$$

$$\left(\frac{\omega_{-3\,\text{dB}}}{\omega_m}\right)^2 = \left(\frac{\omega_p}{\omega_m}\right)^2 + \left[\left(\frac{\omega_p}{\omega_m}\right)^4 + \left(\frac{\omega_r}{\omega_m}\right)^4\right]^{1/2} \tag{11.14}$$

$$M_p^2 = 1 \Big/ \left[\left(\frac{\omega_r}{\omega_m}\right)^2 - \frac{1}{4}\left(\frac{\omega_r}{\omega_m}\right)^4\right]$$

Note that the frequencies ω_p and $\omega_{-3\,dB}$ are normalized to ω_m. In this representation, M_p is independent of laser parameter and is characteristic of all semiconductor lasers.

Normalized resonance frequency curves are plotted in Figure 11.5 as a function of ω_r/ω_m, which scales as the square root of the steady-state output power. At low and moderate output power levels ($\omega_r/\omega_m < 0.4$), ω_p/ω_m rises linearly with a slope close to unity. At higher power levels, ω_p becomes smaller than ω_r and reaches a maximum value of $1/\sqrt{2}$ at $\omega_r/\omega_m = 1$. For $\omega_r/\omega_m > 1$, ω_p falls rapidly and reaches zero at $\omega_r/\omega_m = \sqrt{2}$. The -3 dB frequency, on the other hand, is slightly greater than ω_r for ω_r/ω_m up to $\sqrt{2}$. At this point, $\omega_{-3\,dB}$ saturates and then slowly decreases with increasing ω_r. The maximum achievable -3 dB bandwidth is therefore $\sqrt{2}\omega_m$. The resonance height M_p is relatively large at low output powers and decreases with increasing output power. The value of M_p falls to 0 dB at $\omega_r/\omega_m = \sqrt{2}$.

The above treatment clearly indicates that the ultimate IM bandwidth of a semiconductor laser is affected not only by the value of ω_r, but also by the damping characteristics as defined by ω_m. If a relatively flat response is required with strong damping of the resonance peak, it would be desirable to tailor ω_m to a predetermined value other than the maximum value of ω_m.

11.3.2 Frequency Modulation

Frequency Modulation (FM) or chirp is caused by modulation induced variations in the carrier density. These variations give rise to changes in the

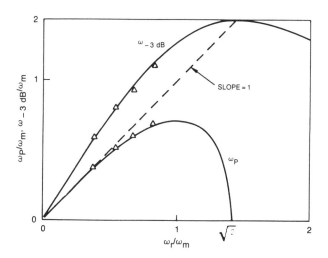

FIGURE 11.5 Normalized resonance frequency response of intensity modulation of semiconductor lasers. (Ref. 11.1. Reprinted with permission of IEEE.)

refractive index of the active region, which in turn, lead to variations in optical frequency. The small signal FM transfer function is defined as

$$F(i\omega) = \frac{\delta\nu(i\omega)}{I(i\omega)} \tag{11.15}$$

where $\delta\nu(i\omega)$ is the small signal component of frequency shift that is proportional to a change of electron density ΔN in the active layer as given by $\alpha\Gamma G\,\Delta N/4\pi$. A more convenient measure of FM is the chirp to modulated power ratio CPR, defined as

$$\text{CPR} = \frac{\delta\nu(i\omega)}{\rho(i\omega)}$$

Based on a gain compression model to solve the rate equations, Tucker (Ref. 11.1) obtained a simple expression for CPR as given by

$$\text{CPR} = \frac{\alpha}{4\pi P_0}\left[i\omega + \frac{\omega_r^2}{\omega_m}\right]$$

where α was the ratio of the real to the imaginary part of a change in active region refractive index due to a change in electron density. For both AlGaAs and InGaAsP, $\alpha \simeq 5$. P_0 is the steady-state output power. At low frequencies, the magnitude of CPR is relatively independent of ω. At frequencies well above ω_r^2/ω_m, the magnitude rises linearly with ω and the phase shift approaches an asymptotic value of $\pi/2$ rad. The normalized peak frequency ω_f/ω_m and height F_p of the FM response are given by

$$\left(\frac{\omega_f}{\omega_m}\right)^2 = \left(\frac{\omega_r}{\omega_m}\right)^2\left[\left(1 + 2\frac{\omega_r^2}{\omega_m^2}\right)^{1/2} - \left(\frac{\omega_r}{\omega_m}\right)^2\right]$$

$$F_p^2 = 1 \bigg/ \left[2\left(\frac{\omega_f}{\omega_m}\right)^2 - 2\left(\frac{\omega_r}{\omega_m}\right)^2 + \left(\frac{\omega_r}{\omega_m}\right)^4\right] \tag{11.16}$$

The shape of the ω_f/ω_m and F_p curves (see Figure 11.3) are similar to the corresponding IM curves, except that the peak height is much larger at low values of ω_r/ω_m. This large resonance-like peak causes ringing in the transient frequency chirp. In coherent communication systems employing phase-shift keying, the peak severely limits the effective modulation bandwidth to a value well below ω_f. The FM resonance frequency ω_f is close to ω_r for low values of ω_r/ω_m, but deviates from ω_r at higher values of ω_r/ω_m, and eventually falls to zero at $\omega_r/\omega_m = 1.54$. At the same point, F_p falls to 0 dB.

11.3.3 Large Signal Dynamic Response

Under large impulse signals, the dynamic response of a semiconductor laser can be quite complex due to highly nonlinear properties of the device, which cause harmonic and intermodulation distortion and in some circumstances, parametric generation. Large signal behavior has been investigated for a variety of modulation schemes, including short pulse generation by gain-switching and pulse code modulation for high-bit-rate data communication. These problems have been analyzed by solving a set of nonlinear differential equations (Ref. 11.3) and only the results are summarized here.

Periodic gain-switching of semiconductor lasers can be used to increase the data carrying capacity of a PCM optical communication system. Results indicate that three distinct advantages in using a higher value of J or a higher J/J_{th} ratio exist:

1. The optical pulse occurs in a shorter time following the application of electrical modulation, which means that higher repetition rates are possible.

2. The duration of the generated optical pulse is shorter, thus providing an opportunity for more pulses to be incorporated within a sampling period by time-division multiplexing, which also gives rise to higher system capacity.

3. The optical pulse has a higher peak power, which helps to increase the signal-to-noise ratio.

The high value chosen for J puts a constraint on the duration of the electrical modulation pulse. The burden is on the pulse current circuit design to generate high amplitude, short duration pulses at high repetition rates. For example, an avalanche pulser system using a step-recovery diode can generate current pulses of 150 ps in duration and peak amplitudes of 600 mA, which are close to the required $10J_{th}$ current density for a stripe-geometry laser with a width of 5 μm. The amplitude of the DC bias current density J_b to be injected into gain-switching diode lasers is also of importance. For optimum performance, J_b should be slightly lower than J_{th} (e.g., $0.9J_{th} < J_b < J_{th}$). A J_b value greater than J_{th} cannot be used because such a situation could lead to self-pulsations. Within the above range of J_b values, a slightly higher J_b/J_{th} ratio is required for a lower J/J_{th} ratio.

Very high repetition rates can be obtained from a gain switched diode laser. If time-division multiplexing is used (e.g., using the outputs of several similar lasers), according to the analysis (Ref. 11.3), approximately 3.5×10^{10} pulses can be transmitted per second. This rate corresponds to a system capacity of 35 G bits/s. As far as PCM communication applications are concerned, a gain-switched diode laser is more advantageous than a mode-locked laser. To mode-lock a diode laser, it is necessary to employ an external cavity that, in turn, introduces a synchronization problem. Furthermore, it is much easier to modulate a laser directly by generating, for example, "on" and "off" pulses in a binary code which can be converted directly to an optical PCM signal. Experimentally, optical pulses as short as 10 ps with peak powers exceeding 120 mW have been generated (Ref. 11.4).

11.4 BIAS AND CONTROL CIRCUITS

The light output and bias voltage versus injection current for a typical LED and LD are shown in Figure 11.6. In the case of LEDs, the light sources can be modulated either by applying current pulses directly through a simple circuit as shown in Figure 11.7(a), or by using an external electro-optic modulator, which is discussed in Section 11.6. In the case of direct modulation of an LED, the spontaneous recombination time τ_{sp} is primarily the limiting factor on the speed of response or the bandwidth of the device. The distorted waveform is equivalent to a filter version of the input current $I(t)$ with a bandwidth B of $1/2\pi\tau_{sp}$. The output $P(t)$ can be expressed as

$$P(t) = \frac{\eta h \nu}{e} \int I(t')e^{-(t-t')/\tau_{sp}} \, dt' \tag{11.17}$$

where η is the quantum efficiency, $h\nu$ is the photon energy, and e is the electronic charge. In the case of LDs, the spontaneous recombination time τ_{sp} is replaced with the photon lifetime τ_p in Equation (11.17).

The use of a simple pulse network to switch a LED usually requires large currents. The currents can cause problems in the circuit design associated with the feedback noise that can interfere with the performance of other system components. Alternatively, one can use a biasing circuit as shown in Figure 11.7(b), where the LED is biased at a constant voltage, but

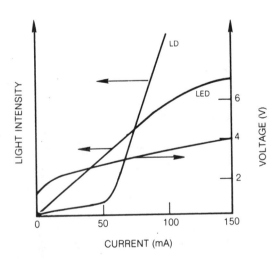

FIGURE 11.6   Typical L and V versus I curves for LEDs and LDs.

and equalization techniques must be employed, especially for analog systems. Figure 11.7(c) shows simple feedback control. In this circuit, a small portion of the output is captured by a local photodetector. The sampled signal is then amplified and compared with the input drive signal. With this technique it is important to assure single mode operation, because the $L-I$ characteristic is not necessarily the same for different modes. For this reason it is easier to adopt this technique for lasers than for LEDs.

One major difference between a laser and a LED is the threshold behavior as indicated by the light versus current curves. For this reason it is desirable to bias a laser near the threshold current to avoid the time delay necessary for building the current density from zero. In the case of a LED, the switching current can be made much lower than the bias current so that high-speed modulation drive circuits can readily be made available.

One problem associated with the pre-bias circuit for laser feedback control is the sudden surge in currents when the modulation signal is momentarily removed. As the modulation signal resumes, a temporary high level of bias voltage could cause a catastrophic failure or burnout. To avoid this problem, a more complex circuitry involves the use of two comparators, as shown in Figure 11.8. One comparator monitors the modulation signal level and the other regulates the feedback control loop. If the drive voltage is off, the average value of the signal voltage goes to zero and simultaneously an adjustment is made in the bias circuit to reduce the laser output. Such a circuit requires not only a large number of components but also very delicate balancing and calibration among various offsets, which are necessary for compensating for the difference in modulation signal levels and in the local photodetector output waveforms. These offsets are interactive, resulting in a complex alignment procedure.

11.5 DIGITAL AND ANALOG CODES

Either digital or analog signals can be used to code an optical carrier for optical fiber systems. Digital coding involves a variety of Pulse-Code Modulation (PCM) formats and can be decoded simply by means of direct detection. For example, in a simple binary pulse code (0 or 1), the only requirement imposed on the receiver is to determine whether a signal is above or below threshold. This case is not true of an analog system, for which the receiver must reproduce as closely as possible the waveform, the frequency, or the phase of an input signal. Although direct detection can also be used for simple analog systems [e.g., the intensity modulation (IM)], heterodyne detection is often employed in systems where the modulation involves either a frequency or phase-shift key. Generally speaking, analog systems require a higher degree of spectral purity and system linearity than do digital systems. Furthermore, system components required for a heterodyne receiver are considerably more complex than those used in a direct detection system.

This section considers only the simplest digital and analog systems—the binary code and the IM—both of which can be regarded as some sort of intensity modulation. The former is usually clocked and the latter is unclocked. Because of the nature of their error occurrence, the fidelity of the transmitted signal for the former is dictated by the Bit Error Rate (BER) and for the latter it is determined by the Signal-to-Noise (S/N) ratio. These topics are discussed in detail in Chapter 13.

A binary signal waveform is shown in Figure 11.9(a), where a well-defined time slot Δt is assigned to a pulse that can be either present or absent. This pulse represents a clocked binary signal that is synchronized with a constant periodic reference signal. If the pulse width is less than Δt, the signal is said to be "Return to Zero (RZ)." Otherwise, the signal is called "Not Return to Zero (NRZ)."

In practice, signals are always degraded and contain errors. For example, within a time slot, a false alarm results in the presence of an output that is not supposed to be there. These errors are introduced as a result of transmitter and receiver noise, fiber dispersion, and imperfections in electronic circuitry. The latter can result in time jitter [Figure 11.9(b)], waveform distortion [Figure 11.9(c)], baseline wander [Figure 11.9(d)], and so on. The rate at which these errors occur is the BER. One way to reduce BER is to use a coder, whose function is to transform the input waveform into a form more suitable for transmission. In this case a decoder is required in the receiver system to convert the signal back to its original waveform.

One example is the bipolar coder, illustrated in Figure 11.10. This coding scheme can alleviate the problem of baseline wander commonly existing in ac coupled filters, through which rectangular pulses are distorted with long tails of opposite polarity. The bipolar coder converts the input pulses alternately into positive and negative pulses. As a result, this coding scheme creates a cancellation in the tail of ac coupled pulses with opposite polarity. It is rather easy to decode this format at the receiving end. Many other coding schemes besides PCM are used [e.g., pulse position modulation (PPM), Manchester code, etc.]. The details and trade-offs of these schemes can be found in the reference. For many low-data-rate applications, unclocked PCM signals are found to be adequate.

The basic requirement for a PCM system is the time synchronization of transmitted signals. This synchronization is usually done by providing timing information in the transmitted signals. However, such timing information can easily be lost, if, for example, the signal contains a long series of zeros. For this reason ternary line codes rather than binary are often used. The simplest code that can ease the clock extraction is the so-called "Alternative Mark Inversion" AMI code. In this code, a power level P represents zero, while $2P$ and 0 represent $+1$ and -1, respectively. A typical

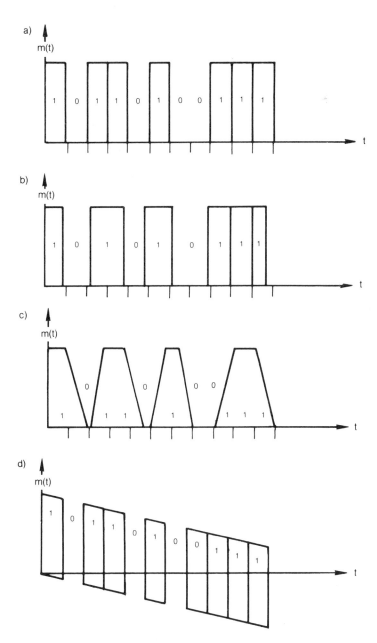

FIGURE 11.9 (a) Clocked binary signal; (b) binary signal with time jitter; (c) distorted waveforms; (d) waveform with baseline wander.

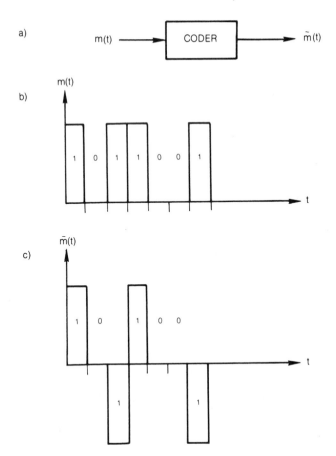

FIGURE 11.10 (a) Function of a bipolar coder that transforms $m(t)$ to $m(t)$; (b) input waveform $m(t)$; (c) output waveform $m(t)$.

AMI signal waveform is shown in Figure 11.11. The main disadvantage of this coding is the relatively stringent stability requirement on transmitter power.

The next level of complexity in coding is the 1B2B (one bit represented by two bits) code. Examples of this code are the "Complemented Mark Inversion" (CMI) and the "Bi-polar Phase" (BP), shown in Figure 11.12. All these codes are inherently well-balanced in such a way that they have no long runs of zeros or ones, as illustrated by these waveforms. The penalty of these codes is the doubling of the modulation rate, which raises the system bandwidth requirement.

The simplest analog modulation format is direct IM. The analog input $m(t)$ is used to modulate the source with an output $P_m(t)$ as given by

$$P_m(t) = P_0[1 + \gamma m(t)] \tag{11.18}$$

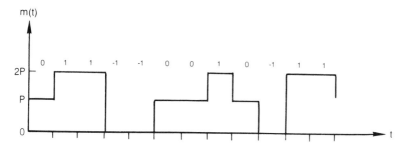

FIGURE 11.11   Typical alternative mark inversion AMI signal waveforms.

where P_0 is the average source output power and γ is the modulation index. A typical analog transmission link is shown in Figure 11.13. The modulation signal $m(t)$ is used as the input to drive the transmitter. The simplest way is to switch the light source directly. A more complicated way is to obtain $P_m(t)$ by using an external modulator, which can be either electro-optic, acousto-optic, or magneto-optic. The output of this transmission link typically contains the modulated signal plus an error signal caused by the noise and nonlinearities of the system. Therefore, system fidelity is measured by the S/N ratio.

The rate for direct modulation depends on the response time of the

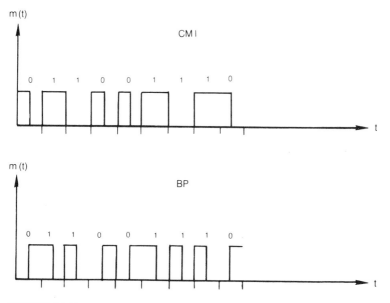

FIGURE 11.12   Typical complemented mark inversion CMI and bipolar BP waveforms.

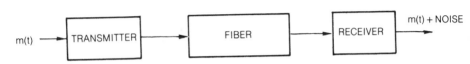

FIGURE 11.13   Typical analog transmission link.

source and also on the time constant of the circuit that provides the drive current. In the case of LEDs, a modulation rate up to 20 megabits per second (Mb/s) is considered to be routine. With careful selection of LED devices and circuits, a modulation rate as high as 100 Mb/s can be achieved; however, it is ultimately limited by the spontaneous lifetime or recombination time, as discussed in Section 11.4. In the case of LDs, modulation speeds beyond 10 GHz have been obtained. The capability of very-high-data-rate communication systems with LDs over LEDs is derived from a faster time constant, which is a fundamental characteristic of stimulated emission over spontaneous emission.

11.6 EXTERNAL ELECTRO-OPTIC MODULATION

The use of an external modulator in a transmitter has a distinct advantage over direct modulation because it relieves the burden imposed on the light source. By passing a CW source through a modulator, a subcarrier that contains information can be generated by applying a voltage on the modulator crystals, which can be electro-optic, acousto-optic, or magneto-optic. This section only discusses the electro-optic effect, which is the most widely used technique to produce intensity, frequency, and phase modulation. Many electro-optic crystals are useful in the wavelength region of interest: for example, potassium dihydrogen phosphate (KDP), ferroelectric peroskites such as $LiNbO_3$ and $LiTaO_3$, and cubic crystals such as GaAs and CdTe. For more information on electro-optic crystals, the review paper by Kaminow and Turner (Ref. 11.5) is very informative.

Electro-optic modulation of light is based on a linear electro-optic effect in crystals whose refractive index is changed upon the application of an electric field. In general, the optical properties of a crystal are described in terms of an index ellipsoid or indicatrix. The equation of indicatrix along the principal axis x_i is

$$\sum_{i=1}^{3} \frac{x_i^2}{n_i^2} = 1 \qquad (11.19)$$

where n_i is the principal refractive index. For example, if an E field is applied along x_1 and the optical carrier is propagating along x_3, a change in refractive index occurs for both n_1 and n_2, depending on the direction of the polarization vector of the carrier. The new indices $(n_1 + \Delta n_1)$ and $(n_2 + \Delta n_2)$ become

$$\frac{1}{(n_1 + \Delta n_1)^2} = \frac{1}{n_1^2} + \gamma_{11}E_1 \qquad (11.20)$$

$$\frac{1}{(n_2 + \Delta n_2)^2} = \frac{1}{n_2^2} + \gamma_{21}E_1 \qquad (11.21)$$

where γ_{11} and γ_{21} are electro-optic coefficients associated with the orientation of the interacting fields. If $\Delta n \ll n$, Equations (11.20) and (11.21) reduce to

$$\Delta n = -\tfrac{1}{2}n_1^3\gamma_{11}E_1 \qquad (11.22)$$

$$\Delta n_2 = -\tfrac{1}{2}n_2^3\gamma_{21}E_1 \qquad (11.23)$$

The situation becomes more complex if the directions of the optical polarization and the E field are not oriented along the principal axes of the crystal. In this case the equation of indicatrix must be expressed in terms of a generalized quadratic electro-optic tensor γ_{ijk}, as given by

$$\sum_{i,j,k} \left(\frac{1}{n_{ij}^2} + \gamma_{ijk}E_k \right) x_i x_j = 1 \qquad (11.24)$$

By the usual contraction, $\gamma_{ijk} \leftrightarrow \gamma_{lm}$, where $l = 1, \ldots, 6$, and $m = 1, 2, 3$. i, j are related to l as follows: 1, 1 $\leftrightarrow$ 1; 2, 2 $\leftrightarrow$ 2; 3, 3 $\leftrightarrow$ 3; 2, 3 $\leftrightarrow$ 4; 3, 1 $\leftrightarrow$ 5; and 1, 2 $\leftrightarrow$ 6. Therefore, in general a total of 18 electro-optic coefficients exist. Because of crystal symmetry, the number of electro-optic coefficients can be greatly reduced.

For KDP (tetragonal class $\bar{4}$2m), all coefficients are zero with the exception of γ_{41}, γ_{52}, and γ_{63}. In the case of cubic crystals such as GaAs and CdTe ($\bar{4}$3m), the nonvanishing coefficients are $\gamma_{41} = \gamma_{52} = \gamma_{63}$. For LiNbO$_3$ and LiTaO$_3$ (trigonal class, 3m), the electro-optic tensor is more complex and has a total of eight nonvanishing tensor components: $\gamma_{11} = -\gamma_{12} = -\gamma_{62}$, $\gamma_{51} = \gamma_{42}$, $\gamma_{13} = \gamma_{23}$, and γ_{33}. Table 11.1 gives the values of refractive indices and nonvanishing electro-optic coefficients of some crystals.

For KDP the equation of indicatrix in the presence of an E field, $\mathbf{E} =$

TABLE 11.1   Refractive Indices and Electro-optic Coefficients of Some Crystals

Material	$\lambda(\mu m)$	n_e	n_0	$\gamma_{ij}(10^{-12} \text{ m/V})$
KDP	0.5	1.472	1.514	$\gamma_{41} = \gamma_{52} = 8.6$, $\gamma_{63} = 9.5$
LiNbO$_3$	0.5	2.245	2.344	$\gamma_{13} = \gamma_{23} = 9.0$, $\gamma_{22} = -\gamma_{12} = -\gamma_{\gamma61} = 6.6$
				$\gamma_{42} = \gamma_{51} = \gamma_{33} = 30$
GaAs	0.8–10	3.6–3.3		$\gamma_{41} = \gamma_{52} = \gamma_{63} = 1.2$
Quartz	0.6	1.553	1.544	$\gamma_{11} = -\gamma_{21} = -\gamma_{62} = -0.47$
				$\gamma_{41} = -\gamma_{52} = -0.2$

$E_1 i + E_2 j + E_3 k_3$, is

$$\frac{x_1^2}{n_0^2} + \frac{x_2^2}{n_0^2} + \frac{x_3^2}{n_e^2} + 2\gamma_{52}E_1 x_2 x_3 + 2\gamma_{52}E_2 x_1 x_3 + 2\gamma_{63}E_3 x_1 x_2 = 1 \quad (11.25)$$

where $n_0 = n_1 = n_2$ and $n_e = n_3$ are the ordinary and extraordinary indices of this uniaxial crystal. Equation (11.25) can be simplified by a proper choice of the coordinate system (x_1', x_2', x_3') for crystal orientation such that the cross terms $x_i x_j$ vanish.

For simplicity, if the E field is applied along x_3 axis, Equation (11.25) reduces to

$$\frac{x_1^2 + x_2^2}{n_0^2} + \frac{x_3^2}{n_c^2} + 2E\gamma_{63} x_1 x_2 = 1 \quad (11.26)$$

By a 45° rotation of the plane formed by x_1 and x_2, a new coordinate system (x_1', x_2', x_3) emerges using the following transformation:

$$x_1 = \frac{\sqrt{2}}{2}(x_1' - x_2')$$

$$x_2 = \frac{\sqrt{2}}{2}(x_1' + x_2') \quad (11.27)$$

Substituting Equation (11.27) into (11.26) yields a simple equation of indicatrix as

$$\frac{x_1'^2}{n_1'^2} + \frac{x_2'^2}{n_2'^2} + \frac{x_3^2}{n_e^2} = 1 \quad (11.28)$$

which is an ellipsoid in its principal axes in the presence of the E field, provided that

$$\frac{1}{n_1'^2} = \frac{1}{n_0^2} + \gamma_{63}E \quad (11.29)$$

$$\frac{1}{n_2'^2} = \frac{1}{n_0^2} - \gamma_{63}E \quad (11.30)$$

Equation (11.29) can be rewritten as follows:

$$n_1' = n_0(1 + n_0^2\gamma_{63}E)^{-1/2}$$
$$= n_0[1 - \tfrac{1}{2}n_0^2\gamma_{63}E + O(\gamma_{63}^2)] \quad (11.31)$$

Therefore, the change of refractive indexes Δn_1 can be approximated by

$$\Delta n_1 = -\tfrac{1}{2}n_0^3\gamma_{63}E \quad (11.32)$$

and similarly,

$$\Delta n_2 = \tfrac{1}{2}n_0^3\gamma_{63}E \quad (11.33)$$

Figure 11.14 shows amplitude modulators that use in part (a) a KDP and in part (b) a GaAs electro-optic crystal. In both cases, the amplitude modulator consists of a properly oriented crystal placed between two crossed polarizers, with the E field applied along the x_3 axis. When a linearly polarized optical field enters the modulator as shown in Figure 11.14(a), it resolves into two orthogonal components along x_1' and x_3. Because of the difference in refractive index between these two components, a phase difference ϕ is developed as they propagate along the length L of the modulator. This phase difference contains two terms as given by

$$\phi = \frac{2\pi}{\lambda} L \left[(n_0 - n_e) - \frac{n_0^3}{2} \gamma_{63} \frac{V}{d} \right] \tag{11.34}$$

in which the first term $(n_0 - n_e)$ is independent of the E field and is due to

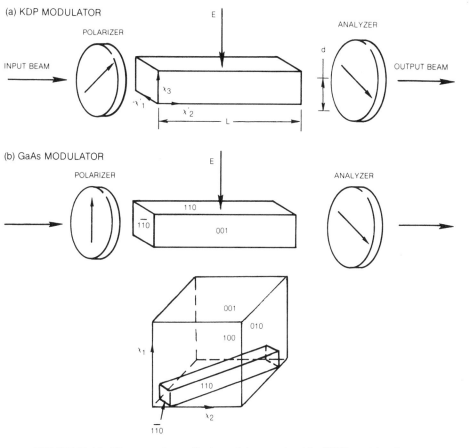

FIGURE 11.14   Electro-optic amplitude modulators using (a) a KDP crystal and (b) a GaAs crystal.

the natural birefringence of the crystal, and the second term is E-field-dependent. The depth of modulation increases with increasing L and decreasing d. As these two components emerge from the modulator, they will interfere with each other because of a phase difference ϕ existing between them. If

$$E_1 = E_0 \quad \text{and} \quad E_2 = E_0 e^{-i\phi}$$

the resultant of these two complex fields is

$$E = \frac{E_0}{\sqrt{2}} (e^{-i\phi} - 1) \tag{11.35}$$

and the corresponding output intensity is

$$I_m \propto EE^* = 2E_0^2 \sin^2 \frac{\phi}{2} \tag{11.36}$$

The fractional intensity transmitted through the modulator is

$$\frac{I_m}{I_0} = \sin^2 \frac{\phi}{2} \tag{11.37}$$

Equation (11.37) defines a half-wave voltage V_π, which is the amount of voltage required to generate a 180° phase shift, so that $I_m = I_0$. Figure 11.14(b) shows one of the crystal orientations for GaAs that yields the maximum phase retardation between two orthogonal components along x_1' and x_2' by the amount

$$\phi_{max} = \frac{2\pi}{\lambda} Ln^3 \gamma_{41} \frac{V}{d} \tag{11.38}$$

At low voltages, the depth of modulation is very low because of the $\sin^2(\phi/2)$ behavior for the transmission. Therefore, it is advisable to bias the modulator with a quarter-wave plate at the output, as shown in Figure 11.15. This $\lambda/4$ plate introduces a fixed retardation of $\pi/2$. The percent transmission, as shown in Figure 11.15(b), is translated to the steepest portion of the $\sin^2(\phi/2)$ curve. In this case, a small sinusoidal voltage $V_m(t)$ can produce a large sinusoidally modulated output as given by

$$\frac{I_m}{I_0} = \sin^2 \left(\frac{\pi}{4} + V_0 \sin \omega_m t \right)$$
$$= \tfrac{1}{2} + \tfrac{1}{2} \sin[V_0 \sin \omega_m t] \tag{11.39}$$

For small V_m, Equation (11.39) can be approximated by

$$\frac{I_m}{I_0} \simeq \frac{1}{2} + \frac{1}{2} V_0 \sin \omega_m t \tag{11.40}$$

To generate a phase shift in the light, it is only necessary to align the polarization vector along one of the induced birefringent axes, instead of

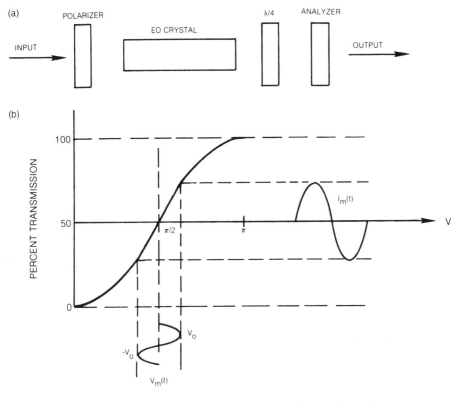

FIGURE 11.15 (a) Electro-optic amplitude modulator biased with a λ/4 plate; (b) percent transmission versus applied voltage for an electro-optic modulator biased with a π/2 phase retardation.

producing two equal components as for the case of amplitude modulation. The phase shift $\Delta\phi$ is directly proportional to the change in refractive index Δn as

$$\Delta\phi = \frac{2\pi}{\lambda} L \, \Delta n \qquad (11.41)$$

With a sinusoidal modulation $\Delta\phi \sin \omega_m t$, the output will contain both the upper and lower sidebands, which can be expressed in terms of Bessel functions of $\Delta\phi$ as

$$E_{\text{out}} = E_0 \cos(\omega t + \Delta\phi \sin \omega_m t)$$

$$= E_0 \left[J_0(\Delta\phi) \cos \omega t + \sum_{n=1}^{\infty} J_n(\Delta\phi) \cos (\omega + n\omega_m)t \right. \qquad (11.42)$$

$$\left. + \sum_{n=1}^{\infty} J_n(\Delta\phi) \cos (\omega - n\omega_m)t \right]$$

The output spectrum is very complex and must be analyzed using either a turnable filter or heterodyne receiver. If a circularly polarized light is used, it is possible to generate a single sideband with very-high-power conversion efficiency by a phase-synchronization with a circularly polarized modulating field. The mathematics, however, is rather involved and is therefore omitted.

For a reactive load, the power per unit bandwidth required to derive an electro-optic modulator is given by

$$\frac{P}{B} = \frac{1}{2} \left(\frac{1}{2} \epsilon E_0^2 \right) d^2 L \tag{11.43}$$

where the first factor $\frac{1}{2}$ is a result of taking the mean-square energy of a harmonic field, the second factor (in parentheses) is a term representing the energy density of the peak field strength E_0, and $d^2 L$ is the volume of the modulator.

The GaAs amplitude modulator will serve as an example to calculate the power required for producing maximum phase retardation. Substituting Equation (11.38) into (11.43) gives

$$\frac{P}{B} = \frac{\pi \epsilon}{2} \frac{d^2 \phi_m^2 \lambda^2}{L n^6 \gamma_{41}^2} \tag{11.44}$$

To make the modulator more efficient, it is necessary to collimate the beam with two lenses so that the beam passes the modulator crystal in a confocal configuration, as shown in Figure 11.16. In this case a Gaussian beam diameter at the center of the modulator rod is $2\omega_0$ and at the ends it is $2\sqrt{2}\omega_0$, where

$$\omega_0^2 = \frac{\lambda L}{2\pi n} \tag{11.45}$$

To minimize the difficulty of beam alignment, it is necessary to introduce a safety factor S such that

$$d = S\sqrt{8}\omega_0 \tag{11.46}$$

Combining Equations (11.45) and (11.46) yields

$$\frac{d^2}{L} = \frac{S^2 4\lambda}{n\pi} \tag{11.47}$$

Substituting Equation (11.47) into (11.44) gives

$$\frac{P}{B} = \epsilon \frac{S^2 \lambda^3 \phi_m^2}{n^2 \gamma_{41}^2} \tag{11.48}$$

Equation (11.48) indicates that modulator driving power increases rapidly with increasing wavelength. For long-wavelength sources, the high power requirement for the modulator may be an important factor that limits system performance.

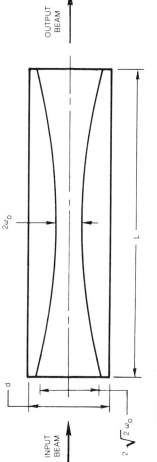

FIGURE 11.16 Optimum configuration for a Gaussian beam in an electro-optic modulator crystal.

The modulation bandwidth of a reactive load can be calculated by treating the modulator as a lump element in the equivalent circuit shown in Figure 11.17. The electro-optic crystal with a dielectric constant ϵ is placed between two electrodes to behave as a parallel plate capacitor C. R_s is the internal resistance of the source. L and R_L are the circuit inductance and the load resistance. Usually $R_s < R_L$, therefore, the modulation bandwidth B is essentially limited by the reciprocal $R_L C$ time constant of the circuit. For broadband operation, it is essential to keep the capacitance at a very low value. In this case, modulation bandwidth is limited by the phase variation during the transit-time period $\tau_l = nl/c$, where l is the length of the crystal through which light travels. Assume that an AC modulation field at a frequency ω_m in the form

$$E(t) = E_m \exp i\omega_m t \tag{11.49}$$

is applied to the crystal. The phase retardation $\phi(t)$ can be obtained by integrating the instantaneous phase retardation given by Equation (11.38) over the interaction length l as

$$\phi(t) = \frac{2\pi}{\lambda} n^3 r_{41} \int_0^l E(z)\, dz = \frac{2\pi}{\lambda} n^3 r_{41} \frac{c}{n} \int_{t-\tau_l}^t E(t)\, dt \tag{11.50}$$

Substituting Equation (11.49) into (11.50) gives

$$\phi(t) = \phi_m \exp i\omega_m t \left[\frac{1 - \exp(-i\omega_m \tau_l)}{i\omega_m \tau_l} \right] \tag{11.51}$$

where $\phi_m = (2\pi n^3 \gamma_{41}/\lambda) l E_m$, which is the peak phase retardation as given by Equation (11.38). The reduction factor F in the bracket of Equation (11.51) represents a degradation in peak retardation as a result of a phase mismatch between the light and time-varying E-field as light travels through the crystal during the finite transit time τ_l. The F factor is plotted in Fig. 11.18, as a function of $\omega_m \tau_l$. For $\omega_m \tau_l \ll 1$, no reduction in modulation efficiency exists. At large $\omega_m \tau_l$ values, the modulation efficiency will be degraded severely. Therefore, in the case of a lump element electro-optic modulator, either the

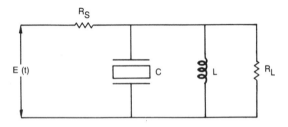

FIGURE 11.17   An equivalent circuit of an electro-optic modulator acting as a parallel plate capacitor (lump element).

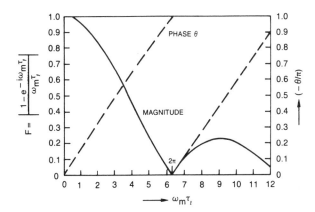

FIGURE 11.18   The magnitude and the phase of the transit-time limiting factor
$F = [1 - \exp(-i\omega_m\tau_l)]/i\omega_m\tau_l$.

modulation frequency ω_m or the length of the modulator must be reduced
to satisfy the transit time limitation.

High-speed and broadband modulation can be achieved by matching
the phase velocity of the electric drive field with that of the lightwave in a
traveling-wave modulator. Modulation frequencies greater than 18 GHz have
been achieved (Ref. 11.6) using GaAs as a traveling-wave modulator. It
exhibits an excellent phase-matching property between the optical and microwave fields, but requires high drive voltage due to its low electro-optic
coefficients as compared with KDP and LiNbO₃ (see Table 11.1). A trade-
off between the drive power and bandwidth can be made using LiNbO₃ as
a traveling-wave waveguide modulator. By careful design and use of thick
low-loss gold electroplated electrodes it has been shown (Ref. 11.7) that a
Ti-diffused LiNbO₃ waveguide directional coupler modulator at $\lambda = 1.32$
microns has a 3 dB bandwidth of 7.2 GHz and requires a drive voltage of
only 4.5 volts, which is equivalent to a power per unit bandwidth of 7.6 mW/
GHz.

A schematic of the Ti:LiNbO₃ traveling-wave directional coupler modulator is shown in Fig. 11.19. This modulator configuration is interesting
because it can provide not only high-speed on-off modulation but also a
switching function by routing the signal from one channel to the adjacent
channel. As a result, this modulator can perform high-speed time-division-
multiplexing/demultiplexing, which is essential for very high-bandwidth lightwave communication systems.

As illustrated in Figure 11.19, the directional coupler modulator consists of two Ti-diffused, z-cut LiNbO₃, y-propagating channel waveguides.
Each guide has a channel width of 6 microns and the channel separation is
6.5 microns. The modulator length is 14.5 mm. On top of this planar structure, an asymmetric stripe electrode is used. It has a characteristic imped-

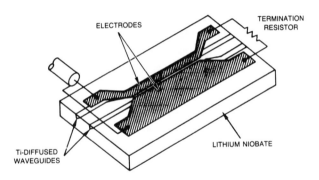

FIGURE 11.19   A schematic of a Ti-diffused LiNbO$_3$ directional coupler traveling-wave switch/modulator. (Ref. 11.7. Reprinted with permission of IEEE.)

ance of 35 Ω, which is determined by the electrode gap (G = 5 microns) to the width (W = 15 microns) ratio, and the effective dielectric constant of LiNbO$_3$ at microwave frequencies. The choice of this electrode width is a compromise between minimizing the impedance mismatch and electrode loss. For a thick (2.8 microns) gold electrode, the microwave insertion loss is about 4 dB at 5 GHz. Using this modulator, a switching voltage of only 4.5 V is required to routine the signal between channels from its maximum to its minimum values with an extinction ratio of 14 dB. The 3 dB modulation bandwidth is 7.2 GHz, which is limited by the phase velocity mismatch in LiNbO$_3$ at a length of 14.5 mm.

11.7 NOISE CHARACTERISTICS

Section 11.5 introduced several types of transmitter waveform distortions that affect the fidelity of a communication system. This section discusses several noise sources that, combined with the waveform distortion, set the fundamental limitation on the information-carrying capability of a transmitter. The noise sources to be treated here are: (1) the intrinsic noise of the laser source, (2) partition noise, and (3) noise resulting from interaction between the source and fiber. Because the noise phenomena are random in nature, statistical methods must be used to deal with communication problems. These methods will be introduced in Section 11.8.

The intrinsic noise properties of semiconductor lasers are strongly affected by the fluctuation of the carrier density. Theory (Ref. 11.8) predicts that the spectral density of FM noise increases due to the carrier effect and that the relaxation resonance appears in FM noise spectra as well as in AM noise spectra. The following analysis assumes that carriers are instantaneously in equilibrium with field intensity. This assumption is true only on a time scale that is long compared to the relaxation oscillation damping time. Because intraband thermalization occurs on a picosecond time scale, whereas the noise phenomena of interest occur on a time scale more than 100 times larger, the above assumption is valid. With this approximation, the fluctuations of the amplitude $a_n(t)$, the phase $\phi_n(t)$ of the field, and the carrier intensity fluctuation Δn obey the following equations (Ref. 11.9):

$$\frac{da_n}{dt} = \frac{a_0}{2} \frac{\partial G}{\partial n} \Delta n + F_r$$

$$\frac{d\phi_n}{dt} = \frac{\partial \omega_c}{\partial n} \Delta n + F_i/a_0 \qquad (11.52)$$

$$\frac{d\,\Delta n}{dt} = -\left(\gamma + a_0^2 \frac{\partial G}{\partial n}\right)\Delta n - 2a_0 a_n G(n_0)$$

where ω_c is the resonant frequency of the cavity, n_0 is the stationary value of n, a_0 is the stationary value of the amplitude, G is the mode gain, γ is the damping constant of the carrier density, and F_r and F_i are the real and imaginary parts of the Langevin force for the field originated from spontaneous emission, respectively. The frequency fluctuation v_n is related to ϕ_n by $(1/2\pi)d\phi_n/dt$. The Fourier transforms of the frequency v_n and the amplitude a_n can be obtained from Equation (11.52) as

$$F_n(v) = \frac{1}{2\pi a_0}\left[\Gamma_i - \frac{\alpha v_r^2 \Gamma_r}{(v_r^2 - v^2) + iv\gamma_e/2\pi}\right] \qquad (11.53)$$

$$A_n(v) = \frac{\Gamma_r}{2\pi} \frac{iv + \gamma_e/2\pi}{(v_r^2 - v^2) + iv\gamma_e/2\pi} \qquad (11.54)$$

where Γ_r and Γ_i are Fourier transforms of F_r and F_i, respectively. The parameters α, v_r, and γ_e are the linewidth enhancement factor, resonance frequency, and damping constant, respectively. They are given by

$$\langle \Gamma_i^2 \rangle = \langle \Gamma_r^2 \rangle = \langle \Gamma^2 \rangle \qquad \text{and} \qquad \langle \Gamma_i \Gamma_r \rangle = 0 \qquad (11.55)$$

$$\alpha = \left(\frac{\partial \omega_c}{\partial n}\right) \Big/ \left(\frac{1}{2}\frac{\partial G}{\partial n}\right) = \frac{\Delta n_r}{\Delta n_i} \qquad (11.56)$$

$$v_r^2 = \left(\frac{1}{2\pi}\right)^2 G(n_0)a_0^2 \frac{\partial G}{\partial n} \qquad (11.57)$$

$$\gamma_e = \gamma + a_0^2 \frac{\partial G}{\partial n} \qquad (11.58)$$

In Equation (11.56), Δn_r and Δn_i are the changes due to a change in carrier density in the real and imaginary parts of the refractive index, respectively. The FM noise spectrum S_{FM}, which is the power spectral density function of the fluctuation of v_n, can be obtained from Equation (11.53) as

$$S_{FM}(v) = \frac{\delta v_l}{\pi} \left[1 + \frac{\alpha^2 v_r^4}{(v_r^2 - v^2)^2 + (\gamma_e/2\pi)^2 v^2} \right]$$

where   (11.59)

$$\delta v_l = \frac{\langle \Gamma^2 \rangle}{4\pi a_0^2}$$

which is equivalent to the previously obtained result given by Equation (8.89). The AM noise spectrum S_{AM}, which is the power spectral density function of the fluctuation of a_n, can be obtained from Equation (11.54) as

$$S_{AM}(v) = \frac{a_0^2 \delta v_l}{\pi} \frac{v^2 + (\gamma_e/2\pi)^2}{(v_r^2 - v^2)^2 + (\gamma_e/2\pi)^2 v^2}$$ (11.60)

The phase noise spectrum $S_{\Delta\phi}$ can be expressed in terms of S_{FM} by using the v_n and ϕ_n relationship as

$$S_{\Delta\phi}(v) = (2\pi\tau)^2 \left(\frac{\sin \pi v \tau}{\pi v \tau} \right)^2 S_{FM}(v)$$ (11.61)

The field spectrum, which is the power spectrum density function of the laser output, can be determined from the FM and AM noise spectra. If AM noise is neglected, the field spectrum can be determined by taking the Fourier transform of the autocorrelation function $R(\tau)$ for the field (see Section 11.8)

$$R(\tau) = \exp(-\langle \delta\phi_n^2(\tau) \rangle / 2)$$ (11.62)

where $\langle \delta\phi_n^2 \rangle$ is the variance of the phase deviation induced in a time interval τ and is given by

$$\langle \delta\phi_n^2 \rangle = 8\pi^2 \int_0^\tau (\tau - t) \, dt \int_0^\infty S_{FM}(t) \cos 2\pi v t \, dv$$ (11.63)

The Fourier transform of $R(\tau)$ given the field spectrum $S(v)$ as

$$S(v) = \frac{\delta v_l}{2\pi[(v - v_l)^2 + (\delta v_l/2)^2]}$$ (11.64)

which has a Lorentzian line shape with a Full Width at the Half Maximum (FWHM) equal to δv_l. v_1 is the laser oscillation frequency.

In coherent optical communication systems (see Chapter 14.8), the FM noises of the transmitter and local oscillator can seriously deteriorate receiver sensitivity. For example, consider a heterodyne receiver in which the instantaneous frequency of the IF signal is

$$v_{IF} = v_{TX} - v_{LO}$$ (11.65)

where ν_{TX} and ν_{LO} are the frequencies of the transmitter and local oscillator lasers, respectively. The autocorrelation function of ν_{IF} can be written in the case of white Gaussian noise as

$$\langle \nu_{IF}(t)\nu_{IF}(t + \tau)\rangle = \frac{1}{2\pi}(\delta\nu_{TX} + \delta\nu_{LO})\,\delta(\tau) \tag{11.66}$$

where $\delta\nu_{TX}$ and $\delta\nu_{LO}$ are the spectral widths of the transmitter and local oscillator lasers, respectively. $\delta(\tau)$ is the Dirac delta function. Equations (11.64) and (11.66) show that the IF signal also has a Lorentzian line shape with an FWHM of $\delta\nu_{TX} + \delta\nu_{LO}$. The power spectrum of the frequency fluctuation of the IF signal is given by

$$\langle \delta\phi_{IF}^2(\nu)\rangle = \frac{\delta\nu_{TX} + \delta\nu_{LO}}{2\pi} \tag{11.67}$$

With coherent detection, a reference signal or delay line in the case of a PSK scheme (see Figure 14.16) is needed to phase-track the IF signal. This reference signal can be generated, for example, by a Phase-Locked Loop (PLL). The phase error $\Delta\phi_n$ between the IF carrier and reference signal depends on the FM noise of the IF signal and the bandwidth of the PLL circuit. The power spectrum of the phase difference between the IF signal and the reference is given by

$$|\Delta\phi_n(i\omega)|^2 = \frac{1}{\nu^2 + \nu_c^2}\langle\delta\phi_{IF}^2(\nu)\rangle \tag{11.68}$$

where ν_c is the locking bandwidth of the PLL. The variance of the phase error in the steady-state is

$$\sigma_\phi^2 = \frac{1}{2\pi}\int_{-\infty}^{\infty}|\Delta\phi_n(i\omega)|^2\,d\omega \tag{11.69}$$

Equations (11.68) and (11.69) give the variance of the phase error $\Delta\phi_n$ between the IF carrier and reference signal as

$$\sigma_\phi^2 = \frac{1}{2\nu_c}(\delta\nu_{TX} + \delta\nu_{LO}) \tag{11.70}$$

For coherent detection, the phase error σ_ϕ must be kept below 0.2 rad Ref. 11.9) to prevent the degradation of the BER performance due to phase noise. For example, to suppress the receiver sensitivity degradation below 3 dB at BER = 10^{-9}, $\sigma_\phi^2 < 0.04$ rad^2. Assuming that $\delta\nu_{TX} = \delta\nu_{LO} = \delta\nu$ and $\sigma_\phi < 0.2$ rad, Equation (11.70) indicates that

$$\delta\nu < 0.04\nu_c$$

The typical spectral width $\delta\nu_l$ of a GaAlAs laser is about 10 MHz operating at a high output power level of 10 mW. In this case, the locking bandwidth v_c must be greater than 250 MHz.

For Differential Phase-Shift-Keying (DPSK) coherent systems, the required $2v_c = v_s$ (the signal bandwidth, $2\pi/T$) must be greater than 500 MHz which leads to a minimum data transmission rate of $1/T > 3$ Gbits/s. For such systems, the intermediate frequency must be comparable to or greater than 10 GHz to satisfy the condition $\sigma_\phi < 0.2$ rad. This requirement is difficult to achieve with existing components. Therefore, it is extremely important to reduce the laser linewidth as much as possible. One technique (Ref. 11.10) used to reduce phase error variance is the addition of an automatic frequency control AFC loop as shown in Figure 11.20. The frequency of the semiconductor laser is discriminated by an optical frequency discriminator (e.g., a Fabry–Perot interferometer) and is detected by a photodetector. The output voltage of the amplifier is proportional to the frequency error from a prescribed value of the optical frequency. Laser frequency is controlled by the output voltage of the amplifier fed to the bias current terminal. The AFC circuit changes the FM-noise spectrum of the laser. Two parameters, the loop gain G and the loop bandwidth B, can describe the characteristics of noise suppression. To achieve an effective linewidth reduction by the AFC scheme, the AFC bandwidth B must be wider than the linewidth δv_l of the free running laser.

Figure 11.21(a) shows the lineshape of a free-running semiconductor laser and Figure 11.21(b) shows the reduced linewidth of an AFC laser where $G = 10$ dB and $B = 2\delta v_l$. These calculated spectra indicate that the linewidth of the central part of the spectrum can be reduced by a factor of G. This reduction is due to the line shape around the central part and is mainly determined by the FM noise spectrum, which can be suppressed by a factor of G. Because $B > \delta v_l$, the linewidth reduction ratio is independent of the loop bandwidth. Figure 11.22 shows the phase error variance as a function

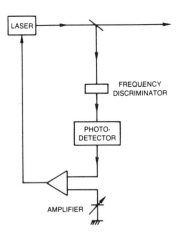

FIGURE 11.20 Block diagram of an automatic frequency control loop. (Ref. 11.10. Reprinted with permission of IEEE.)

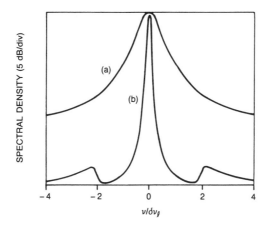

FIGURE 11.21 Field spectrum of semiconductor lasers (a) in a free-running state, and (b) in the AFC state, where $G = 10$ dB and $B = 2\,\delta v_1$. (Ref. 11.10. Reprinted with permission of IEEE.)

of bit duration T with the loop bandwidth B as a parameter. This calculation assumes that $\delta v_l = 1$ MHz and $G = 10$ dB. The lower and upper cutoff frequencies are 0 Hz and $2/T$ Hz, respectively. The results indicate that the phase error variance is always below 0.04 rad^2 by using a 50 MHz loop bandwidth for a free running laser having a 1 MHz linewidth.

For direct detection or incoherent systems, the AM noise that is caused by the fluctuation in carrier density is the dominant factor affecting receiver sensitivity. The measured S/N ratios by direct detection of laser light are shown in Figure 11.23 as a function of injection current at a fixed frequency of 50 MHz and with a receiver bandwidth $B = 10$ MHz for an index-guided CSP laser and a gain-guided V-groove laser. The S/N ratio reaches a minimum value at a point slightly above threshold, indicating that this minimum point represents the state of maximum noise occurring at or near the thresh-

FIGURE 11.22 Phase error variance σ_ϕ^2 as a function of the bit duration T. The free-running laser linewidth is 1 MHz and the AFC gain is 10 dB. (Ref. 11.10. Reprinted with permission of IEEE.)

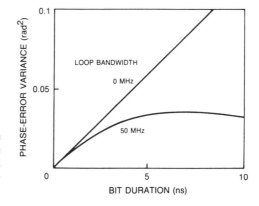

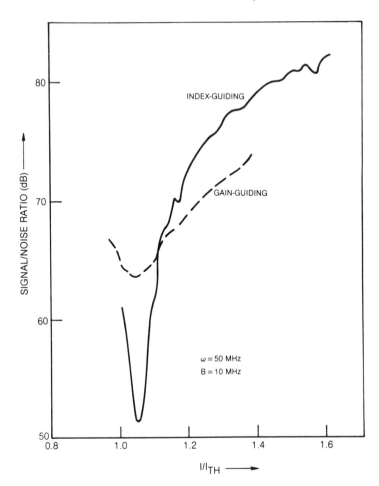

FIGURE 11.23   Measured signal-to-noise ratio (S/N) as a function of normalized injection current I/I_{th} for both index-guided and gain-guided semiconductor lasers. (From Ref. 11.11. Reprinted with permission of IEEE. © 1982.)

old of laser oscillation. The smoother the transition from the nonlasing to the lasing state, the lower the noise at threshold. For this reason, the noise is lower in a gain-guided laser than in an index-guided laser near threshold. For injection currents sufficiently above threshold ($I/I_{th} > 1.2$), the index-guided laser exhibits a better S/N ratio than does the gain-guided laser. In either case S/N is better than 70 dB for a noise bandwidth at 10 MHz. In actual systems, the S/N is lower by a certain amount, which is determined by the modulation index.

Besides intrinsic noise and waveform distortions, other problems, such as partition noise and power fluctuation are due to the interaction between the laser and fiber. Partition noise is due to fluctuations among different lasing modes. Clearly, no partition noise would exist if the laser would oscillate only in a single longitudinal mode. If a laser has two or three modes, the partition noise is greatest, because the total photon density is shared among these modes, and a slight perturbation can cause a significant amount of energy transfer between modes. On the other hand, if the laser output contains a large number of longitudinal modes, the partition photon density is relatively small in each mode. In this case, the spontaneous emission factor for each mode becomes relatively large, and can provide a natural damping of the modal fluctuation. However, a situation could occur in which a single mode laser, when modulated by injection currents, could be forced to oscillate in multimodes, thus introducing partition noise.

Noise can also be induced by feedback as a result of reflection from the end-face of a fiber. This problem can be treated by analyzing two coupled-cavities. One of the two cavities is the laser itself and the other is the external cavity formed between the output facet of the laser and the end-face of the fiber. The interaction between these two cavities can cause a considerable change in the emission spectra. A distinction must be made between the reflections from the near end and those from the far end. In the case of near-end reflection, a low-frequency noise usually occurs and is caused by fluctuations of either the laser cavity or external cavity. The power spectrum of these fluctuations typically extends up to several kilohertz. In the case of far-end reflection, the submode spacing corresponding to the long external cavity is very narrow so that a single longitudinal mode of the laser could break up into a number of submodes, which is a phenomenon commonly known as mode locking. Due to this self-modulation, the spectral envelope of the submodes is considerably broadened. Self-pulsation can also occur if the inverse round-trip time is approximately equal to the relaxation frequency. The amount of reflection that a system can tolerate depends critically on a laser's degree of coherence. For a laser cavity of 400-μm length, the round-trip time is on the order of 10 to 15 ps. If the front-end of the fiber is located very close to the laser output facet, so that the cavity mode of a single-longitudinal-mode laser coincides with the external cavity mode, the natural line width could be significantly narrowed. However, this situation is not very stable because a slight change in cavity length of one of the cavities can result in a significant change in the wavelength of the laser. In both cases, interaction occurs if the end reflection R is greater than the amplitude of the spontaneous emission within the laser cavity. A typical value for the power reflection coefficient R to produce laser instability is about 10^{-4} for an index-guided laser and about 3×10^{-3} for a gain-guided laser. To avoid these complications, an optical isolator is required for high-data-rate optical communication systems.

11.8 ASPECTS OF COMMUNICATION THEORY

The inability to predict the exact frequency and amplitude of an output from a receiving channel requires a best estimate of the message generated by the transmitter. Statistical averaging using probability theory is therefore a necessary process for many situations involving randomly varying phenomena such as random waveforms and noise problems in a communication system. In this section some basic conceptual aspects of communication theory are introduced. A more comprehensive discussion of this subject can be found in many texts (e.g., Ref. 11.12).

This section first introduces the concept of a probability density function $p(x)$. In the real world an average value can be obtained by performing either a series of measurements on a system over a long period of time or a simultaneous set of measurements on a large number of identical systems. (If these experiments are performed carefully, they should lead to the same result.) The latter is known as the ensemble method and provides an elegant basis for establishing a mathematical model. This method avoids the extended time required to complete the experiment. To assure that the two processes noted above are equivalent, new statistics involving an ergodic process must be considered. This matter is of theoretical interest and will not be treated here.

Let x represent a typical measured value that defines a point at a corresponding distance from a fixed reference point on a straight line. If this line is divided into many small equal intervals of length Δx and the number of points that fall in each interval are counted, a probability density function $p(x)$ emerges, which is defined by

$$p(x) = \lim_{\Delta x \to 0} \frac{\text{number of points in } \Delta x \text{ at } x}{N \, \Delta x}$$

where N is the total number of points.

The probability that the value falls within a specified range, say x_1 to x_2, is

$$P(x_1 < x < x_2) = \int_{x_1}^{x_2} p(x) \, dx$$

The probability that the value is less than a specified value, x, is

$$P(x) = \int_{-\infty}^{x} p(x) \, dx \tag{11.71}$$

The average value of a physical quantity $F(x)$ that satisfies the distribution function $p(x)$ is

$$\langle F(x) \rangle = \int_{-\infty}^{\infty} F(x)p(x) \, dx \tag{11.72}$$

The average value of x^n, which is known as the nth moment of the distribution, is defined by the expression

$$\langle x^n \rangle = \int_{-\infty}^{\infty} x^n p(x)\, dx \tag{11.73}$$

Many types of probability density functions describe the distribution of the ensemble of interests. The two distributions used most frequently in noise theory are the Poisson and Gaussian distribution functions. The probability density function of a Gaussian distribution is of the form

$$p(x) = \frac{1}{\sigma\sqrt{2\pi}} \exp\left[-\frac{(x - x_0)^2}{2\sigma^2} \right] \tag{11.74}$$

In Equation (11.74), the parameters have been adjusted such that $P(x)$ is normalized to give

$$\int_{-\infty}^{\infty} p(x)\, dx = 1 \tag{11.75}$$

The mean value $\langle x \rangle$ is x_0 and the variance

$$\sigma^2 = \langle x^2 \rangle - x_0^2 \tag{11.76}$$

Figure 11.24(a) and (b) show the nature of Gaussian functions for $p(x)$ and $P(x)$, respectively. Substituting Equation (11.74) into (11.71) gives

$$P(x) = \tfrac{1}{2}[1 + F(z)] \tag{11.77}$$

where $F(z)$ is defined by

$$F(z) = \frac{2}{\sqrt{\pi}} \int_0^z e^{-t^2}\, dt \qquad \text{and} \qquad z = \frac{x - x_0}{\sqrt{2}\sigma} \tag{11.78}$$

An important theorem in statistics, called the central limit theorem, shows in a very general way that the distribution of an ensemble of infinitely large numbers of independently and randomly distributed quantities must approach the Gaussian distribution. Almost all physical phenomena involving noise obey this theorem. A proof of this theorem is beyond the scope of this book.

The number of photons detected in a certain period of time is discrete, even though the light intensity is a continuous quantity. The probability $p(n)$ of detecting n photons is statistically independent of the number of photons previously detected and over a given time period obeys the Poisson distribution, which is

$$p(n) = \frac{\langle n \rangle^n \exp(-\langle n \rangle)}{n!} \tag{11.79}$$

where $\langle n \rangle$ is the mean value of n. More discussion on Poisson's distribution

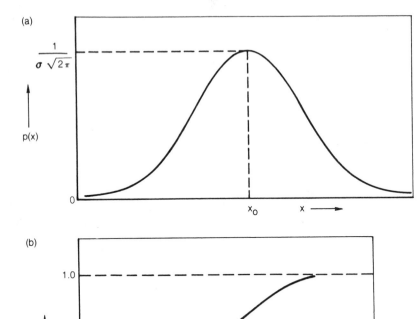

FIGURE 11.24 (a) Probability density function $p(x)$ of a Gaussian distribution; (b) probability function $P(x)$ of the corresponding distribution.

is found given in Chapter 12. Only one special case is considered here. When $n \gg 1$, use Stirling's approximation,

$$n! = \sqrt{2\pi n}\, n^n e^{-n} \qquad (11.80)$$

Substituting Equation (11.80) into (11.79) and letting $n = \langle n \rangle + \delta n$ gives

$$p(n) = \frac{1}{\sqrt{2\pi\langle n\rangle}}\left(1 + \frac{\delta n}{\langle n\rangle}\right)^{-(\langle n\rangle + \delta n + 1/2)} e^{\delta n} \qquad (11.81)$$

Because

$$\lim_{\langle n\rangle \to \infty}\left(1 + \frac{\delta n}{\langle n\rangle}\right)^{-(\langle n\rangle + \delta n + 1/2)} = e^{-\delta n - (\delta n)^2/2\langle n\rangle} \qquad (11.82)$$

Equation (11.81), in the limit that $\langle n \rangle$ is very large, becomes

$$p(\langle n \rangle) = \frac{1}{\sqrt{2\pi\langle n \rangle}} e^{-(\delta n)^2/2\langle n \rangle} \tag{11.83}$$

Equation (11.83) indicates that when the average number of photoelectrons is large, the fluctuations approach a Gaussian distribution about the mean with $\langle n \rangle = \sigma^2$. This example is representative of the central limit theorem. The next topic to be covered is the concept of convolution integral. Many physical situations require the relation of the measured sum of components in a system to the statistics of the individual contributions. If $p_1(x_1)$ and $p_2(x_2)$ are the probability density functions of two independent quantities x_1 and x_2, the probability density function of x_3, which is the sum of the two quantities (e.g., $x_3 = x_1 + x_2$), is

$$p_3(x_3) = \int_{-\infty}^{\infty} p_1(x_1)p_2(x_3 - x_1)\, dx_1 \tag{11.84}$$

Equation (11.84) is commonly known as the convolution integral, and has the familiar form of the Fourier integral in linear network analysis.

The probability density function $p(x_3)$ for the sum of two independent variables is the inverse Fourier transform of the product of the Fourier transforms of the individual probability density functions. In statistics this function is called the characteristic function. If

$$P_1(t) = \int_{-\infty}^{\infty} p_1(x_1)e^{ix_1 t}\, dx_1 \tag{11.85}$$

and

$$P_2(t) = \int_{-\infty}^{\infty} p_2(x_2)e^{ix_2 t}\, dx_2 \tag{11.86}$$

the characteristic function of the variable x_3 is equal to the product of the two Fourier transforms as

$$P_3(t) = P_1(t)P_2(t) \tag{11.87}$$

Extending this process to the sum of n variables yields the following for the characteristic function of x_{n+1};

$$P_{n+1}(t) = P_1(t)P_2(t) \cdots P_n(t) \tag{11.88}$$

The inverse of P_{n+1} is

$$p_{n+1}(x_{n+1}) = \frac{1}{2\pi} \int_{-\infty}^{\infty} P_1(t)P_2(t) \cdots P_n(t)e^{-it(x_1 + \cdots + x_n)}\, dt \tag{11.89}$$

An example of the result above is to identify the sum of noise caused by the superposition of n independently occurring sinusoidal distributions, such as the outputs of n nonsynchronous oscillators which can be represented by the zeroth-order Bessel function of the first kind, J_0. The probability density function for the sum of n independent sinusoidal sources with peak value $\alpha_1 \ldots \alpha_n$ can be written in accordance with Equation (11.89) as

$$P_{n+1}(x_{n+1}) = \frac{1}{2\pi} \int_{-\infty}^{\infty} J_0(\alpha_1 t) J_0(\alpha_2 t) \cdots J_0(\alpha_n t) e^{-ix_{n+1}t} \, dt \qquad (11.90)$$

For small $\alpha_n t$ only the first two terms of power-series expansion of $J_0(\alpha_n t)$ are retained as

$$J_0(\alpha_n t) = 1 - \frac{(\alpha_n t)^2}{4} \qquad (11.91)$$

Substituting Equation (11.91) into (11.90) yields

$$p_{n+1}(x_{n+1}) \simeq \int_{-\infty}^{\infty} e^{-ix_{n+1}t - (\alpha_1 + \alpha_2 + \cdots + \alpha_n)t/4} \, dt = \frac{1}{\sigma\sqrt{2\pi}} e^{-x_{n+1}^2/2\sigma} \qquad (11.92)$$

where

$$\sigma^2 = \tfrac{1}{2}(\alpha_1^2 + \cdots + \alpha_n^2) \qquad (11.93)$$

The result above is rather remarkable and indicates that the resultant distribution is again in a Gaussian form with a standard deviation equal to the square root of the sum of the mean squares of each individual distribution. This example is another representation of the central limit theorem.

So far, this section has dealt with the probability density function involving one variable. Similar techniques can also be applied to the analysis of physical quantities involving two or more random variables. Expressions for multivariate probability functions are rather complex and are beyond the scope of this text. However, an important application of bivariate statistics is the autocorrelation of two measurements that can be made for either a single noise source at two specified instances of time or two noise sources. As for the single-noise-source case, let the two measured values be $x = f(t)$ and $y = f(t + \tau)$. The autocorrelation function $R(\tau)$ is defined by the expression

$$R(\tau) = \lim_{T \to \infty} \frac{1}{T} \int_0^T f(t) f(t + \tau) \, dt$$

$$= \int_{-\infty}^{\infty} \int_{-\infty}^{\infty} xy p(x, y) \, dx \, dy \qquad (11.94)$$

where $p(x, y)$ is a two-dimensional probability density function. The corresponding power spectrum $W(\omega)$ can be expressed in terms of $R(\tau)$ by Fourier transforms as

$$W(\omega) = \frac{1}{2\pi} \int_{-\infty}^{\infty} R(\tau) e^{i\omega t} \, d\tau$$

$$= \lim_{T \to \infty} \frac{F_x(-\omega) F_y(\omega)}{2\pi T} \qquad (11.95)$$

where

$$F_x(\omega) = \int_0^T f(t)e^{-i\omega t}\,dt$$

$$F_y(\omega) = \int_0^T f(t+\tau)e^{-i\omega t}\,dt \tag{11.96}$$

The autocorrelation of two noise sources is not discussed here. The concepts and methods introduced above provide only the background necessary to treat the problems associated with the design of an optimum receiver, discussed in Chapter 13.

The remainder of this section introduces the most important concept concerning the information capacity of a transmitting channel. It was first formulated by Shannon (Ref. 11.13). The theorem of channel capacity states that a maximum information-carrying capacity C for any communication channel exists if the channel is constrained in power. To operate at a higher rate than C, the system is subjected to a higher probability of error. The following states this theorem without offering the proof.

Theorem.   There exists a maximum capacity C given by

$$C = B\log_2\left(1 + \frac{S}{N}\right) \tag{11.97}$$

for any communication channel having an information bandwidth B and a given S/N ratio.

It follows from Equation (11.97) that as S/N approaches infinity, the information-carrying capacity also approaches infinity. Obviously, this relationship is impossible, because every electronic system has noise that cannot be less than the fundamental limit set by the quantum noise. It has been shown by Gordon (Ref. 11.14) that in the quantum noise-limiting case, the classical channel capacity as given by Equation (11.97) is no longer valid. Gordon showed that for a single mode system having an output power P given by

$$P = \langle n\rangle h\nu B \tag{11.98}$$

where $\langle n\rangle$ is the average number of photons per mode, the maximum information-carrying capacity C is given by

$$C = B\log\left(1 + \frac{S}{N + h\nu B}\right) + \frac{S+N}{h\nu}\log\left(1 + \frac{h\nu B}{S+N}\right)$$
$$- \frac{N}{h\nu}\log\left(1 + \frac{h\nu B}{N}\right) \tag{11.99}$$

For example, consider a resistor, having a thermal noise N, that can be expressed by the blackbody radiation formula at a temperature T as given by

$$N = h\nu B(e^{h\nu/kT} - 1)^{-1} \qquad (11.100)$$

At room temperature, $N \gg h\nu B$. Substituting this condition into Equation (11.99) provides

$$C \simeq B\left[\log\left(1 + \frac{S}{N}\right) - \frac{h\nu BS}{2N(S + N)}\log e\right] \qquad (11.101)$$

Because the second term in Equation (11.101) is very small compared with the first term in parentheses, the quantum mechanical result approaches the classical result in the limit as $N \gg h\nu B$.

The capacity theory does not state how to achieve the maximum capacity. To achieve a reasonably high rate, a signal must be coded in such a way as to have a statistically predictable noise. Several basic coding schemes were introduced in previous sections. Still more efficient coding schemes, which are variations of the basic coding schemes are available. Readers are urged to pursue additional information from books specializing in this subject.

PROBLEMS

11.1. In the approximation that $\epsilon = 0$ and $\beta = 0$, calculate the time-dependent electron density N_1, photon density ρ_1, and injection current I_1 of semiconductor lasers.

11.2. If the E field is applied along the (111) direction [e.g., $\mathbf{E} = E(i + j + k)/\sqrt{3}$], show that the equation of indicatrix of GaAs crystal is

$$\frac{1}{n^2}(x_1^2 + x_2^2 + x_3^2) + \frac{2}{\sqrt{3}}\gamma_{41}E(x_1x_2 + x_2x_3 + x_3x_1) = 1$$

11.3. Using the coordinate transformation

$$x_1' = \frac{\sqrt{2}}{2}(x_2 - x_3)$$

$$x_2' = \frac{\sqrt{6}}{6}(-2x_1 + x_2 + x_3)$$

$$x_3' = \frac{1}{\sqrt{3}}(x_1 + x_2 + x_3)$$

show that the equation of indicatrix as given by Problem 11.2 can be reduced to an ellipsoid in principal axes, with indices given by

$$n_{x_1'} = n_{x_2'} = n + \frac{1}{2\sqrt{3}}n^3\gamma_{41}E$$

$$n_{x_3'} = n - \frac{1}{\sqrt{3}}n^3\gamma_{41}E$$

11.4. Let x_1 and x_2 be two independent random variables with the probability density functions

$$p_1(x_1) = \begin{cases} (\pi\sqrt{1 - x_1^2})^{-1} & -1 \le x_1 \le 1 \\ 0 & \text{elsewhere} \end{cases}$$

$$p_2(x_2) = \begin{cases} x_2 e^{-x_2^2/2} & x_2 \ge 0 \\ 0 & \text{elsewhere} \end{cases}$$

Show that the product $x_3 = x_1 x_2$ has a Gaussian density function.

11.5. Let x be a random variable with the probability density function

$$p_1(x_1) = \tfrac{1}{2} e^{-|x_2|}$$

Establish the probability density function $p_2(x_2)$ of a random variable

$$x_2 = e^{x_1}$$

11.6. A random variable x, which takes on only an integer value, is said to be "Poisson" distributed if

$$p(n) = \delta(\langle n \rangle - n) \frac{\langle n \rangle^n e^{\langle n \rangle}}{n!}$$

(a) Plot $p(n)$ for $\langle n \rangle = 2$, $n \le 5$.
(b) Find $\langle n \rangle$ and σ^2.

11.7. Let x be a sinusoidal function of t:

$$x(t) = A \cos(\omega t + \varphi)$$

where A, ω, and φ are statistically independent random variables.
(a) Show that the correlation function $R(\tau)$ is

$$R(\tau) = e^{-|\tau|}$$

(b) The power spectrum $F(\omega)$ is

$$F(\omega) = \frac{2}{1 + \omega^2}$$

REFERENCES

11.1. R. S. Tucker, *IEEE J. Lightwave Tech.*, LT-3, 1180 (1985).

11.2. H. Kressel and J. K. Butler, *Semiconductor Lasers and Heterojunction LED's*, Academic Press, Inc., New York, 1977, Sect. 17.5.

11.3. M. S. Demokan and A. Nacaroglu, *IEEE J. Quant Elect.*, QE-20, 1016 (1984).

11.4. G. J. Aspin and J. E. Carroll, *IEEE Proc. 1, Solid State Elect. Dev.*, 129, 283 (1982).

11.5. I. P. Kaminow and E. H. Turner, *Proc. IEEE*, 54, 1374 (1966).

11.6. P. K. Cheo, *IEEE J. Quant. Elect.*, QE-20, 700 (1984).

11.7.   R. C. Alferness, C. H. Joyner, L. L. Buhl, and S. K. Korotky, *IEEE J. Quant. Elect.*, *QE-19*, 1339 (1983).

11.8.   K. Vahala and A. Yariv, *IEEE J. Quant. Elect.*, *QE-19*, 1096–1109 (1983).

11.9.   K. Kikuchi and T. Okoshi, *IEEE J. Quant. Elect.*, *QE-21*, 1814 (1985).

11.10. K. Kikuchi and T. P. Lee, *IEEE J. Lightwave Tech.*, *LT-5*, 1273 (1987).

11.11. K. Peterman and G. Arnold, *IEEE J. Quant. Elect.*, *QE-18*, 543 (1982).

11.12. J. M. Wozencraft and I. M. Jacobs, *Principles of Communication Engineering*, John Wiley & Sons, Inc., New York, 1965.

11.13. C. E. Shannon, *Proc. IRE, 37*, 10 (1949).

11.14. J. P. Gordon, *Proc. IRE, 50*, 1898 (1962).

12

Photodetectors

12.1 INTRODUCTION

Photodetectors are devices that convert optical signals into identical electrical waveforms. Many types of detectors operate on the basis of pyroelectric, thermoelectric, or photoelectric effects. Detectors for optical fiber systems usually are photodiodes, which are photoelectric devices. Photodiodes convert incoming photons into electron-hole pairs in a regeneration time on the order of 10^{-10} s. Physical processes responsible for these devices are exactly the reverse of light emission in a semiconductor and were discussed in Chapter 8. This chapter discusses the device aspect of photodiodes and describes the structural configurations. Then, device responsivity and quantum gain are analyzed.

It is important to have a working knowledge of the operating mechanisms of photodiodes. Because of the statistical nature of the electron-hole pair-production and multiplication processes, these devices are inherently noisy. This chapter examines various sources of noise and analyzes the effect of input signal waveform on detector noise output. Results indicate that detector noise increases with increasing bandwidth or device's frequency response, and is dependent on the transfer functions of the detector as well as the input signal waveform. By minimizing these noise sources, it is possible to achieve a high fidelity optical data transmission system at a low BER with minimal signal power.

12.2 *pn* AND *pin* PHOTODIODES

Photodiodes are made of semiconductor *pn* junctions under reverse bias. When a photodiode is illuminated by photons of a given frequency, electrons are excited from the valence band into the conduction band. Thus electron-hole pairs are generated within the depletion region. These electron-hole pairs drift toward the *n*- and *p*-regions, respectively, under the influence of the internal field, which builds up from the internal space-charge density as a result of the difference in Fermi level between the *n*- and *p*-materials. At zero bias, this drifting current flowing through the junction is balanced by an opposite current due to diffusion of majority carriers. If an external reverse bias is applied across the junction, as shown in Figure 12.1, the diffusion of majority carriers will be greatly reduced, resulting in a net current flow. The photoexcited carriers can migrate across the junction. If the loss due to recombination in the depletion region is ignored, the photocurrent can be written as

$$I_0 = \frac{e\eta P_0}{h\nu} \tag{12.1}$$

where e, P_0, and $h\nu$ are the electronic charge, optical power, and photon energy, respectively. η is the conversion efficiency. If electron-hole pairs are produced outside the depletion region, the probability for them to recombine in the diffusion region is very great, so that the conversion efficiency will be reduced.

Let ρ_n and ρ_p be the charge densities and W_n and W_p be the depletion widths of the *n*- and *p*-materials, respectively. From Gauss's law, an expression for the field E is given by

$$E = \frac{\rho_n W_n}{\epsilon} = \frac{\rho_p W_p}{\epsilon} \tag{12.2}$$

The bias voltage V_b across the junction can be obtained using the relationship between W and the bias voltage for a *pn* junction as discussed in Chapter 9:

$$V_b = \frac{\rho_n W_n^2 + \rho_p W_p^2}{2\epsilon} \tag{12.3}$$

If $\rho_p \ll \rho_n$, $W_n \ll W_p$, then

$$W_p \simeq \sqrt{\frac{2\epsilon V_b}{\rho_p}} \tag{12.4}$$

Substituting Equation (12.4) into (12.2) yields an approximate expression for

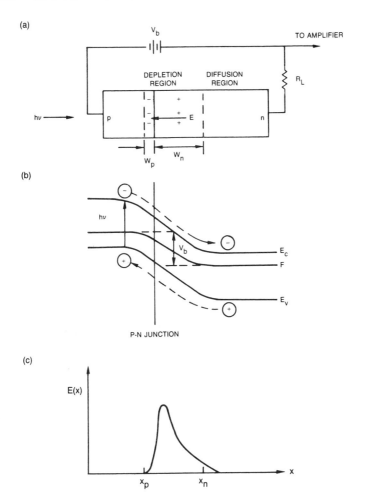

FIGURE 12.1   (a) External reverse-biased *pn* photodiode showing the depletion region in *p*- and *n*-materials. (b) Energy band diagram of a reverse-biased *pn* junction. (c) Electric field $E(x)$ in a reverse-biased *pn* photodiode.

the field at or near the junction:

$$E \simeq \sqrt{\frac{2V_b \rho_p}{\epsilon}} \qquad (12.5)$$

The quantum efficiency of a simple *pn*-junction device is usually very low because a large fraction of optical power is not efficiently utilized. In the case of a silicon *pn*-junction photodiode, the depletion width of the junction is much shorter than the absorption length because the optical absorption coefficient is very small, so that most of the optical power is absorbed inside the diffusion region. Only a very small fraction of the carriers can be generated in the depletion region to provide a displacement current. Clearly, such a device is not only inefficient, but also has a relatively slow response time, as a result of a random diffusion process. These difficulties can be avoided by adding an intrinsic (undoped) and high-resistivity semi-insulating layer within the *pn*-junction to form a *pin* structure, as shown in Figure 12.2. The width of the *i*-layer must be made many times wider than the absorption length. Because the field strength is high in the *i*-layer, electron-hole pairs can be quickly swept out of the *i*- toward the *n*- or *p*-region, respectively. The wider the thickness of the *i*-layer, the higher the quantum efficiency. However, a penalty of reducing the speed of response is taken by using a very thick insulating layer between the junction. For relatively low-data-rate systems, pin diodes have been widely used. The typical quantum efficiency of pin diodes lies in the range of 0.5 to 0.9 and the responsivity varies from 0.4 to 0.65 A/W. For Si-based pin diodes, the response time is typically on the order of several nanoseconds and the spectral response covers a range from ultraviolet to near infrared with a responsivity peak at around 0.9 μm. The typical range for the bias voltage is between 20 and 100 V.

(a) **pin** PHOTODIODE CONFIGURATION

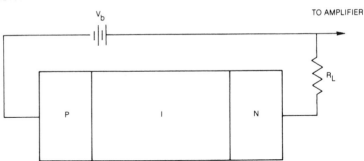

(b) E FIELD IN THE DEPLETION REGION

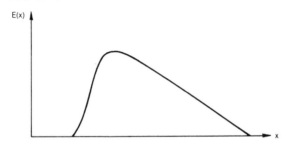

FIGURE 12.2 (a) External reverse-biased *pin* photodiode; (b) *E* field in the depletion region of a *pin* photodiode.

Materials that are commonly used to fabricate photodiodes are Si, Ge, GaAs, InAs, and InGaAs. In general, indirect bandgap materials, such as Si and Ge, are preferred over direct bandgap materials, primarily because the surface recombination in a direct bandgap material can lead to a substantial loss of carriers as a result of trapping by surface states, without producing photocurrents of significant magnitude. On the other hand, indirect absorption requires the help of a third body such as a phonon to conserve the momentum in the transfer process. This requirement makes the transition probability less likely to occur than the process involving only two bodies in a collision. However, it is possible to avoid this difficulty by making a pin configuration to provide a longer absorption length. Figure 12.3 shows the absorption coefficients of Si and Ge as a function of wavelength. For comparison purposes, the absorption coefficient of GaAs is also presented. The absorption begins at the band edge and increases rapidly with increasing photon energy; however, the increase is more gradual for indirect materials than for direct materials. The threshold for direct absorption in silicon occurs at 4.1 eV, which lies in the ultraviolet, whereas in GaAs, direct absorption occurs at 1.45 eV. The results shown in Figure 12.3 indicate that an absorption length of about 50 μm is required to absorb all the light at 0.83 μm in Si, whereas in GaAs only 1 μm is required. For wavelengths below 1 μm, Si photodiodes are considered to be ideal for use in optical fiber systems, but they are not very sensitive when the wavelength exceeds 1 μm. For wavelengths above 1 μm, Ge photodiodes are often used. At 1.5

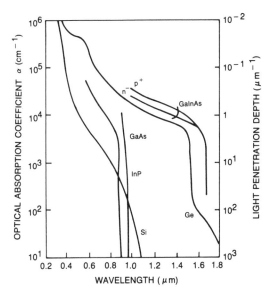

FIGURE 12.3 Absorption coefficient versus wavelength for several semiconducting materials. (Ref. 12.1. Reprinted with permission of IEEE.)

μm a rapid increase in the absorption coefficient occurs in Ge, as shown in Figure 12.3. This increase corresponds to a direct transition in Ge from its band structure. As previously mentioned, surface recombination becomes a problem at this wavelength. A maximum efficiency of 50% can be obtained by using a Ge photodiode and 30% for an InAs diode at about 1.3 μm. Other photodiodes such as AlGaAsSb and InGaAs are made by an alloy process of binary into ternary or quaternary III–V compounds. High quantum efficiency and high-speed photodiodes for detection of light at longer wavelengths could become commercially available using these alloys. In addition to its material dependence, quantum efficiency is also dependent on the reflectivities and geometry of a device, the absorption coefficient $\alpha(\lambda)$, and the operating temperature. At room temperature, the quantum efficiency of a Si pin diode can be as high as 90% at 0.9 μm. The value decreases rapidly with increasing λ and reduces to about 20% at 1.06 μm. At longer wavelengths, materials other than Si must be considered. More recently results (Ref. 12.1) indicate that η values as high as 80% can be obtained for wavelengths ranging from 1.8 to 2.5 μm by using $Al_xGa_{1-x}AsSb$ photodiodes.

12.3 AVALANCHE PHOTODIODES

Another way to improve the performance of a pn-junction photodiode is to produce an avalanche gain in these devices by means of impact ionization. The avalanche condition can be created by increasing the reverse-biased voltage just slightly below the breakdown value of the semiconductor so that a very high electric field ($>10^5$ V/cm) is established across the junction. Electrons and holes traversing under such a high field can acquire sufficient kinetic energy to produce additional electron-hole pairs through inelastic collisions. The energy transfer in this process can be sufficient to bring an electron from the valence band into the conduction band. This process can be multiplied rapidly with a multiplication factor M that is usually represented by an exponential function of the bias voltage.

Figure 12.4(a) shows a typical avalanche photodiode structure that separates the depletion region into two different regions. The first region is a wide drift i- or π-region, in which photons are absorbed. The second region is a narrow p-region, in which photogenerated carriers are multiplied. When a sufficiently high reverse-biased voltage is applied so that the depletion layer of the diode just reaches through into the low-concentration i-region, the electric field at the junction is about 5 to 10% less than that required to cause avalanche breakdown. A slight increase in applied voltage could cause the depletion layer to extend rapidly to the p^+-contact. The resistivity of the i-region of the device is typically about 5000 Ω-cm and its thickness can be as much as 200 μm at a bias voltage of slightly less than that required to achieve reach-through. Usually, this device operates in a fully depleted mode so that all carriers are collected by drift alone.

a) AVALANCHE PHOTODIODE CONFIGURATION

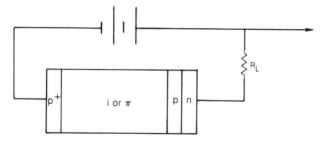

b) ELECTRIC FIELD PROFILE CROSSING THE DEVICE

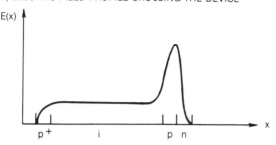

FIGURE 12.4   (a) External reverse-biased avalanche photodiode; (b) E-field profile extended across the device configuration.

The electric E-field profile for this diode is shown in Figure 12.4(b). Avalanche multiplication occurs in the high-field region near the pn-junction with a typical width of about 2 μm. The field in the i-region is much lower but sufficient to assure that the carriers gain enough kinetic energy to reach saturation. The transition time required for reach-through is in the nanosecond range, which is the typical response time of a diode. It can be estimated by the expression W/v_e, where v_e is the drift velocity of the electrons. The transit-time multiplication is much shorter than the time required for electrons to go across the depletion layer and can be considered to occur instantaneously.

Both avalanche gain and excess noise are related to an important parameter k, which is the ratio of the ionization rates α and β for electrons and holes, respectively. Clearly, the rates at which electrons and holes are created in the semiconductor are different and also structurally dependent. Figure 12.5 shows the measured α and β values for silicon. The results of Figure 12.5 are only qualitative, because the information on the orientation of the electric field for these measurements is unknown. At low fields, ionization rates differ significantly and are usually structurally dependent. At high fields, the rates become relatively independent of field orientation and crystal structure, because the number of collisions is so large that the ion-

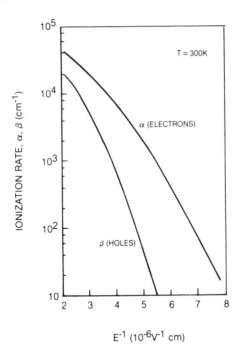

FIGURE 12.5 Ionization rates for electrons α and holes β in silicon as a function of $1/E$ field at 300°K. (From Optical Fibre Communications, ed. M. I. Howe and D. V. Morgan. Reprinted with permission of John Wiley and Sons, Inc., New York. © 1980.)

ization rates become isotropic. It is reasonable to expect that at very high fields the k value approaches unity.

The multiplication factor M and excess noise factor F of avalanche photodiodes have been analyzed in great detail by McIntyre and co-workers (Ref. 12.2). Only a few special cases that can provide a reasonable under standing of the operating mechanism of typical devices will be examined here. Consider that additional electron-hole pairs are generated within the depletion layer at x under a sufficiently high field. In traversing a distance dx, the electron and the hole will produce, on the average, $\alpha\,dx$ and $\beta\,dx$ ionizing collisions, respectively. These additional carriers, in turn, can gain enough energy from the field to cause further ionization, until an avalanche of carriers has been created.

Over a long path W, the multiplication $M(x)$ can be expressed by the expression

$$M(x) = 1 + \int_0^x \alpha(x)M(x)\,dx + \int_x^W \beta(x)M(x)\,dx \qquad (12.6)$$

Differentiating Equation (12.6) gives

$$\frac{dM(x)}{dx} = (\alpha - \beta)M(x) \qquad (12.7)$$

Equation (12.7) has a solution of the form

$$M(x) = M(0) \exp\left[\int_0^x (\alpha - \beta) \, dx \right] = M(W) \exp\left[- \int_0^W (\alpha - \beta) \, dx \right]$$

(12.8)

Substituting Equation (12.8) into (12.6) yields

$$\frac{1}{M(W)} = 1 - \int_0^W \alpha \, dx \exp\left[- \int_x^W (\alpha - \beta) \, dx \right]$$

(12.9)

Substituting Equation (12.9) into (12.8) provides

$$M(x) = \frac{\exp\left[- \int_x^W (\alpha - \beta) \, dx \right]}{1 - \int_0^W \alpha \, dx \exp\left[- \int_x^W (\alpha - \beta) \, dx \right]}$$

(12.10)

For simplicity consider the case for which electron and hole ionization coefficients are equal, e.g., $\alpha = \beta$. Equation (12.10) indicates that M is no longer a function of x and is given by

$$\frac{1}{M} = 1 - \int_0^W \alpha \, dx = 1 - \delta$$

(12.11)

In reality, $\alpha \neq \beta$. The average of ionization rate ratio is defined as

$$k = \frac{\beta}{\alpha}$$

(12.12)

In terms of k, Equation (12.10) can be approximated by

$$M = \frac{1 - k}{\exp[-(1 - k)\delta] - k}$$

(12.13)

If $\beta \ll \alpha$ or k approaches zero, then Equation (12.13) becomes

$$M \simeq e^\delta$$

(12.14)

In this case, carrier multiplication increases exponentially with δ; however, it remains finite as long as $\beta = 0$. Thus such a diode would never reach the avalanche breakdown. Figure 12.6 shows carrier multiplication as a function of electric field and δ for various values of k in an avalanche photodiode. Results of Figure 12.6 indicate that for k values other than zero, the multiplication goes to infinity at some finite values of electric field. Therefore, the choice of k values is very important in the design of a fiber optic receiver system.

Another important consideration in choosing an avalanche photodiode

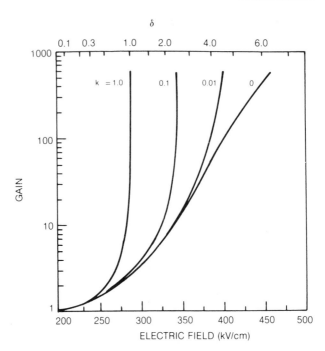

FIGURE 12.6   Electronic gain in an avalanche photodiode as a function of E field at various k values. (From Ref. 12.2. Reprinted with permission of IEEE.)

is the excess noise factor F. Assuming that the diode is shot-noise limited, the noise current I generated in W is determined by the mean square of the fluctuation in photocurrent generated in the same space. It is given by

$$\langle (I - \langle I \rangle)^2 \rangle = 2eI_0B\langle M^2 \rangle \qquad (12.15)$$

where I_0 is given by Equation (12.1) and B is the detector bandwidth. The result of Equation (12.15) will be derived in Section 12.4. The mean square of M is given by M^2F, where F is the excess noise factor. Therefore Equation (12.15) can be written as

$$N_S^2 = 2eI_0BM^2F \qquad (12.16)$$

where I_0 is the average current of the photodiode, and F is defined as the ratio of the actual noise to the noise of an ideal device if the multiplication process is noiseless. Expressions for the excess noise factor have been derived (Ref. 12.3) for both electrons and holes as given by

$$F_e = \frac{k_2 - k_1^2}{1 - k_2}M_e + 2\left[1 - \frac{k_1(1 - k_1)}{1 - k_2}\right] - \frac{(1 - k_1)^2}{M_e(1 - k_2)} \qquad (12.17)$$

$$F_h = \frac{k_2 - k_1^2}{k_1^2(1 - k_2)}M_h - 2\left[\frac{k_2(1 - k_1)}{k_1^2(1 - k_1)} - 1\right] + \frac{(1 - k_1)^2k_2}{k_1^2(1 - k_2)M_h} \qquad (12.18)$$

where k_1 and k_2 are two different weighted averages of k as defined by

$$k_1 = \frac{\int_x^W \beta M \, dx}{\int_0^W aM \, dx} \tag{12.19}$$

$$k_2 = \frac{\int_x^W \beta M^2 \, dx}{\int_0^x \alpha M^2 \, dx} \tag{12.20}$$

For all practical purposes, F_e and F_h can be simplified in the forms

$$F_e = k_2 M_e + \left(2 - \frac{1}{M_e}\right)(1 - k_2) \tag{12.21}$$

$$F_h = \frac{k_2 M_h}{k_1^2} - \left(2 - \frac{1}{M_h}\right)\left(\frac{k_2}{k_1^2} - 1\right) \tag{12.22}$$

Figure 12.7 shows the functional relationship between F_e and M_e for various values of k. For low-noise operation, it is desirable to keep k at a relatively low value. The value of k depends not only on material but also on device structure. For silicon, k is usually very small at low fields but becomes larger with increasing field strength.

Silicon is a favorable material for making avalanche photodiodes, be-

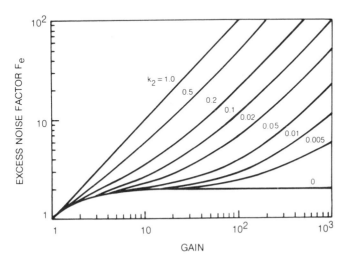

FIGURE 12.7   Excess electronic noise factor of an avalanche photodiode as a function of gain at various k values. (From Ref. 12.2. Reprinted with permission of IEEE.)

cause its ionization rate for electrons is 10 to 100 times larger than that for holes in the spectral region 0.8 to 0.9 μm. In first-generation optical fiber systems, Si APDs were used extensively and provided excellent performance, with a quantum efficiency reaching nearly 100% and a response time around 1 ns. These devices usually provided a current gain of 100 with an excess noise factor of about 5. In the longer-wavelength region (1 to 1.6 μm), silicon becomes transparent, so that other materials with narrower bandgaps must be used. Again similar to the case of pin photodiodes, germanium can be used for making APDs. However, the dark current in Ge i much larger than in Si (about 10^{-7} A), and nearly equal carrier ionization rates result in high excess noise. In spite of these shortcomings, recently developed Ge APDs have exhibited an excess noise factor of 7 at a gain of 10, a bandwidth of 500 MHz, and a quantum efficiency of ~0.8, at λ = 1.2 μm.

Other materials, such as InGaAsP and InGaAs, have also been used to make APDs sensitive at longer wavelengths. Figure 12.8 depicts the construction of two types of APDs employing a narrow-bandgap InGaAs or InGaAsP layer for absorption and a wide-bandgap InP layer for multipli

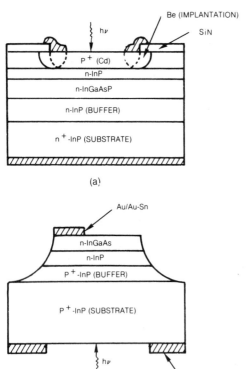

(a)

(b)

FIGURE 12.8 InGaAs/InP avalanche photodiodes having separate regions for absorption (InGaAs) and multiplication (InP): (a) planar type; (b) mesa type. (From Ref. 12.3. Reprinted with permission of IEEE, © 1983.)

cation. This scheme has been shown to minimize the tunneling dark current. For example, the mesa-type APD has a primary dark current of 3 nA and an excess noise factor $F = 10$ at a gain of 20. The planar structure, on the other hand, usually is more noisy than the mesa type because of the dark current problem. More discussion on long wavelength photodetectors can be found in Section 12.6.

12.4 NOISE IN PHOTODIODES

Many different types of noise sources exist in a photodetector. If no proper precaution is taken, they can degrade receiver performance. This section shall review these noise sources which consist of (1) background noise, (2) dark noise, (3) shot noise, and (4) quantum noise. Most of the noise sources are the result of the randomness associated with the rate of arrival of photons at the detector and the statistical nature of the generation of primary electron-hole pairs. Background noise is generated by the ambient light other than the signal. Dark noise that occurs in the absence of light is produced primarily by high-energy radiation or is due to imperfections in the detector. Shot noise is signal-dependent and therefore, is proportional to the signal power. Quantum noise represents the ultimate limit of an ideal detector whose performance is limited only by the statistical uncertainty in predicting a quanta of energy $h\nu$ in a given state. At low signal levels, the probability of generating exactly n electron-hole pairs is characterized by a Poisson distribution as given by

$$p(n) = \langle n \rangle^n e^{-\langle n \rangle}/n! \tag{12.23}$$

where $\langle n \rangle$ is the average number of photons in the detector's output pulse in response to an incoming signal pulse $P_0(t)$. In an ideal detector with no dark current, the average number in a given time period τ is given by the integral as follows:

$$\langle n \rangle = \int_{t_0}^{t_0 + \tau} \frac{\eta}{h\nu} p_0(t) \, dt \tag{12.24}$$

Statistically, the fluctuation about the mean δn is defined by the expression

$$\delta n = n - \langle n \rangle \tag{12.25}$$

and the mean square of the fluctuation, or the variance, is defined by $\langle (\delta n)^2 \rangle$. Equation (12.25) provides

$$\langle (\delta n)^2 \rangle = \langle n^2 - 2n\langle n \rangle + \langle n \rangle^2 \rangle = \langle n^2 \rangle - \langle n \rangle^2 \tag{12.26}$$

The quantity $\langle n^2 \rangle$ in Equation (12.26) can be shown to be

$$\langle n^2 \rangle = \langle n \rangle + \langle n \rangle^2 \tag{12.27}$$

Substituting Equation (12.27) into Equation (12.26) gives

$$\langle (\delta n)^2 \rangle = \langle n \rangle \tag{12.28}$$

which is one of the remarkable results of Poisson distribution. It indicates that the mean-square fluctuation of this distribution is equal to its mean value. This result will be used in the calculation of the variance of various photodiode noises in fiber optic receiver circuits.

Photodiodes are reverse-biased devices that absorb incident photons and convert them into electron-hole pairs. These photogenerated electron-hole pairs will drift to detector electrodes thus producing a photocurrent in the external circuit. An equivalent circuit of a detector is shown in Figure 12.9, representing a current source shunted by the depletion capacitance C_d of the detector. Because of the pair-production process the output voltage will be characterized by an average value $\langle n \rangle$ or the mean square of the fluctuations from this average. Because of the random arrival time of the photons, the output voltage of the detector is characterized by a frequency response function, which corresponds to a filtering process with a limited bandwidth. To include the frequency dependence of the device, the average output voltage is

$$\langle V \rangle = e\langle M \rangle \int \frac{\eta}{h\nu} P_0(t') Z_T(t - t') \, dt' \tag{12.29}$$

where $Z_T(t)$ is the impulse response of the detector with a gain multiplication $\langle M \rangle$ and is defined by

$$Z_T = Rh_T \tag{12.30}$$

For a linear channel, R is a constant with dimensions of resistance and h_T is a normalized impulse response function related to the input and output pulse shape function as

$$h_T = \frac{h_{out}}{h_{in}} \tag{12.31}$$

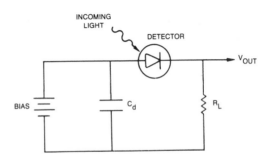

FIGURE 12.9   An equivalent circuit of a photodetector.

At any time, the mean square deviation or variance of the voltage from the average value $\langle V \rangle$ is a measure of the noise. Using the result of Equation (12.28), the following can be written:

$$\langle (\delta V)^2 \rangle = e \langle M^2 \rangle \int \left(\frac{e\eta}{h\nu} \right) P_0(t') Z_T^2(t - t') \, dt' \tag{12.32}$$

In terms of the variance of noise current,

$$\langle (\delta I)^2 \rangle = e \langle M^2 \rangle \int \left(\frac{e\eta}{h\nu} \right) P_0(t') h_T^2(t - t') \, dt' \tag{12.33}$$

Qualitatively, the impulse response function of the photodiode can be illustrated by arbitrarily choosing an incoming optical pulse shape as shown in Figure 12.10(a). Each pulse contains n number of photons. The detector, in response randomly to the ith photon at t_i, produces a displacement current $I_i(t)$ as shown in Figure 12.10(b). The composite response from all of the displacement currents, as shown in Figure 12.10(c) should resemble the signal waveform superimposed with various noises.

The impulse response function can be expressed either in time-domain or in frequency-domain by means of the following Fourier transforms:

$$Z(t) = \frac{1}{2\pi} \int_{-\infty}^{\infty} Z(\omega) e^{i\omega t} \, d\omega$$

The inverse transform is

$$Z(\omega) = \int_{-\infty}^{\infty} Z(t) e^{-i\omega t} \, dt$$

It can be shown that

$$\int_{-\infty}^{\infty} Z^2(t) \, dt = \frac{1}{2\pi} \int_{-\infty}^{\infty} Z(\omega) Z^*(\omega) \, d\omega = 2 \int_{0}^{\infty} Z(\omega)^2 \, d\nu \tag{12.34}$$

Using the above relationships, the mean square output noise current generated by the signal current $e\eta P_0/h\nu$, and also the dark current I_d either in time-domain or in frequency-domain is as follows:

$$N_{\text{det}}^2 = \langle (\delta I)^2 \rangle_{\text{det}} = e \langle M^2 \rangle \int_{-\infty}^{\infty} \left[\frac{e\eta}{h\nu} P_0(t) + I_d(t) \right] h_T^2(t) \, dt$$

$$= 2e \langle M^2 \rangle \int_{0}^{\infty} \left[\frac{e\eta}{h\nu} P_0(\omega) + I_d(\omega) \right] |H_T(\omega)|^2 \, d\nu \tag{12.35}$$

where the quantity in the brackets represents the mean value of the noise

(a)

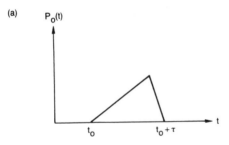

(b)

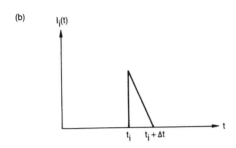

(c)

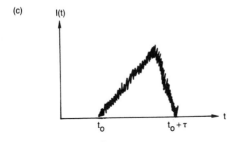

FIGURE 12.10 Detector input and output waveforms. (a) An arbitrary input waveform to the detector. (b) The displacement current resulting from the ith electron-hole pair generation. (c) The composite pulse response waveform resulting from all electron-hole pairs.

current spectral density. $H_T(\omega)$ is defined as

$$H_T(\omega) = \frac{H_{out}(\omega)}{H_{in}(\omega)} \qquad (12.36)$$

where $H_{out}(\omega)$ and $H_{in}(\omega)$ are the Fourier transforms of $h_{out}(t)$ and $h_{in}(t)$, respectively.

To evaluate the above integrals for a simple case in which a photodiode is illuminated by a light at a constant power level P_0, the detector response function $Z(t)$ can be determined by considering the time-dependent passage of an electron through the photodiode. For a transit-time τ as defined by the width W of the depletion region divided by the average drift velocity of the electron under a constant electric field, the time-dependent current can be expressed as

$$i(t) = ev(t)/W = e(vt/\tau)/(v\tau/2) = 2et/\tau^2$$

where v is the final drift velocity at a time τ so that the average drift velocity is just one-half of v. Therefore, the pulse response function in this case is

$$Z(t) = 2t/\tau^2 \tag{12.37}$$

Substituting Equation (12.37) into Equation (12.34) yields the transfer function

$$Z(\omega) = \frac{2}{\tau^2} \int_0^\tau te^{-i\omega t}\, dt$$

$$= \frac{2}{(\omega\tau)^2} [(\omega\tau \sin \omega\tau + \cos \omega\tau - 1) - i(\sin \omega\tau - \omega\tau \cos \omega\tau)]$$

The product of $Z(\omega)Z^*(\omega)$ is

$$|Z(\omega)|^2 = \frac{4}{(\omega\tau)^4} [(\omega\tau)^2 + 2(1 - \cos \omega\tau - \omega\tau \sin \omega\tau)]$$

For $\omega\tau \ll 1$, that is if the frequency response is not limited by the transit-time effects, this approximation gives

$$|Z(\omega)|^2 = 1 \tag{12.38}$$

Substituting Equation (12.38) into Equation (12.35) and integrating over the entire bandwidth provides

$$N^2_{\text{detector}} = 2e\langle M^2 \rangle(I_0 + I_d)B \tag{12.39}$$

where $I_0 = e\eta P_0/h\nu$ and B is the detector's bandwidth. The first term of Equation (12.39) represents the variance of the shot-noise of a photodiode under constant illumination, which is directly proportional to the incident light power. The second term represents the variance of the dark current noise of a photodetector, which is proportional to the dark current in the absence of light. For APD, the higher the multiplication factor, the higher the detector's noise as indicated by Equation (12.39). For a pulse-code signal, the pulse response function, or transfer function is more complicated than the expression given by Equation (12.37) or (12.38) and depends on the relative shapes of the signals and the width of the time slot associated with the code. This discussion will be expanded in the next section.

Now consider other noise sources encountered by the detector. The most common sources are thermal background noise, quantum noise, and Johnson noise. The level of the thermal spectral density S_{th} is often given in terms of kT watts per Hertz, where k is Boltzmann's constant and T is the absolute temperature. The familiar expression for S_{th} is given by

$$S_{th} = \frac{h\nu}{\exp(h\nu/kT) - 1} \tag{12.40}$$

At optical frequency (e.g., $\lambda = 1$ micron), $h\nu \gg kT$ and the photon energy has a value on the order of 2×10^{-19} J, as compared to the kT value of 4×10^{-21} J at room temperature. Therefore, at optical frequency thermal background noise is usually insignificant. Quantum noise, on the other hand, manifests itself in several possible ways, depending on the detection techniques. For direct detection as usually is the case for many optical fiber systems, quantum noise N_Q arises primarily from the statistical counting of photons. If a detector is relatively free of dark current noise, no error will be detected in the absence of an optical pulse. The probability of an error in the presence of an optical pulse of energy E can be estimated from Equation (12.23) by letting $n = 0$. Thus, the probability that no pairs are produced is given by

$$P(E) = e^{-E/h\nu} \tag{12.41}$$

Equation (12.41) shows that in a quantum-noise-limited case, the energy E must be greater than 21 $h\nu$ for an error probability less than 1 in a billion (10^{-9}). In other words, only 1 out of a billion events will generate no electron-hole pairs, if the number of photons in a pulse is greater than 21 for quantum-noise-limited detection. The equivalent minimum detectable power for a 10^{-9} bit error rate is

$$P_{min} = \frac{21h\nu B}{2} \tag{12.42}$$

In Equation (12.42), a factor of two is introduced because of the use of binary code. For example, for a system bandwidth B of 10 Mb/s, the P_{min} is about 2×10^{-11} W or -77 dBm for an incident light wave at 1 micron. This estimate assumes an error rate of 10^{-9} and the receiver is assumed to be quantum-noise-limited.

The output from a photodiode is usually fed into an amplifier, which is an integral part of the receiver system. For simplicity, assume that amplifier noise is dominated by Johnson noise that is generated in the 50 ohm input resistance. This assumption is reasonable because the admittance of the combined diode capacitance C_d and the amplifier input capacitance C_a is usually very small compared to that of the 50 ohm resistance. More details can be found in Chapter 13.

A resistor R at a temperature T emits noise power according to Equation (12.40), which can be rewritten as

$$S_J = \frac{h\nu}{1 + h\nu/kT + \frac{1}{2}(h\nu/kT)^2 + \cdots - 1}$$

At radio frequencies (e.g., $hv \ll kT$), an expression for the mean value of Johnson noise spectral density is given by $S_J = kT$. The mean square fluctuation of the photocurrent generated by the shunt resistor R can be expressed by

$$N_{shunt}^2 = \langle (\delta I)^2 \rangle_{shunt} = \frac{2S_J}{R} \int_0^\infty | H_T(\omega) |^2 \, d\omega \qquad (12.43)$$

where $H_T(\omega)$ is the Fourier transform of the impulse response of the RC network. For a total of two pass bands, the integral yields a value of $2B$, where B is the filter bandwidth. Therefore, the mean square fluctuation of the thermal noise current or the Johnson noise in the load resistor R can be expressed as

$$N_{shunt}^2 = 4kTB/R \qquad (12.44)$$

Or, the mean square fluctuation in voltage developed across the shunt resistor R can be expressed as

$$\langle (\delta V) \rangle^2 = N_{shunt}^2 R^2 = 4kTBR \qquad (12.45)$$

where B is the bandwidth of the RC network. The expressions above indicate that the thermal or Johnson noise from a resistor can contribute significantly to the excess noise power.

Because photodiodes are modeled as a current source in parallel with a capacitor as shown in Figure 12.9, essentially no thermal noise is associated with these sources. All thermal noises come from the input impedance of the amplifier. It is possible to reduce Johnson noise by using a high-input impedance, however, to maintain a large bandwidth, a low-capacitance must be structured by bonding the diode to a substrate containing either a Si or GaAs FET amplifier of comparable capacitance. Alternatively, it is also possible to use a bipolar transistor with a hgh-impedance front-end. Operating at high signal levels, transimpedance amplifiers may be good candidates. More details on amplifiers can be found in Chapter 13.

12.5 EFFECTS OF SIGNAL WAVEFORMS

This section deals with various noise sources of a photodiode when illuminated with time-varying pulse signals in a binary code at a signaling rate B and a time slot of width $T = B^{-1}$. In each time slot, signal power is assumed to take on one of two discrete values with a pulse shape function $h_{in}(t)$. The optical power impinging on the detector is then of the form

$$P_0(t) = \sum_{k=-\infty}^{\infty} b_k h_{in}(t - kT) \qquad (12.46)$$

where k is a variable denoting the time slot and b_k takes on one of two discrete values, $b(0)$ or $b(1)$. The current generated by the photodiode in response to $P_0(t)$ consists of a series of pulses each of which corresponds to an average number of electron-hole pairs. For a given time slot, this number is a random variable obeying a Poisson distribution. If the detector is an APD, this number will be multiplied by the avalanche gain M, which is also a random variable with a mean value $\langle M \rangle$ and a mean square value $\langle M^2 \rangle$. For p-i-n detectors, $\langle M \rangle = \langle M^2 \rangle = 1$. Following the treatment of Smith and Personik (Ref. 12.4), this section presents the signal-dependent mean square noise currents. Assuming that the input shape function $h_{\text{in}}(t)$ is normalized such that

$$\frac{1}{T} \int_{-\infty}^{\infty} h_{\text{in}}(t) \, dt = 1$$

the energy associated with a given pulse is $b_k T$ where $b_k = b(0)$ or $b(1)$. The quantity b has the dimensions of power with this normalization. The output pulse shape function $h_{\text{out}}(t)$ is chosen such that its maximum value occurs at $t = 0$, therefore, $h_{\text{out}}(0) = 1$.

Because various transfer functions depend explicitly on the width of the time slot T, it is convenient to introduce a normalized frequency variable y such that

$$y = \frac{\omega}{2\pi B} = \frac{\omega T}{2\pi} \tag{12.47}$$

With this change of variables, two new functions $H_{\text{in}}(y)$ and $H_{\text{out}}(y)$ are obtained. They are expressed as

$$H'_{\text{in}}(y) = \frac{1}{T} H_{\text{in}}(\omega)$$

and

$$H'_{\text{out}}(y) = \frac{1}{T} H_{\text{out}}(\omega) \tag{12.48}$$

Substituting Equation (12.48) into (12.36) shows that

$$H_T(\omega) = H'_T(y) \tag{12.49}$$

These transfer functions are useful to express the variance of various noise currents. The second term of Equation (12.35) shows the variance of dark current I_d to be

$$N^2_{\text{dark current}} = 2eI_d\langle M^2 \rangle BI_2 \tag{12.50}$$

where I_d is assumed to be independent of ω and

$$I_2 = \int_0^{\infty} |H'_T(y)|^2 \, dy \tag{12.51}$$

Using Equation (12.43), the variance of shunt noise is

$$N^2_{\text{shunt}} = \frac{4kT}{R} BI_2 \tag{12.52}$$

Shot noise, on the other hand, is dependent on the incoming signal format. To calculate the shot noise, consider a digital signal with a binary code. The quantity b_k in Equation (12.46) takes on two discrete values b_k = $b(0)$ or $b(1)$. If state 0 is transmitted with probability $P(0)$ and state 1 with probability $P(1)$, then the average optical power is given by

$$\langle P_0 \rangle = b(0)P(0) + b(1)P(1)$$

If the two states are equally probable, e.g., $P(0) = P(1) = \frac{1}{2}$, as is generally the case (more discussion can be found in Chapter 13), then

$$\langle P_0 \rangle = \frac{1}{2}[b(0) + b(1)]$$

If $b(0) = 0$, then

$$\langle P_0 \rangle = \frac{1}{2}b(1) \tag{12.53}$$

Using the above result, the first term of Equation (12.35) can be written as (Ref. 12.4),

$$N^2_{\text{signal}} = e \left(\frac{e\eta}{h\nu} \right) \langle M^2 \rangle B \sum_k b_k I_1 \tag{12.54}$$

where

$$I_1 = \int_{-\infty}^{\infty} \exp[i2\pi y(t - kT)/T] H'_{\text{in}}(y) \mid H'_T(y) \mid^2 dy \tag{12.55}$$

All of the above noise sources are proportional to B. In the case of the signal noise source, Equation (12.55) shows that it depends in a detailed manner not only on the value of the output pulse within the time slot T of interest, but it contains contributions from all other time slots weighted by the term in the bracket of Equation (12.55). It is convenient to separate the summation in Equation (12.54) into two parts. The first part is due to the contribution from the time slot in question and the second part is due to the signal contribution from all other time slots. To evaluate the first part let $t = 0$ at which a maximum signal is obtained as given by

$$N^2_s(0) = 2e \left(\frac{e\eta}{h\nu} \right) \langle M^2 \rangle BbI_1 \tag{12.56}$$

where b takes one of two discrete values $b(0)$ or $b(1)$ and

$$I_1 = \text{Re} \int_0^\infty H'_{\text{in}}(y) \mid H'_T(y) \mid^2 dy \qquad (12.57)$$

The remaining part, which results from the contributions over all other time slots, depends on the values of all the b_k. For the worst case, in which a maximum value of b is assumed in all other time slots, the noise term can be evaluated at $t = 0$ by using the expression

$$N_s^2(0) = 2e \left(\frac{e\eta}{h\nu}\right) \langle M^2 \rangle B b_{\text{max}}(\textstyle\sum - I_1) \qquad (12.58)$$

where

$$\sum = \frac{1}{2} \sum_{k=-\infty}^{\infty} H'_{\text{in}}(k) \mid H'_T(k) \mid^2 \qquad (12.59)$$

In choosing the shape of the output pulse, it is desirable to select the coordinate system so that the maximum value is located at the center of the time slot and zero at the center of all other time slots. In other words, the output signal, but not necessarily the noise, at the center of the kth time slot is due only to the input power in the kth time slot. Therefore, the intersymbol interference is zero with the assumed pulse shape. One class of functions that satisfies the above criterion is the "raised cosine" family (Ref. 12.4) as given by

$$h_{\text{out}}(t) = \frac{\sin(\pi t/T) \cos(\pi\beta t/T)}{(\pi t/T)[1 - (2\beta t/T)^2]} \qquad (12.60)$$

where β is a parameter and takes on values from zero to unity. The Fourier transform of $h_{\text{out}}(t)$ is given by

$$H_{\text{out}}(y) = \begin{cases} 1 & 0 < y < \dfrac{1-\beta}{2} \\[2mm] \dfrac{1}{2}\left[1 - \sin\left(\dfrac{\pi y}{\beta} - \dfrac{\pi}{2\beta}\right)\right] & \dfrac{1-\beta}{2} < y < \dfrac{1+\beta}{2} \\[2mm] 0 & \text{elsewhere} \end{cases} \qquad (12.61)$$

Figure 12.11 shows the pulse-shape function, the Fourier transform, and the eye diagram for $\beta = 0.1, 0.3$, and 1.0, respectively.

The commonly used pulse-shape functions for input pulses are: (1

(a) RAISED COSINE FUNCTION

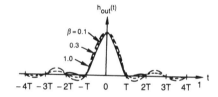

(b) FOURIER TRANSFORM OF h_{out} (t)

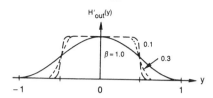

(c) WORST CASE EYE-DIAGRAM

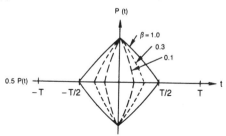

FIGURE 12.11 The most commonly used output pulse waveforms, (a) in time domain, (b) in frequency domain, and (c) of the worst case eye diagram for the raised cosine family. (Ref. 12.5. Reprinted with permission of American Telephone and Telegraph Company.)

rectangular, (2) Gaussian, and (3) exponential. The parameter α represents a fraction of time slot. For a rectangular pulse-shape,

$$h_{in}(t) = \frac{1}{\alpha} \qquad -\frac{\alpha T}{2} < t < \frac{\alpha T}{2} \qquad (12.62)$$

For a Gaussian pulse-shape,

$$h_{in}(t) = \frac{1}{\sqrt{2\pi}} \exp(-t^2/2\alpha^2 T^2) \qquad (12.63)$$

(a) RECTANGULAR INPUT PULSE SHAPE FUNCTION

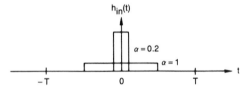

(b) GAUSSIAN INPUT PULSE SHAPE FUNCTION

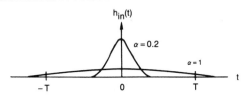

(c) EXPONENTIAL INPUT PULSE SHAPE FUNCTION

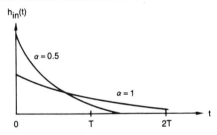

FIGURE 12.12 (a) Rectangular input pulse shape function, (b) Gaussian input pulse shape function, and (c) Exponential input pulse shape function. (Ref. 12.5. Reprinted with permission of American Telephone and Telegraph Company.)

And, for an exponential pulse-shape,

$$h_{in}(t) = \frac{1}{\alpha} \exp(-t/\alpha T) \qquad (12.64)$$

For dispersion-free systems where input pulses are undistorted by the transmission medium, the rectangular pulse-shape function is often used. The Gaussian shaped function is commonly used for multimode fibers that have strong mode coupling or mode mixing. For a multimode fiber with very weak mode coupling, the exponential pulse-shape function may be used. Figure 12.12 shows the input pulse-shape functions for these three types. The normalized transfer function $H_T(y)$ is shown in Figure 12.13. The values of I_1

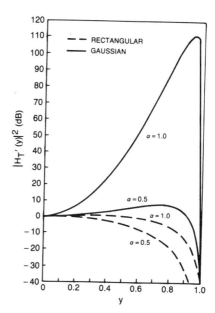

FIGURE 12.13 Normalized transfer functions for rectangular and Gaussian input pulse shape functions. (Ref. 12.4. Reprinted with permission of Springer-Verlag.)

I_2, and $\sum$ for various input pulse-shape functions can be found in Refs. (12.4) or (12.5).

12.6 FREQUENCY RESPONSE OF SILICON PHOTODIODES

For light wavelengths below one micron (see Figure 12.3), silicon photodiode is an excellent detector. The frequency response of Si photodiodes is determined primarily by three factors: (1) the carrier collection time τ_c, which is the time required for the electric field to sweep out the photoexcited carriers within the depletion region; (2) the RC time τ_{RC}, which is the time required to discharge the junction capacitance C_d through a combination of internal and external resistances; and (3) the diffusion time τ_d, which is the time required for these carriers to diffuse into the depletion region. The total photodiode rise time τ_r, defined as the time of response from 10% to 90% of a pulse height is essentially equal to the largest of the three mentioned above. To a good approximation,

$$\tau_r = (\tau_c^2 + \tau_{RC}^2 + \tau_d^2)^{1/2} \tag{12.65}$$

The expression for τ_c is given by

$$\tau_c \simeq \frac{W}{2} \frac{W}{\mu V_b} \tag{12.66}$$

where W, μ, and V_b are the width of the depletion region, carrier mobility, and reverse-bias voltage, respectively. For silicon photodiodes, τ_c can be estimated to a good approximation by using the expression

$$\tau_c \simeq \frac{\rho_n}{400} \quad \text{or} \quad \frac{\rho_p}{1000} \quad \text{ns} \tag{12.67}$$

where ρ_n and ρ_p are the resistivities in ohm-cm of n-type or p-type materials. The circuit defining the RC time constant of a silicon photodiode is shown in Figure 12.9, in which the total resistance R includes, in addition to R_L, R_s the internal resistance, and R_j the junction resistance. C_d is the junction capacitance, and I the photocurrent generated by the incident light. R_s is comprised of three parts: the resistance of the undepleted region of the silicon chip, the contact resistance, and the collection resistance associated with the resistivity of the front surface generated by diffusion. The total capacitance of the diode could exceed the C_d value by an additional capacitance associated with external packaging and wiring connections. Taking these factors into account, the following can be written:

$$\tau_{RC} = 2.2(R_S + R_j + R_L)(C_d + \sum C_n) \tag{12.68}$$

Typical R values range from a few ohms to a few hundred ohms and C_d can be estimated by

$$C_d = \frac{19{,}200A}{\rho_n^{1/2} V_b^{1/2}} \quad \text{pF} \tag{12.69}$$

where A is the photodiode active area. The C_d value could be increased by several picofarads due to additional external capacitance.

The diffusion time depends critically on the photon absorption depth, which is strong function of λ. If the absorption depth exceeds the depletion depth, the diffusion time can be estimated for the p-on-n structure by the expression

$$\tau_d \simeq \frac{1}{13} \left(\frac{3}{\alpha} - 0.54 \rho_n^{1/2} V_b^{1/2} \times 10^{-4} \right)^2 \tag{12.70}$$

where α is the absorption coefficient, and for the n-on-p structure

$$\tau_d = \frac{1}{36} \left(\frac{3}{\alpha} - 0.32 \rho_p^{1/2} V_b^{1/2} \times 10^{-4} \right)^2 \tag{12.71}$$

Risetime values for commercially available pin silicon photodiodes range from about 0.5 ns for small-area devices up to several microseconds for very large-area devices operating in a nonbias mode. For the latter case, the expressions above involving V_b must be modified by replacing V_b with V_0, which is the self-depletion bias voltage. Fall-time values are usually longer than τ_r, for partially depleted devices but are about equal to τ for fully depleted structures. The τ_r for avalanche silicon photodiodes is significantly shorter than that of pin diodes; however, APDs require very careful control of operating bias voltage and temperatures and they are available only in very small sizes.

Assuming that τ_c and τ_d can be greatly reduced by optimizing the device configuration, the upper cutoff frequency f_c is then determined by the RC time constant alone as

$$f_c = \frac{1}{2\pi RC} \tag{12.72}$$

Combining Equations (12.72) and (12.65) enables the determination of a relation between the 10% to 90% rise time τ_r, to the cutoff frequency f_c by the expression

$$\tau_r \simeq \frac{0.35}{f_c} \tag{12.73}$$

where the cutoff frequency f_c is often regarded as the bandwidth B of the system.

12.7 HIGH-SPEED AND LONG-WAVELENGTH PHOTODETECTORS

The primary wavelengths of interest at present for optical fiber communication are in the region of 1.3 to 1.6 microns, where both optical attenuation and material dispersion can be minimized. GaInAs, whose bandgap energy is 0.8 eV, is ideal material for making heterojunction photodetectors in this wavelength region. In this section, very high-speed devices of both p-i-n and APD photodetectors operating in the long wavelength region are discussed. Presently, the internal gain of the most broadband APDs permits the most sensitive receivers to operate frequencies up to 7 GHz. For higher-speed applications, p-i-n/FET receivers offer higher sensitivity than APD detectors with responsivity extending well above 30 GHz. The major limitations on the speed of heterojunction long-wavelength photodetectors are the same for silicon detectors. In addition, long-wavelength photodetectors suffer from charge trapping at the interface of the heterojunction.

The transit time of both electrons and holes is an important parameter to be considered in the design of a high-speed device. It depends on carrier velocities and thus on the electric field. Figure 12.14 shows the velocity-impedance on electric field for GaAs and GaInAs. For high-speed operation, the intrinsic layer of a p-i-n device should be completely depleted and the field throughout the intrinsic layer should be above 50 kV/cm for the case of GaInAs (Ref. 12.19). At this field strength, carriers travel at the saturation velocity at the value given in Figure 12.14 during most of the their transit.

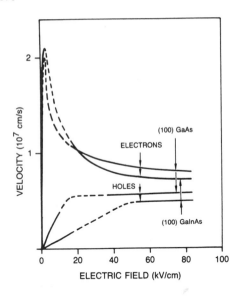

FIGURE 12.14 Carrier velocity versus electric field for GaInAs and GaAs. (Ref. 12.1. Reprinted with permission of IEEE.)

At doping levels in an intrinsic layer of approximately 10^{16} cm^{-3} or less, the electric field and carrier velocities are quite constant across most of the detector for a bias of at least a few volts below the breakdown voltage. To reduce the transit time limitation, very thin intrinsic layers are used. However, as intrinsic layer thickness is decreased, capacitance is increased, and quantum efficiency decreases. Consequently, each given detector area has an optimum intrinsic layer thickness that yields the maximum bandwidth. This relationship is illustrated in Figure 12.15 for a GaInAs detector designed for 1.3 microns (α = 1.16 μm^{-1}) operation. It is a design trade-off plot between the speed of response and the quantum efficiency of a device. The calibrations (Figure 12.15) assume the sum of the load resistance and the detector resistance to be 50 Ω. In general, the largest bandwidth is obtained with the smallest detector area at reduced quantum efficiency. As shown in Figure 12.15, a 200 GHz GaInAs detector could have, at best, a quantum efficiency of 16%.

For high-speed detectors with $\alpha L \ll 1$, the quantum efficiency can be estimated by the formula:

$$\eta = (1 - r)(1 - e^{-\alpha L}) = (1 - r)\alpha L \qquad (12.74)$$

where r is the Fresnel reflectivity, α is the optical absorption coefficient (see Figure 12.3) and L is the thickness of the intrinsic layer. Quantum efficiency can be increased by collecting light parallel to the junction plane as shown in Figure 12.16. Two versions of this structure exist. The first, which is an edge-absorbing detector labeled "A," can be made very small and therefore can have a very high bandwidth. The major drawback of this device configuration is that the light spreads as it travels through the device. As a result

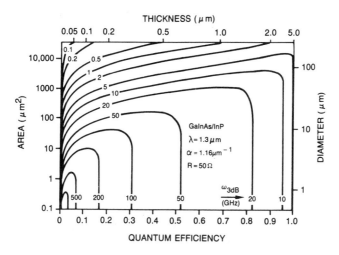

FIGURE 12.15   Contours of constant 3 dB bandwidth in the detector area, depletion layer thickness, quantum efficiency plane. (Ref. 12.1. Reprinted with permission of IEEE.)

the quantum efficiency is strongly dependent on the characteristics of the incoming beam. The second version, which is a waveguide structure, eliminates this difficulty. The structure is made by surrounding the absorbing layer with a transparent waveguide to confine the light to the region of the absorbing junction.

For long-wavelength operation (1.3 − 1.55 μm), the waveguide structure can be made of GaInAs ($n = 3.5$) cladded with InP ($n = 3.2$), which is transparent in the 1 to 1.6 micron region. For a 0.2 micron intrinsic layer thickness, a waveguide of 1 micron width, 10 micron length and with an AR-

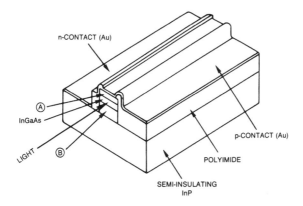

FIGURE 12.16   Schematic diagram of transverse detectors: an edge absorbing detector (A-p^+ GaInAs; b-n^+ GaInAs or InP) and a waveguide detector (A-p^+ InP; B-n^+ InP). (Ref. 12.1. Reprinted with permission of IEEE.)

coated facet, has a quantum efficiency of 49%, assuming only 50% of the light is coupled into the waveguide. The bandwidth, which is basically limited by the transit time because of extremely low parasitics, is expected to be 150 GHz. Unfortunately, this waveguide detector suffers from charge trapping at the p-InP/GaInAs interface as illustrated in Figure 12.17. Charge trapping can be avoided by using a continuously graded layer grown via LPE or super-lattices, grown by Chemical Beam Epitaxy (CBE), which has been found to reduce charge trapping at p-InP heterojunctions. Vapor-phase epitaxy is found to produce layers with very abrupt interfaces that cause charge trapping and produce a long tail in the response.

The most commonly used high-speed p-i-n photodiodes are back-illuminated etched-mesas as shown in Figure 12.8(b). The p- and n-contacts are conventional Au-Zn and Au-Sn electroplated and alloyed layers. Mesas of typically 10 microns in height are etched in a bromine-methanol solution and the edges are coated with nonconducting epoxy. The contact to the mesa is made via a gold mesh tape and conductive epoxy. The performances of two mesa photodiodes are shown in Figure 12.18. For the higher-speed GaInAs device, the diameter of the mesa is approximately 20 microns and the depletion layer thickness is 0.5 microns. The lower-speed device has a diameter of 70 microns and a depletion layer thickness of 2.7 microns. The quantum efficiencies of these two devices versus wavelength are plotted in Figure 12.18(b). The higher quantum efficiency and reduced wavelength dependence is consistent with the wavelength dependent of the absorption coefficient.

The most advanced high-speed and long-wavelength APDs are made

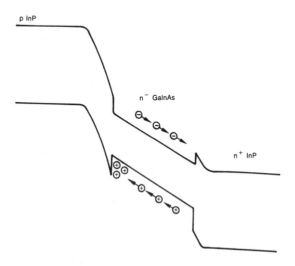

FIGURE 12.17 Band diagram of a p-i-n heterojunction photodetector showing interfacial trapping. (Ref. 12.1. Reprinted with permission of IEEE.)

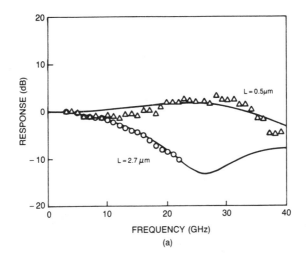

(a)

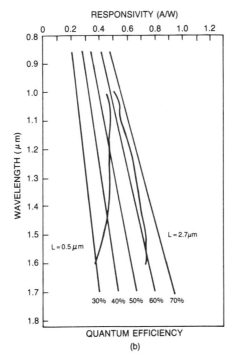

(b)

FIGURE 12.18 (a) Frequency response of two GaInAs detectors: One has a depletion layer thickness $L = 0.5$ μm; the other has a L-value of 2.7 μm. (b) Wavelength dependence of the responsivity of the two detectors in (a). (Ref. 12.1. Reprinted with permission of IEEE.)

of InP/InGaAsP/InGaAs grown by chemical beam epitaxy (CBE). They exhibit low dark current (<50 mA at 90% breakdown), good external quantum efficiency (>90% at λ = 1.3 μm), and high avalanche gain (G = 40). In the low gain region (G < 10), bandwidths as high as 8 GHz can be achieved (Ref. 12.6). In the high gain region (G > 15), the gain-bandwidth product is measured to be 70 GHz. A schematic cross-section of such a back-illuminated mesa structure is shown in Figure 12.19. It consists of seven epitaxial layers grown on an (100) oriented Zn-doped (1×10^{19} cm^{-3}) InP substrate. The layer parameters are as follows: the first layer is a 0.25 μm thick Be-doped (10^{19} cm^{-3}) InP buffer layer; the second layer is a 0.6–0.7 μm thick Sn-doped (4.8×10^{16} cm^{-3}) InP multiplication region; the transition region consists of three thin (<700 Å) InGaAsP (E_g = 1.13, 0.95 and 0.80 eV) layers with a carrier concentration of 2×10^{16} cm^{-3}; and, the InGaAs absorbing layer is a n-type (<5×10^{15} cm^{-3}) epitaxial material with a thickness in the range of 1.8–2.8 microns. This entire structure is capped with a n^+ InP (>2 $\times 10^{17}$ cm^{-3}) contact layer. The p-(Au-In-Zn) and n-(Au-Su) type contacts are deposited by pulse electroplating. The mesa is etched in a dilute solution of Br methanol. A window is etched in the p-type contact to permit illumination through the substrate. The chip is then mounted in a low capacitance pill package. If the absorbing layer denoted by "a" is completely depleted, the electric field profile shown in Figure 12.20 results. The maximum E-field is about 6.2×10^5 V/cm and the field at the InGaAsP/InGaAs interface is 1×10^5 V/cm. At these field values equivalent to 90% of breakdown field strength, the total dark current is typically in the range of 5–150 nA and the primary dark current is only about 1.5–6 nA. Figure 12.21 shows the dark current and measured gain versus bias voltage. Multiplication values as high as 40 can be obtained before breakdown. The solid curve represents the gain values calculated from Equation (12.10), using the following ioni-

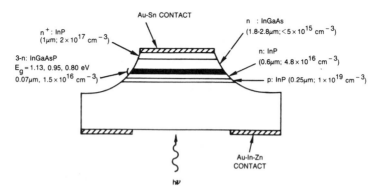

FIGURE 12.19 Cross-sectional diagram of an InP/InGaAsP/InGaAs back illuminated mesa-structure APD. (Ref. 12.6. Reprinted with permission of IEEE.)

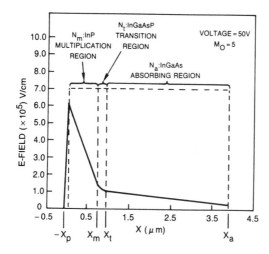

FIGURE 12.20   Electric field profile in an InP/InGaAsP/InGaAs APD biased at 90% of breakdown voltage. (Ref. 12.6. Reprinted with permission of IEEE.)

zation coefficients for various regions:

$$\alpha_m = 3.45 \times 10^5 \exp[-(1.04 \times 10^6/E)^{1.54}] \text{ cm}^{-1}$$

$$\beta_m = 3.80 \times 10^5 \exp[-(1.01 \times 10^6/E)^{1.46}] \text{ cm}^{-1}$$

$$\alpha_t = 3.37 \times 10^6 \exp(-2.29 \times 10^6/E) \text{ cm}^{-1}$$

$$\beta_t = 2.94 \times 10^6 \exp(-2.40 \times 10^6/E) \text{ cm}^{-1}$$

$$\alpha_a = 2.27 \times 10^6 \exp(-1.13 \times 10^6/E) \text{ cm}^{-1}$$

$$\beta_a = 3.95 \times 10^6 \exp(-1.45 \times 10^6/E) \text{ cm}^{-1}$$

where the subscripts m, t, and a refer to the multiplication, transition and absorption regions. These ionization coefficients are obtained by adjusting the carrier concentration in the multiplication region to provide the best fit with the measured values. The discrepancies fall well within the range of experimental error from the sample measurements.

The responsivity of an APD with a 3 μm thick absorbing layer and an AR coating is shown in Figure 12.22. Because of extremely high absorption of the InP substrate, the quantum efficiency of the back-illuminated APD device is strongly dependent on the thickness of the substrate. The results of Figure 12.22 were obtained for a device with an InP substrate thinned to a thickness of 25 microns, and coated with an AR film of 1.3 microns. This APD photodetector has a quantum efficiency greater than 90% in the wavelength range of 1.2 to 1.43 microns.

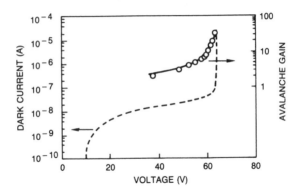

FIGURE 12.21 Dark current (dashed curve) and avalanche gain (data points) versus bias voltage. The solid curve is the calculated gain. (Ref. 12.6. Reprinted with permission of IEEE.)

The frequency response versus gain for the same device is shown in Figure 12.23. These measurements were made using a high-speed laser (λ = 1.3 μm) with direct modulation. The output of the APD was analyzed with a network analyzer. In the high gain region ($G > 15$), the measured gain-bandwidth product is 70 GHz. This limitation is attributed to the transit time of the carriers through the depletion region. The transit time for an APD structure is usually longer than that of a *p-i-n* photodiode because the multiplication and absorption regions are not coincidental. This attribute adds length to the drift region. In addition, a transit time is associated with the secondary electrons that are injected from the multiplication region into the absorbing layer. In fact, for large multiplication, these secondary electrons dominate the transit time. The frequency response of an APD for the

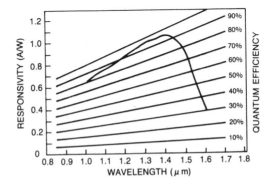

FIGURE 12.22 Responsivity and quantum efficiency of an AR-coated InP/InGaAsP/InGaAs APD versus wavelength. (Ref. 12.6. Reprinted with permission of IEEE.)

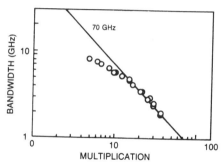

FIGURE 12.23   Measured frequency response of an InP/InGaAsP/InGaAs APD at 1.3 μm versus gain multiplications. (Ref. 12.6. Reprinted with permission of IEEE.)

general case of photogeneration on both sides of a *p-n* junction with both electrons and holes injecting into the multiplication region has been analyzed by Campbell et al. (Ref. 12.6). Using the one-dimensional model shown in Figure 12.20, the frequency dependence of the gain $M(\omega)$ is given by

$$\frac{M(\omega)}{M(0)} = \frac{\sin(\omega W/2v)}{\omega W/2v} \cdot [1 + (\omega\tau_{RC})^2]^{-1/2} \cdot [1 + (\omega/R_h)^2]^{-1/2}$$
$$\cdot [1 + (\omega\tau_B M(0))^2]^{-1/2} \tag{12.75}$$

where $M(0)$ is the DC gain, v is the saturated carrier drift velocity, W is the depletion width, R is the total resistance, C is the diode capacitance, R_h is the emission rate for holes trapped at the heterojunction interface, and τ_B is the effective transit time through the multiplication region or the build-up time in the avalanche process. Equation (12.75) is derived under the following assumptions: (1) the light is absorbed only in the depleted portion of the InGaAs layer ($x_t < x < W$), therefore, the diffusion effect is ignored; (2) only holes are injected into the multiplication region at $x = x_m$; and, (3) drift velocities for electrons and holes are equal.

The first term in Equation (12.75) is the common expression for the frequency response of a transit time-limited device with a total deletion width W and equal drift velocity for electrons and holes. At low gain the frequency response of an APD is very close to that of a *p-i-n* photodiode with comparable depletion width. At high gain, the transit time effect of the secondary electrons significantly reduces the APD bandwidth. For *p-i-n* photodiodes, higher bandwidths can be achieved by decreasing the width of the absorption layer but at the cost of lower quantum efficiency. In any case, the transit time is not usually a limiting factor at frequencies less than 10 GHz.

The second term in Equation (12.75) is a *RC* time-limited frequency response. With mesa structures, the typical device capacitance can be made as small as 0.1 to 0.2 pF. For a total resistance in the range between 10 to 50 ohms, the minimum and maximum spread in bandwidths limited by the *RC* time constant are 8 and 26 GHz.

The third term in Equation (12.75) is due to charge accumulation at the heterojunction interface. The tunneling process is the dominant mechanism responsible for the penetration of holes through the interfacial barrier at low temperature; but, at high temperature, the thermionic emission becomes the dominant mechanism. The problem of interfacial trapping can be eliminated by introducing either a transition layer or continuously grading the bandgap energy between the InP and InGaAs. With these approaches, bandwidths greater than 8 GHz have been obtained.

The last term in Equation (12.75) is due to the APD's avalanche build-up time. Owing to its regenerative nature, the higher the gain the longer it takes the avalanche process to build up. This process gives rise to a constant gain-bandwidth product, which is often the most restrictive term. One of the factors that determines the avalanche build-up time and hence, the gain bandwidth product, is the ratio of the electron and hole ionization coefficient k. In general, the greater the difference in α and β, the higher the bandwidth that can be attained. In InP the convergence of α and β values at high electric fields degrades the gain bandwidth product. This reduction can be compensated for by decreasing the width of the avalanche region. By reducing the multiplication layer thickness from 1.5 μm to less than 0.5 μm, the gain bandwidth product of a back-illuminated mesa structure InP/InGaAsP/ InGaAs APD can be increased from 18 to 70 GHz.

Figure 12.23 shows a gain bandwidth product of 72 GHz, which corresponds to a calculated effective transit time of about 2.2 ps, which is in agreement with the theory. The 3 dB rolloff frequency due to the RC effect is 11.5 GHz for a 0.2 pF capacitance and 70 Ω device-plus-load resistance. Because the absorption region is completely depleted and the heterojunction interfaces are graded over several lattice constants, the diffusion and interfacial trapping effects are insignificant and therefore can be ignored.

PROBLEMS

12.1. With 50 V of reverse-biased voltage, calculate the doping level ρ_p required to fully deplete the photoexcited carriers within a width $W = 0.1$ mm. What is the maximum electric field inside this device?

12.2. Show that the second moment of Poisson's distribution is

$$\langle n^2 \rangle = \langle n \rangle + \langle n \rangle^2$$

12.3. Calculate the Johnson noise of a 50 Ω resistor for a bandwidth $B = 10$ MHz.

12.4. If the diffusion time is negligible, calculate the rise time of a silicon photodiode having an active area of 0.1 cm^2 and a thickness of 0.04 cm. The resistivity is 400 Ω-cm. The device is reverse-biased at 50 V and is connected to an external load of 50 Ω.

12.5. The absorption depth of a silicon photodiode at 0.9 microns is about 75 microns. Assuming the same parameters as in Problem 12.4, what would be the depletion depth of the diode? Calculate the diffusion time if the reverse-biased voltage of the diode is reduced to 5 V.

12.6. For an InGaAs detector having an intrinsic layer thickness of one micron, estimate its quantum efficiency for operating at 1.3 microns.

12.7. Calculate the Fourier transform of a single rectangular pulse of duration τ, centered at $t = 0$ with an amplitude a_m. Plot the result as a function of $\omega\tau/2$.

12.8. Calculate the gain or multiplication factor M for the APD shown in Figure 12.19 using the electric field $E(x)$ shown in Figure 12.20.

REFERENCES

12.1. J. E. Bowers and C. A. Burrus, Jr., *IEEE J. Lightwave Tech.*, *LT-5*, 1339 (1987).

12.2. R. J. McIntyre, *IEEE Trans. Electron Devices ED-13*, 164 (1966) and *ED-19*, 703 (1972); P. P. Webb, R. T. McIntyre, and J. Conradi, *RCA Rev.*, *35* 234 (1974); J. Conradi, *IEEE Trans. Elect. Devices, ED-19*, 173 (1972).

12.3. T. Li, *IEEE J. Select. Areas Comm. SAC-1*, 356 (1983).

12.4. *Semiconductor Devices for Optical Communications*, ed. by H. Kressel, Springer-Verlag, 1980, Ch. 4.

12.5. S. D. Personick, *Bell Syst. Tech. J. 52*, 843 (1973).

12.6. J. C. Campbell, W. T. Tsang, G. J. Qua, and B. C. Johnson, *IEEE J. Quant. Elect., QE-24*, 496 (1988).

13

Optical
Receivers

13.1 INTRODUCTION

The interplay between detector output, the front-end of a preamplifier, the post amplifier, an equalizer, and a filter is very delicate and has a profound effect on the sensitivity of an optical receiver. In this chapter, various components of an optical receiver are presented. Receiver circuit noise is analyzed with special emphasis on the optimized design of low-noise FET and bipolar amplifier front-ends. Receiver sensitivity can be evaluated using state-of-the-art components. The fundamental goal in the design of an optical receiver is to minimize the amount of optical power, which must be regenerated by the receiver to maintain a desired fidelity measured in terms of a given BER for digital systems or S/N ratio for analog systems. The gain versus bandwidth and the signal power versus BER are just some of the important trade-off parameters in the design of a receiver for optical fiber systems.

13.2 RECEIVER CIRCUITS

A typical receiver consists of a photodetector, low noise preamplifier, post amplifier, equalizer, and filter. The output of a photodetector usually needs extensive amplification to be useful for further signal processing. As a result, additional noises are often introduced into the system. The post amplifier provides further power amplification of a signal and may also provide other functions such as automatic level control and clamping circuitry by keeping a signal to a fixed reference level. The role of an equalizer is to remove signal waveform distortions introduced by the amplifier and other components such as fiber transmission lines and to regenerate a reasonable output pulse shape. The main function of a filter is to maximize the S/N ratio while preserving essential features of the signal. Especially at long wavelengths, InGaAs photodetectors are relatively more noisy and less sensitive than their counterparts, silicon photodiodes operating at short wavelengths. The reason is that at longer wavelengths the quantum efficiency of photodetectors is usually lower, and so is the gain in the case of avalanche photodiodes. Therefore, a very careful design of the receiver for long-wavelength systems is needed. This consideration is particularly true for systems requiring very high bit rates $B > 1$ Gbit/s, because the preamplifier noise of the receiver increases rapidly with the bit rate.

This section deals with various noise sources of the components in a receiver. Figure 13.1 shows the front-end of a receiver, which is assumed to be a linear channel, consisting of various noise sources. The signal $I_s(t)$ is the photocurrent of the detector generated by the incident light $P_0(t)$. C_d is the depletion capacitance of the detector, and C_s is the stray capacitance associated with the interconnection of the detector to the input of the amplifiers. R_L is the load resistance in the bias circuit. The signal generators I_d, I_b, I_a, and V_a characterize, respectively, the dark noise current of the photodetector, thermal noise associated with the biasing resistor, noise current, and noise voltage of the preamplifier. Noise sources associated with the remaining portion of the linear channel are assumed to be negligibly small. Current noise sources, such as I_s, I_d, and I_b have been previously discussed.

The noise sources of the remaining portion of the circuit are analyzed here following the treatment by Smith and Personick (Ref. 13.1). As shown in Figure 13.1, I_a and V_a are the shunt and series noises of the preamplifier whose admittance is denoted by Y_a. The total input admittance is determined

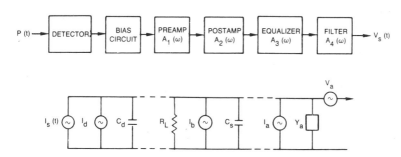

FIGURE 13.1   An equivalent circuit of the front-end of an optical receiver.

by

$$Y_I = \frac{1}{R_I} + i\omega C_T \qquad (13.1)$$

where

$$\frac{1}{R_I} = Y_a(\omega) + \frac{1}{R_L} \qquad (13.2)$$

$$C_T = C_d + C_s \qquad (13.3)$$

Let the transfer functions of the preamplifier and post amplifier be $A_1(\omega)$ and $A_2(\omega)$, and those of the equalizer and filter be $A_3(\omega)$ and $A_4(\omega)$, respectively. ω is the angular frequency, e.g., $\omega = 2\pi\nu$. The signal voltage at the output of a linear channel $V_s(\omega)$ is then given by

$$V_s(\omega) = Z_T(\omega)I_s(\omega) \qquad (13.4)$$

where $Z_T(\omega)$ is the transfer function of the system as given by

$$Z_T(\omega) = A_1(\omega)A_2(\omega)A_3(\omega)A_4(\omega)/Y_I(\omega) \qquad (13.5)$$

To evaluate various circuit noises, two types of noise generators in the receiver circuit must be examined. The first is the shunt noise source, which is due to noise current generators the same as those in a photodiode. The second is the series noise source, which is due to the series noise voltage generators. Series noise is measured in terms of the mean square output voltage instead of current. For convenience, the transfer function is defined as

$$Z_T(\omega) = R_T H_T(\omega) \qquad (13.6)$$

where R_T is a constant with dimensions of resistance and $H_T(\omega)$ contains the frequency-dependence of $Z_T(\omega)$. $H_T(\omega)$ will be used to calculate the shunt or current noise sources while $Z_T(\omega)$ will be used to calculate the series or voltage noise sources. The shunt current generator associated with the bias circuit and amplifier can be lumped together into an equivalent shunt current generator with a noise spectral density given by

$$\frac{d}{d\nu}\langle(\delta I)^2\rangle = \frac{4kT}{R} + \frac{d}{d\nu}\langle(\delta I)_a^2\rangle \qquad (13.7)$$

where $\nu = \omega/2\pi$ so that the variance resulting from this source can be expressed in the form

$$N_{\text{shunt}}^2 = \int_0^\infty \frac{d}{d\nu}\langle(\delta I)^2\rangle \mid H_T(y)\mid^2 d\nu$$

$$= \left[\frac{4kT}{R} + \frac{d}{d\nu}\langle(\delta I)_a^2\rangle\right] BI_2 \qquad (13.8)$$

where I_2 is given by Equation (12.51).

The output noise due to the series voltage generator with a noise spectral density $d\langle(\delta V)_a^2\rangle/dv$ can be expressed in the form

$$N_{series}^2 = \int_0^\infty \frac{d}{dv}\langle(\delta V)_a^2\rangle \mid Y_I(\omega)\mid^2 \mid Z_T(\omega)\mid^2 dv \qquad (13.9)$$

Substituting Equation (13.1) into (13.9) yields

$$N_{series}^2 = \frac{1}{R_I^2}\frac{d}{dv}\langle(\delta V)_a^2\rangle \int_0^\infty \mid Z_T(\omega)\mid^2 dv$$
$$+ (2\pi C_T)^2\frac{d}{dv}\langle(\delta V)_a^2\rangle \int_0^\infty v^2\mid Z_T(\omega)\mid^2 dv \qquad (13.10)$$

Substituting Equation (13.6) into Equation (13.10) yields

$$N_{series}^2 = \frac{d}{dv}\langle(\delta V)_a^2\rangle\left[\left(\frac{R_T^2}{R_I}\right)BI_2 + (2\pi C_T)^2 R_T^2 B^3 I_3\right] \qquad (13.11)$$

where I_2 is given by Equation (12.51) and

$$I_3 = \int_0^\infty \mid H_T'(y)\mid^2 y^2\, dy \qquad (13.12)$$

The mean square noise current can be obtained from the series noise generator by simply dividing Equation (13.12) with R_T^2. Note that the series noise generator contributes two terms to the output noise voltage: one is proportional to the bandwidth of the circuit, and the other is proportional to the cube of the bandwidth. This latter term is the limiting factor in the optimization of optical receivers.

13.3 FET AND BIPOLAR AMPLIFIER NOISE

The common configuration for an input amplifier when using a field effect transistor FET is shown in Figure 13.2 where the resistor R_L represents the parallel combination of resistors used to bias the gate of the transistor and to provide a DC return path for the detector circuit. The input of the FET is a capacitance C_T. The transistor produces an output current proportional

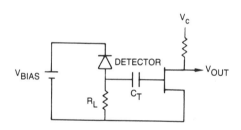

FIGURE 13.2 A schematic of a FET amplifier front-end.

to the input voltage with a transconductance constant g_m. The principal sources of noise in this circuit are the thermal noise of the equivalent resistor R_L, the shot noise associated with the gate leakage current, and the noise associated with the channel conductance. The equivalent input shunt current spectral density is given by

$$\frac{d}{dv} \langle (\delta I)_a^2 \rangle = \frac{4kT}{R_L} + 2eI_{\text{gate}} \qquad (13.13)$$

where I_{gate} is the gate leakage current. The channel conductance contributes a noise current at the output of the FET with a spectral density of $4kT\Gamma g_m$, where Γ is a numerical factor that is about 0.7 for Si devices and approximately 1.1 for GaAs devices. This current source gives rise to a series voltage source at the input of the amplifier with a series voltage spectral density as given by

$$\frac{d}{dv} \langle (\delta V)_a^2 \rangle = \frac{4kT\Gamma}{g_m} \qquad (13.14)$$

The equivalent circuit of a bipolar input amplifier is shown in Figure 13.3. The dominant noise sources in a bipolar transistor front-end are the shot noise associated with the base current I_b and the collector currents I_c, and also the thermal noise of the base resistor R_b. In the common emitter configuration, the shot noise spectral density of the base current I_b is modeled by an equivalent shunt current generator with a noise current spectral density as given by

$$\frac{d}{dv} \langle (\delta I)_a^2 \rangle = 2eI_b \qquad (13.15)$$

The shot noise spectral density of the collector current is given by $2eI_c$, which gives rise to a series noise spectral density as given by

$$\frac{d}{dv} \langle (\delta V)_a^2 \rangle = 2eI_c/g_m^2 \qquad (13.16)$$

where g_m is given by $g_m = eI_c/kT$.

The input admittance of the transistor at the internal base contact is

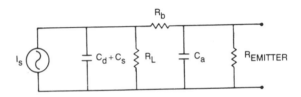

FIGURE 13.3 An equivalent circuit for a bipolar front-end.

given by

$$Y_a = \frac{1}{R_b} + i\omega C_a \qquad (13.17)$$

where C_a includes the base-emitter and base-collector capacitances. The total input admittance is

$$Y_I = \left(\frac{1}{R_L} + \frac{1}{R_b}\right) + i\omega C_T \qquad (13.18)$$

where

$$C_T = C_d + C_s + C_a \qquad (13.19)$$

The noise in receiver circuits involving GaAs MESFET and bipolar junction transistors can now be evaluated. Using the results obtained for the noise sources associated with the shunt and series circuit generators as given by Equations (13.13) and (13.14), variance of the FET amplifier can be written as follows:

$$N_{\text{FET}}^2 = \frac{4kT}{R_L}I_2B + 2eI_{\text{gate}}I_2B + \frac{4kT\Gamma}{g_m}[(2\pi C_T)^2B^3I_3 + BI_2/R_L^2] \qquad (13.20)$$

where the first term represents the shunt current noise source and the second term corresponds to the shot noise associated with gate leakage current. In Equation (13.20), g_m is the transconductance of the device, Γ is a channel noise factor whose value is 0.7 for Si devices and 1.1 for GaAs FETs, I_{gate} is the FET gate leakage current, and C_T is the total input capacitance. The values of the integrals I_2 and I_3 depend on input and output pulse-shapes. For NRZ input and output pulses with a full raised-cosine spectrum that results in zero intersymbol interference, the corresponding integral values are $I_2 = 0.562$ and $I_3 = 0.0868$ (Ref. 13.2). For RZ input pulses with a 50% duty cycle, the integral values are $I_2 = 0.402$ and $I_3 = 0.0361$. The total capacitance C_T is given by

$$C_T = C_d + C_{gs} + C_{gd} + C_s \qquad (13.21)$$

where C_d is the detector capacitance, C_{gs} is the FET gate-source capacitance, C_{gd} is the FET gate-drain capacitance, and C_s is the stray capacitance that includes the load resistor capacitance.

The last term in Equation (13.20) is the series thermal noise associated with the FET channel. This noise is inherent in FET operation and represents the dominant receiver noise mechanism. This noise can be minimized by making g_m/C_T as large as possible. This optimization of the FET gate width implies that the following condition is satisfied:

$$C_{gs} + C_{gd} = C_d + C_s \qquad (13.22)$$

A figure of merit for an optimized FET receiver is defined by

$$\text{FET figure of merit} = \frac{f_T}{\Gamma(C_d + C_s)} \tag{13.23}$$

where

$$f_T = \frac{g_m}{2\pi(C_{gs} + C_{gd})} \tag{13.24}$$

In other words, an ideal FET preamplifier must have an f_T value as high as possible and an Γ value as low as possible.

The results of Equations (13.8) and (13.11) show the variance of the equivalent current noise for a bipolar front-end to be

$$N_{BP}^2 = \left(\frac{4kT}{R_L} + 2eI_b\right) I_2 B + \left[\frac{2eI_c}{g_m^2} C_T^2 + 2kTR_b(C_d + C_s)^2\right] 4\pi^2 I_3 B^3 \tag{13.25}$$

In Equation (13.25), the total capacitance C_T is also given by

$$C_T = C_d + C_s + C_a \tag{13.26}$$

where C_a includes base emitter and base collector capacitances. The noise in a bipolar amplifier and also in the FET amplifier can be minimized by choosing the highest possible f_T value, by minimizing C_d and C_s, and by making R_L as large as possible.

In general, the noise of silicon bipolar transistors is somewhat higher than that of GaAs MESFETs, although future III–V bipolar transistors may have performances comparable or superior to those of present FETs. A comparison of equivalent input-noise-current variance for a state-of-the-art GaAs MESFET and Si-bipolar transistor is shown in Figure 13.4. The parameters for the MESFET are $f_T = 20$ GHz, C_d and $C_s = 0.5$ pF, $\Gamma = 1.1$, and $I_{gate} = 100$ nA. The parameters for the bipolar transistor are $f_T = 10$ GHz, $R_b = 20\ \Omega$, $I_c/I_b = 100$, C_d and $C_s = 0.5$ pF, and $C_a = 0.8$ pF. The results of Figure 13.4 are calculated using the NRZ input pulse format and neglect the external noise from the resistor R_L. The slope of the bipolar noise curve changes from 6 dB/octave (e.g., $\propto B^2$) at 0.1 Gbit/s to 9 dB/octave (e.g., $\propto B^3$) at 20 Gbit/s as a result of the saturation of the optimum collector current caused by the increase in emitter diffusion capacitance with I_c. The slope of the FET curve, on the other hand, is a constant 9 dB/octave, except at the low frequency where $1/f$ noise ($f < 50$ MHz) and gate leakage current begin to play an important role. With the best available preamplifier, the noise for a GaAs MESFET is lower than that for a Si-bipolar transistor

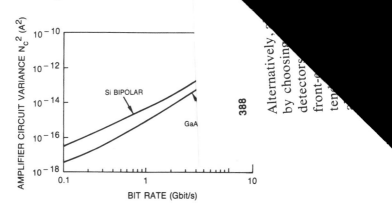

FIGURE 13.4   Calculated noise current variance N_c^2 of optimized silicon bipolar and GaAs MESFET preamplifiers as a function of the bit rate B. (Ref. 13.2. Reprinted with permission of IEEE.)

by about 10 dB at 0.1 Gbit/s to 5 dB at 20 Gbit/s. This noise level corresponds to a receiver sensitivity of 7.5 − 5 dB advantage for the MESFET.

13.4 RECEIVER DESIGN

The design goal of an optical receiver is to minimize the amount of optical power required to achieve highest receiver sensitivity at a given BER. Previous sections described various noise sources in a receiver in detail. It is apparent that a low-noise, broadband optical receiver relies on an optimum design of the front-end. In this section, alternatives for front-end designs are discussed. They are the high impedance or integrating front-ends and the transimpedance or current-to-voltage converter. The choice of input termination is very important to provide a desired bandwidth while keeping the input noise level to a minimum.

The most straightforward but not necessarily the optimum method of designing a front-end is to terminate the input to the preamplifier with a load resistor R_L so that it satisfies the following condition:

$$R_L \leq 1/2\pi B C_T$$

In other words, the bandwidth of the input admittance must be equal to or greater than the desired bit rate B. The ideal situation is one where a standard 50 ohm termination is small enough to satisfy this condition, otherwise a small resistance R_L must be used. In either case, the signal would be amplified requiring little or no equalization. However, the output must be filtered to regenerate its original pulse shape. The penalty of this simple approach is an increase of the circuit noise over that potentially achievable.

...all sources of noise can be reduced to their absolute minimum ... extremely low input capacitance, and using low dark current ... In this case a high input impedance R_L can be tolerated for the ...nd. If R_L is large, the input admittance is dominated by C_T, which ...s to integrate the signal current. Thus, the high impedance front-end is ...lso commonly referred to as integrating front-end.

In a high impedance front-end, the equalizer plays the critical role of restoring the pulse-shape, which otherwise is distorted due to the limited bandwidth of the input admittance. Often the equalizer takes the form of a simple differentiator, which attenuates the low frequency components of the signal and restores a flat transfer function to the system. However, some difficulties are associated with this approach. First, with this approach the zero position of the equalizer must be calibrated to compensate for the zero of the input admittance. Because the location of the input zero depends on the values of C_T and R_L as well as other device parameters, the zero position may vary from device to device. The precision with which equalization must be achieved also depends on the amount of the signal spectrum near the "cut-in" frequency of the equalizer. Another problem with a high impedance front-end is that it has a relatively low dynamic range (I_{max}/I_{min}). The loss of dynamic range occurs because the charge on the input capacitance can build up over the duration of the time slots. As a result, high input signals can produce premature saturation of the amplifier.

 Another approach commonly used in front-end design is to use a transimpedance or shunt feedback amplifier. It is basically a current-to-voltage converter as shown schematically in Figure 13.5. In the limit of large loop gain, the relation between the output voltage and input current is given by

$$V_{out} = -Z_F I_{in} \qquad (13.27)$$

where Z_F is the effective feedback impedance from the output to the input of the amplifier. The transimpedance amplifier is widely used because it has wide bandwidths and large dynamic range. The frequency response of this device is governed by the feedback resistor R_F and a capacitor C_F, which can be made exceedingly small compared with the input capacitance. The dynamic range of a transimpedance amplifier is greater than that of a high-impedance front-end. The reason for this difference is that in a transimpedance amplifier the attenuation of the low frequency components is accomplished through negative feedback, and hence the low frequency components are amplified by the closed-loop gain, instead of the open-loop as in other types of amplifiers. Therefore, the improvement in dynamic range is approximately given by the ratio of the open- and closed-loop gains. The noise sources of this amplifier consist of contributions from detector leakage, the input transistor, and the feedback resistor. It can be determined from the expression for FET or bipolar transistors by replacing R_L with R_F. In practice, the noise performance of a transimpedance amplifier is not as good as

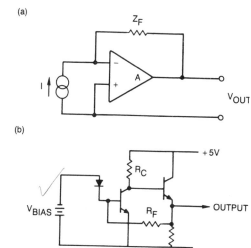

FIGURE 13.5 Schematics of a transimpedance amplifier. (a) Simplified representation and (b) a circuit diagram.

that achieved with the high-impedance approach, therefore, the transimpedance amplifier should not be used if an extremely low-noise receiver system is required.

13.5 SIGNAL TO NOISE RATIO AND ERROR PROBABILITY

The S/N and BER are major parameters for determining the fidelity of an optical receiver system. S/N is defined by the ratio of the signal power to the sum of the noise power. Therefore, for a linear channel, it is convenient to express S/N in terms of the mean square of the signal current to the sum of the mean square of the noise currents. For example, the S/N at the output of an APD having an internal gain M and an external load resistor R_L can be written as

$$\frac{S}{N} = \frac{(M\eta P_0/h\nu)^2}{2eI_0M^2FB + 4kTB/R_L} \tag{13.28}$$

where P_0 is the average power of the light signal, and I_0 is given by Equation (12.1). In the denominator, only the shot noise and Johnson noise are considered. Other noise sources, such as thermal background, dark current, leakage current, and quantum fluctuation, should also be included, if they represent a significant portion of the total system noise. One meaningful quantity that determines a detector's figure of merit is the noise equivalent power (NEP), which is defined as the amount of light power impinging on the active detector area that produces an output signal power equivalent to the noise output of the detector. NEP represents the minimum detectable power of a detector and is a function of the bandwidth. Typical NEP values for silicon pin diodes are in the range of 1 to 10 $\times$ 10^{-14} W/Hz$^{1/2}$. More discussion on minimum detectable power is given in the next section.

The S/N ratio in general increases with increasing P_0. For high-fidelity communication systems, $S/N \gg 1$. The higher the S/N, the lower the probability of errors in the system. This section examines the causes of errors and calculates the probability of BER. Exact calculation is very difficult. Only a simple approximation is introduced to estimate the BER for a series of digital Gaussian pulses incident on a photodiode. Because of pulse spreading, adjacent pulses can overlap, as shown in Figure 13.6. This problem is commonly referred to as intersymbol interference. Figure 13.6(a) shows a series of three incident Gaussian pulses having $\sigma = 0.25\tau$, where τ is the spacing between pulses. If $\sigma > 0.25\tau$, intersymbol interference becomes significant. For example, consider a case for which $\sigma = 0.5\tau$, as shown in Figure 13.6(b). The amplitude of a normalized Gaussian pulse is $1/\sqrt{2\pi\sigma}$. The intersymbol interference is represented by the eye formation. The eye opening ϵ is a measure of the interference. In this case $\epsilon = 0.8\tau(1 - 0.27) = 0.584\tau$. To reduce the overlap, an equalizer can be used as a filter. Caution must be used, however, in the implementation of an equalizer because more noise could be introduced into the system. Therefore, great care must be exercised in choosing a filter that matches the input pulse shape precisely.

A comprehensive analysis of error probability for optical communi-

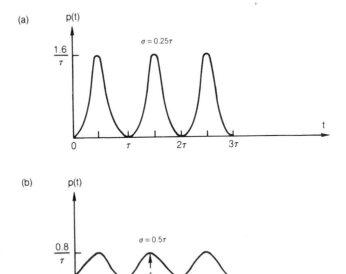

FIGURE 13.6   Series of three normalized Gaussian light pulses showing the effect of intersymbol interference with (a) $\sigma = 0.25\tau$; (b) $\sigma = 0.5\tau$. (From Ref. 13.3. Reprinted with permission of Academic Press, Inc., New York.)

cation systems can be found in the book by Pratt (Ref. 13.4). Two different regimes exist: (1) small signal detection involving photon counting; and (2) large signal detection involving statistical averaging. For both situations, a simple digital PCM intensity modulation with direct detection is used. The carrier signals of a PCM/IM system are represented by ones and zeros. The transmitted signals with a given information bandwidth can be corrected and filtered at the receiver terminal. For example, a bit decision can be made as to whether the photodetector output exceeds a decision threshold during a bit period. Within each bit interval, two probabilities of detection must be evaluated to determine the probability of system detection error. The first one is the probability P_S that the decision threshold will be exceeded by the transmitter signal and detector noise. The second one is the probability P_N that the threshold will be exceeded by the noise alone. The probability of making an error, P_E, depends on the probability that the signal and noise do not exceed the threshold when the transmitter signal is present, and that the noise alone exceeds the threshold when the transmitter signal is absent. For a system transmitting "one" or "zero" bits with a probability P,

$$P_E = P(1 - P_S) + (1 - P)P_N \qquad (13.29)$$

For equal likelihood (e.g., $P = \frac{1}{2}$) the following applies:

$$P_E = \tfrac{1}{2}(1 - P_S + P_N) \qquad (13.30)$$

Case 1: Small signal detection.   The threshold decision is based on the actual counting of the photoelectrons. The number of counts, n, during a bit period must be compared with the threshold number n_{th}. If n is equal to or larger than n_{th}, a "one" bit is registered. If n is less than n_{th}, a "zero" bit is registered. For extremely low level detection when $n < 10^2$, the detection probabilities are assumed to have a Poisson distribution. Therefore,

$$P_S = \sum_{n = n_{th}}^{\infty} (\langle n_S \rangle + \langle n_N \rangle)^n \frac{\exp(-\langle n_S \rangle - \langle n_N \rangle)}{n!} \qquad (13.31)$$

and

$$P_N = \sum_{n = n_{th}}^{\infty} \langle n_N \rangle^n \frac{\exp(-\langle n_N \rangle)}{n!} \qquad (13.32)$$

where $\langle n_S \rangle$ is the average number of photoelectrons per bit due to the signal and $\langle n_N \rangle$ is the average number of photoelectrons per bit due to the noise. The decision rule for judging the presence of a signal is usually followed by a ratio test:

$$\frac{P_S}{P_N} \geq \frac{1 - P}{P} \qquad (13.33)$$

where P is the a priori probability that a signal is transmitted. Solving for the threshold value n_{th} for which equality exists in Equation (13.33) (e.g., $P = \frac{1}{2}$) yields

$$n_{th} = \frac{\langle n_S \rangle}{\ln(1 + \langle n_S \rangle/\langle n_N \rangle)} \tag{13.34}$$

Figure 13.7 is a plot of the likelihood ratio tests thresholds n_{th} as a function of the average number of signal and shot-noise photoelectron counts for a pulse-code intensity modulation system. Figure 13.8 is a plot of the probability of detection error for the shot-noise-limited decision threshold based on the likelihood ratio test. The cusps in the curves are due to integer changes in the detection threshold as $\langle n_s \rangle$ increases. For the case $\langle n_N \rangle = 0$, a value for P_E reaches 10^{-9} when $\langle n_S \rangle = 21$, consistent with the quantum noise-limited case previously discussed. For example, suppose that the system noise produces an average of $n_N = 20$ photoelectrons per bit and there is an average of $n_s = 10$. The threshold for equal probability of error in the two states as given by Equation (13.34) is $n_{th} = 24.7$. If 25 or more photoelectrons are detected when a "0" is transmitted, an error results. Errors can also occur when a "1" is transmitted with less than 25 photoelectrons per bit detected. Raising the threshold creates the likelihood that more errors in "1" and less in "0" will result. On the other hand, decreasing the threshold will decrease the errors in "1" at the expense of the errors in "0". Therefore,

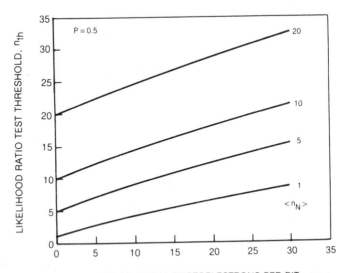

AVERAGE NUMBER OF SIGNAL PHOTOELECTRONS PER BIT, $<n_S>$

FIGURE 13.7 Threshold n_{th} as a function of $\langle n_S \rangle$ and $\langle n_N \rangle$ for a PCM/1 M system with $P = \frac{1}{2}$. (From Ref. 13.4. Reprinted with permission of John Wiley and Sons, Inc., New York. © 1969.)

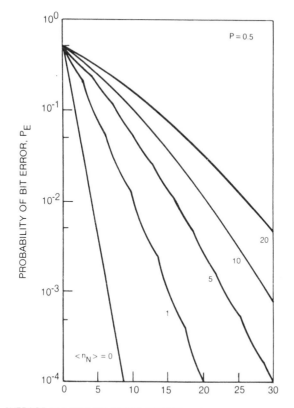

AVERAGE NUMBER OF SIGNAL PHOTOELECTRONS PER BIT, $<n_S>$

FIGURE 13.8   Probability of detection error as a function of $\langle n_S \rangle$ for a shot-noise-limited PCM/1 M system with $P = \frac{1}{2}$. (From Ref. 13.4. Reprinted with permission of John Wiley and Sons, Inc., New York, © 1969.)

a precise determination of the optimum threshold is critical for systems involving low-level signal detection.

Case 2: Large signal detection.   As system noise increases, optical power must be increased accordingly. If photoelectrons contained in one bit become very large ($n > 10^4$), Gaussian statistics can be used to describe the detection probabilities. For most optical fiber systems, this situation turns out to be the case because system operation is usually thermal-noise-limited. The use of Gaussian statistics is relatively straightforward and yields simple analytical solutions. Furthermore, results obtained using Gaussian statistics are in agreement with measurements for the range of parameters encountered in many incoherent optical fiber systems involving direct detection. For the subsequent treatment on receiver sensitivities for incoherent optical PCM communication systems, Gaussian statistics will be used. Let the expected values of the signals in the two transmitted states be $I(0)$ and $I(1)$ as shown in Figure 13.9 and assume the probability density functions of the two states to be Gaussian with variances of σ_0^2 and σ_1^2, respectively. Because the noise in each signal state contains contributions proportional to the signal, σ_0 and σ_1 will not, in general, be equal. Choosing the decision threshold I_{th} to yield equal probability of error in the two states, the probability of making an error P_E is given by

$$P_E = \frac{1}{\sqrt{2\pi}} \int_Q^\infty \exp(-x^2/2)\, dx \equiv \text{Erfc}(Q) \qquad (13.35)$$

where

$$Q = | I_{th} - I_i |/\sigma_i \qquad (13.36)$$

and I_i and σ_i are the expected value and standard deviation of the ith signal level ($i = 0, 1$). For Gaussian statistics, the value of the variance σ^2 is equal to the mean square noise associated with the signal level. Therefore, for all practical purposes, Q is equivalent to the square root of S/N. To a good approximation, Equation (13.35) can be expressed as

$$P_E(Q) = \frac{1}{\sqrt{2\pi}Q} \exp - Q^2/2 \qquad (13.37)$$

Figure 13.10 shows P_E versus Q. This figure shows that for $Q = 6$, $P_E = 10^{-9}$.

The analysis above is oversimplified. In general, receivers contain intersymbol interference, and detectors such as an APD have excess multiplication noise. For these reasons, the Gaussian approximation is not completely satisfactory. However, experimental results indicate that the Gaussian approximation for calculating receiver sensitivity is good to within 1 dB. The shortcoming lies in the fact that the optimum threshold is closer

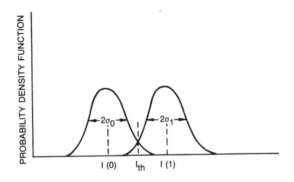

FIGURE 13.9 Probability density functions of two Gaussian states with variance σ_0 and σ_1.

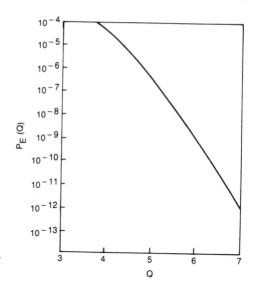

FIGURE 13.10   A plot of the error function.

to the center of the "eye" than the Gaussian approximation predicts. This consequence is due to the highly skewed nature of the probability distribution of randomly multiplied shot noise.

13.6 RECEIVER SENSITIVITY

A receiver's sensitivity is defined in terms of the received optical power required to achieve a desired BER. This power is often measured in dBm where 0 dBm corresponds to 1 mW of optical power at the receiver. One of the goals of receiver design is to minimize the amount of power required to achieve desired system fidelity. In digital systems, information is transmitted as a series of pulses "1" or "0" occurring at a rate B. At the receiver, the optical signal is attenuated from the transmitted signal and may also be distorted. It is the function of the digital regenerator to determine whether a mark designated by "1" or a space designated by "0" was transmitted and to regenerate the information with the least probability of error. The expected values of the transmitted signal are therefore in two states, namely

$$I_s(0) = \left(\frac{\eta e}{hv}\right) b(0) \qquad (13.38)$$

and

$$I_s(1) = \left(\frac{\eta e}{hv}\right) b(1) \qquad (13.39)$$

Mean square noise currents also exist in two states, namely N_0^2 and

N_1^2. Choosing the decision threshold I_{th} to yield equal probability of error in the two states as would be appropriate when $P(0) = P(1)$, the definition of Q indicates that

$$I_{th} - \left(\frac{\eta e}{h\nu}\right) b(0) = QN_0$$

$$\left(\frac{\eta e}{h\nu}\right) b(1) - I_{th} = QN_1 \qquad (13.40)$$

For direct detection systems, the noise associated with a signal is usually negligible compared to circuit noise. Hence, to a good approximation, all noise currents are assumed to be the same as the circuit noise current N_c. With this assumption, Equation (13.40) yields

$$\frac{\eta e}{h\nu} [b(1) - b(0)] = 2QN_c \qquad (13.41)$$

The average power is given by

$$\overline{P} = b(0)P(0) + b(1)P(1) \qquad (13.42)$$

For equal likelihood, $P(0) = P(1) = \frac{1}{2}$, Equation (13.42) gives

$$\overline{P} = \frac{1}{2}[b(0) + b(1)] \qquad (13.43)$$

In the limit $b(0) = 0$, a simple expression for the detectable optical power required to achieve a desired BER that corresponds to a chosen Q value is given by

$$\overline{P} = \left(\frac{h\nu}{e\eta}\right) QN_c \qquad (13.44)$$

If $Q = 1$, $\overline{P} = P_m$ which is the minimum detectable power and is also called the NEP. The above expression is useful for receivers employing a p-i-n photodiode. In the case of APDs, the signal current in the "1" state can introduce significant noise into the system due to avalanche gain. The noise in the "1" state is given by

$$N_1^2 = N_c^2 + 2e \left(\frac{\eta e}{h\nu}\right) \langle M^2 \rangle BI_1 b(1) \qquad (13.45)$$

Using Equations (13.45) and (13.40) and again assuming $b(0) = 0$ gives

$$I_{th} = QN_c \qquad (13.46)$$

$$\left(\frac{e\eta}{h\nu}\right) \langle M \rangle b(1) - I_{th} = Q \left[N_c^2 + 2e \left(\frac{e\eta}{h\nu}\right) \langle M^2 \rangle BI_1 b(1) \right]^{1/2}$$

With Equation (13.46), $b(1)$ can be obtained as

$$b(1) = \left(\frac{hv}{e\eta}\right) Q \left[\frac{2N_c}{\langle M \rangle} + 2eBI_2Q \frac{\langle M^2 \rangle}{\langle M \rangle^2}\right] \qquad (13.47)$$

Substituting Equation (13.47) into (13.43) provides

$$\overline{P} = \left(\frac{hv}{e\eta}\right) Q \left[\frac{N_c}{\langle M \rangle} + eBI_1Q \frac{\langle M^2 \rangle}{\langle M \rangle^2}\right] \qquad (13.48)$$

The first term in Equation (13.48) corresponds to the circuit noise and the second term is the additional noise introduced by the avalanche effect on the signal. The quantity $\langle M \rangle$ is the average avalanche gain and can be simply expressed by the letter M, whereas the mean square of the avalanche gain is given by

$$\langle M^2 \rangle = M^2 F(M) \qquad (13.49)$$

where $F(M)$ is referred to as the excess noise factor, which is given explicitly in Equation (12.21).

Substituting Equation (13.49) into (13.48) yields

$$\overline{P} = \left(\frac{hv}{e\eta}\right) Q \left[\frac{N_c}{M} + eBI_1QF(M)\right] \qquad (13.50)$$

The above expression is derived by assuming zero dark current. If $I_d \neq 0$, the result of Equation (12.50) is added into Equation (13.50) to obtain

$$\overline{P} = \left(\frac{hv}{e\eta}\right) Q \left[eBI_1QF(M) + \left(\frac{N_c^2}{M^2} + 2eI_dI_2BF(M)\right)^{1/2}\right] \qquad (13.51)$$

The calculated receiver sensitivity for a *p-i-n* and APDs with varied k values at bit rates from 0.1 Gbit/s to 20 Gbit/s is shown in Figure 13.11. The sensitivities are calculated for $Q = 6$ or a BER of 10^{-9} and for an NRZ input-pulse at a wavelength of 1.3 microns. The results of Figure 13.11 ex-

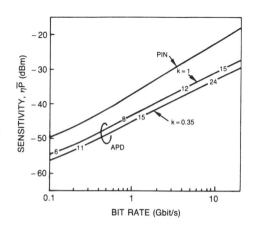

FIGURE 13.11 Calculated sensitivities versus bit rate for an optimized *p-i-n* diode and an APD using a GaAs MESFET preamplifier. The optimum APD gains are indicated at points along the curves. (Ref. 13.2. Reprinted with permission of IEEE.)

TABLE 13.1　The Best Reported Values on Receiver Sensitivities

B (Gbit/s)	(μm)	$\bar{P}$ (dBm)	$\langle n \rangle$ (per bit)	Detector	Amplifier	Pulse-Format	Ref.
1	1.55	−42	471	InGaAs APD	FET	NRZ	a
	1.3	−37.5	1162	InGaAs APD	FET	NRZ	b
2	1.55	−37.6	705	InGaAs APD	FET	RZ	c
	1.3	−33.4	1494	Ge APD	BP	RZ	d
4	1.55	−32.4	1114	InGaAs APD	FET	RZ	e
	1.3	−31.5	1157	InGaAs APD	FET	RZ	e
8	1.3	−25.8	2150	InGaAs APD	FET	NRZ	f

a. Ref. 12.7.

b. J. C. Campbell, et al., *Electron. Lett., 18* 818 (1983).

c. M. Shikada, et al., Proc. 5th Int'l. Conf. Integrated Opt. and Opt. Fiber Comm., Venice, Italy (1985).

d. J. Yamada, et al., *IEEE J. Quant. Elect., QE-18,* 1537 (1982).

e. S. Fujita, et al., Proc. 12th Eur. Conf. Opt. Comm., Barcelona, Spain (1986).

f. B. L. Kasper, et al., *J. Lightwave Tech., LT-5,* 344 (1987).

clude the thermal noise from the load resistor. Other parameters are: $f_T = 20$ GHz, $\Gamma = 1.1$, $I_{\text{gate}} = 100$ nA, $I_d = 0$, $C_d = 0.5$ pF, and $C_s = 0$. Optimum avalanche gains are indicated at points along the curves. Two different k values were used in these calculations. The case of $k = 1$ corresponds to a Ge APD and the case of $k = 0.35$ corresponds to an InGaAs APD. If $I_d \neq 0$, the sensitivity is further degraded by a small amount, e.g., at $I_d = 100$ mA, the sensitivity is decreased by about 0.5 dB at 20 Gbit/s. A summary of measured receiver sensitivities as reported in the literature for bit rates up to 8 Gbit/s is given in Table 13.1.

The general expression (Equation 13.51) for receiver sensitivity shows that system performance is dependent on the shapes of the light pulses arriving in each time slot. If received pulses tend to overlap and cause intersymbol interference, errors will most likely be introduced into the decision-making process. The overlap can be removed by the process of equalization, but this process has two disadvantages: (1) Equalization, in general, requires a knowledge of the input pulse-shape, which depends on the pulse broadening or dispersion properties of the fiber; and, (2) Equalization increases the noise at the linear channel output, which may degrade receiver sensitivity. Therefore, in designing an optimized system, a trade-off exists between fiber dispersion and receiver sensitivity.

PROBLEMS

13.1. Using the Gaussian pulse-shape function (a) plot the curves for $\sigma = 0.25\tau$ and $\sigma = 0.5\tau$, and (b) determine the eye-opening parameter formed by the zero and one pulses with a $\sigma = 0.5\tau$.

13.2. Calculate the minimum optical power required for a 50 Mb/s communication system that has a total system loss of 40 dB, and a $P_E = 10^{-9}$.

13.3. Show that the optimum FET gate width must be chosen to satisfy the condition

$$C_{gs} + C_{gd} = C_d + C_s$$

13.4. To achieve a BER of 10^{-9}, a particular optical receiver requires a minimum input level of 3000 photons per "on" pulse. Assume that "on" pulses correspond to 3000 photons of received energy. Consistent with this assumption, the receiver accommodates a 10% extinction ratio. Assume that in the "off" condition (digital zero sent) the received energy is 300 photons. If the data rate being communicated is 100 Mb/s (half ones and half zeros on the average), at 1.3 microns, what is the average power required at the receiver in watts?

13.5. A 50-Mb/s transimpedance digital receiver has an average optical signal input level of -55 dBm. The transimpedance is 4000 Ω. The detector responsivity is 0.5 A/W with a detector gain M. Calculate the receiver output in millivolts for $M = 50$. Calculate the number of incident photons per pulse for $hv = 2 \times 10^{-19}$, assuming half the input pulses are present and half are not.

REFERENCES

13.1. *Semiconductor Devices for Optical Communications,* ed. by H. Kressel, Springer-Verlag, 1980, Ch. 4.

13.2. B. L. Kasper and J. C. Campbell, *IEEE. J. Lightwave Tech.,* LT-5 1351 (1987).

13.3. Fundamentals of Optical Fiber Communications, ed. M. K. Barnoski, Academic Press, New York, 1976.

13.4. W. K. Pratt, *Laser Communication Systems,* John Wiley & Sons, New York, 1969.

14

Optical
Fiber Systems

14.1 INTRODUCTION

The accumulated knowledge of various optical fiber components imparted in previous chapters enables the analysis of optical fiber systems for which each component has a given set of specifications to be considered for various applications. Based on system requirements and constraints, a designer must first choose the most suitable components and then evaluate overall performance in terms of the total frequency response and power budget of a system. An optimum system is one in which the power margin is at a minimum while system response is faster than that required for system bandwidth and fidelity. To achieve this goal, a trade-off analysis must be performed among various possible component selections. This type of design analysis usually involves an iterative process for which computer-aided design (CAD) is often necessary. This section introduces various techniques for selecting suitable components. To assure system reliability, an allowance for component degradation is often made. Parameters involved in component and system optimization are summarized. Examples are given to illustrate the procedure for making this type of analysis. Descriptions of several practical optical fiber systems are also given. In particular, techniques for avoiding collisions in a congested local area network are described in detail. This chapter also includes a section on the performance of coherent light-wave communication systems.

14.2 PRELIMINARY DESIGN GUIDE

Four basic inputs are required for the design of a communication link: (1) the data rate or bandwidth; (2) the fidelity, indicated by either S/N or BER; (3) the link length and number of terminals; and, (4) the type of data and signal waveforms to be transmitted. Once these requirements are specified by the user, a designer can begin the iterative process for system optimization. It is customary to start an analysis by examining the total dispersion bandwidth of the fiber and source combination for a given terminal spacing. If a pin photodiode is sufficient, there is no reason to use the more sophisticated circuit for operating an APD. If the initial choice for a fiber-source combination cannot meet the required data rate, either an upgrading of the combination or a repeater may be considered without sacrificing transmission fidelity. Because the optical fiber industry is still in a stage of rapid growth and expansion, both the quality and cost of various components will undergo several significant transitions in years to come. Therefore, it is desirable to install the best commercially available fibers into systems because the replacement costs of these items is far greater than those of terminal components such as sources and detectors. It is expected that considerably more reliable sources and efficient couplers will become available at lower costs in the future. If necessary, a system can easily be upgraded by only replacing these terminal components. It is advisable to allow for possible system expansion in data-handling capacity even though the initial investment cost may be higher.

Table 14.1 serves as a guide for the initial choice of various components based on communication length and data-rate requirements. The initial selection is not important because through an iterative procedure these components will be optimized in terms of the total system performance, cost, and availability. Many companies in the United States, Canada, Japan, and Europe have their commercial product lines that are well characterized and performing reliably in accordance with their specifications.

For data transmission between short distances at rates below 2 Mb/s, very reliable systems can be made at a very reasonable cost. For these systems the best choice of source is a LED, because the cost is low and its performance is very reliable, with a typical lifetime greater than 10^6 h. Using large NA values, the coupling efficiency between a LED and a fiber bundle can be made as high as $1 - L_p$, where L_p is the packing fraction loss. For short communication links at low-data-rates, fiber dispersion usually does not present a problem and information can be transmitted using either a pulse code or IM with either transistor logic or a simple IC circuit. A pin photodiode with an integrated FET preamplifier at the front-end is sufficient to use as the receiver. The typical receiver noise for a pin receiver at 2 Mb/s is below -40 dBm for a BER of 10^{-9}. In addition to the considerations

TABLE 14.1 Preliminary Choice of Components for Various System Requirements on Channel Length L and Bandwidth B

Component	$L < 100$m $B < 2$ Mb/s	$L < 1$ km $B < 30$ Mb/s	$L > 1$ km $B < 100$ Mb/s	$L > 1$ km $B > 100$ Mb/s
Source	LED (GaAs)	LED (AlGaAs) LD (AlGaAs)	LD (AlGaAs)	LD (InGaAsP) LD (AlGaAs)
Fiber	Step-index fiber bundle	Step-index fiber or bundle	Graded-index (low loss)	Single mode or graded-index
Detector	pn or pin	Si pin	Si APD GaAs APD	Si APD Ge APD
Amplifier	FET	Trans-impedance	Temperature-compensated bipolar	Temperature-compensated bipolar
Driver	IC or transistor	Prebias	Pre-bias or EO modulator	EO modulator

above, environmental constraints, such as the operating temperature range, humidity, corrosion, radiation, and so on, must be taken into account for the design of system packaging.

For system requirements falling in the medium range (i.e., $L > 1$ km and $B \leq 100$ Mb/s) the component trade-off and selection process become more involved, because many options exist, each of which requires a detailed examination, if the system is to be optimized. In this case, a computer-aided design can be very helpful and will be outlined in the following section. For very high-data-rate and long-distance communication channels, the choice of components is again limited. For these cases inevitably the best component in each category is chosen.

14.3 DESIGN ANALYSIS

For systems with intermediate bandwidth requirements, a detailed analysis must be performed to optimize the system by evaluating its performance with specifications for the transmitter, cable channels, and receiver. In the case of the transmitter, the parameters that need to be considered include the optical power, wavelength, spectral width, beam size and shape, frequency response, output linearity, and modulation or coding format. Other parameters that have an influence on system performance are the lifetime, transmitter noise, and environmental factors such as temperature, vibration, and humidity. In the case of the fiber, critical parameters are the diameters for the core and cladding, index profile, attenuation coefficient, modal and material dispersion, pulling strength, and microbending loss. In the case of the receiver, the noise equivalent power, frequency response, quantum efficiency, amplifier noise, and gain are important parameters that characterize receiver performance.

Even though the system trade-off analysis among these parameters can be rather complex, the system requirements in terms of data rate, distance between terminals, number of terminals, fidelity, and environmental constraints such as temperature range, humidity, vibration, space, and cost usually dictate the choice of certain key components that are the most likely candidates for the system. Once the initial choices are made, a routine iterative process can be followed for the optimization of each component. Figures 14.1, 14.2, and 14.3 are the flowcharts for the design analysis of a transmitter, fiber, and receiver, respectively. The analysis is complicated by the fact that some of the parameters in one subsystem are dependent on parameters in another subsystem. For example, the choice of spectral width and wavelength for the source depends on the dispersion characteristics of the fiber and also on the spectral response of the detector. For this reason, an optimum system may not be unique, and often the choices are made on the basis of availability and cost.

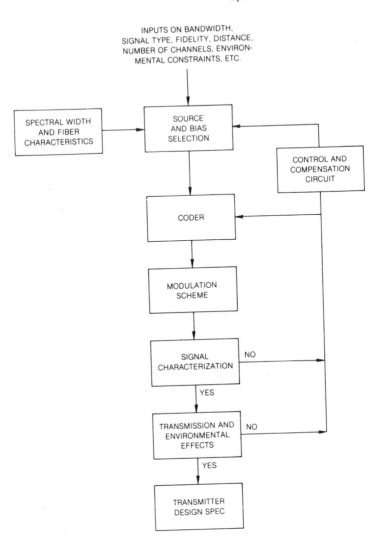

FIGURE 14.1   Design flowchart for optical fiber transmitter.

Once a system configuration is established, two computations must be carried out, one involving the calculation of the total system response and the other related to the power budget. The former gives an estimate for the system's information-handling capability and the latter gives an estimate on the system's reliability, based on available transmitter power. Both of these calculations are necessary and provide an indication of the fidelity of the system. The total system response can be expressed in terms of the rise time for the transmitter τ_l, the fiber τ_f, and the receiver τ_r, as follows:

$$\tau_s = \sqrt{\tau_l^2 + \tau_f^2 + \tau_r^2} \tag{14.1}$$

On the other hand, the amount of information to be transmitted by a system with bandwidth B can also be expressed in terms of a time constant τ_B. Table 14.2 gives τ_B in terms of either bit rate or bandwidth B for various signal types. If $\tau_s < \tau_B$, system response is considered to be adequate.

Power budget can be estimated by calculating (1) the minimum optical power P_d that can be detected by a chosen receiver at a given bandwidth and a signal level required by the S/N ratio, and (2) the maximum signal power P_s that can reach the receiver after allowing for all possible system losses. If $P_s > P_d$, the system chosen is considered to be adequate. The power margin that is the difference between P_s and P_d should be at least 10 dB to allow for component degradation and other unexpected problems. Table 14.3 gives the results of an analysis for power budget and rise times of a multichannel color TV system. The system components consist of an AlGaAs DH laser, a graded-index fiber, and a star coupler, which form a distribution network to 10 receiving terminals at a distance of 0.5 km utilizing an APD receiver.

Another example, given below, describes a typical digital data link involving only one transmitter and one receiver at a bit rate of 20 Mb/s. The

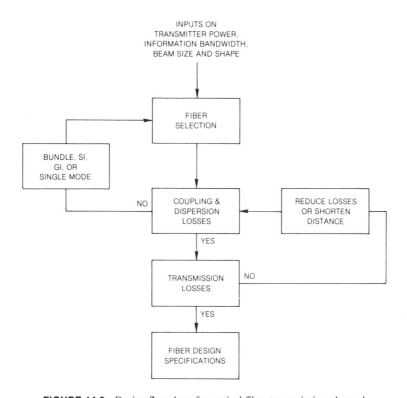

FIGURE 14.2   Design flowchart for optical fiber transmission channel.

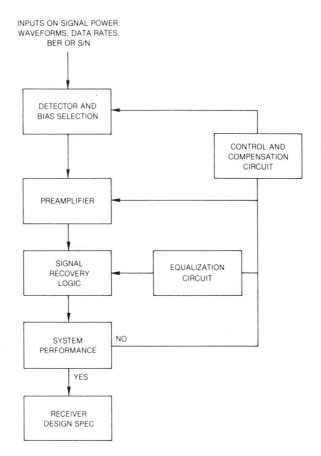

FIGURE 14.3   Design flowchart for optical fiber receiver.

length of this link is assumed to be 8 km. Over this length, four graded-index fiber cables are used with a total of three splicings. A LED is chosen as the source with an average output at 3 dBm. An APD detector is used in this link and has a time constant of 3 ns. Table 14.4 gives the results of the power budget and system rise-time analyses for this data link. The calculated sys-

TABLE 14.2   Rise-Time Estimate for Various Signal Types

Signal	τ_B
NRZ	0.7/bit rate
RZ	0.7/2(bit rate)
IM	0.7/2B
PCM	1/(sampling rate)(bits/sample)(B)

TABLE 14.3   Design Analysis for a Multi-channel Color TV System[a]

Component	Power Budget	Risetime
Transmitter		$\tau_t = 5$ ns
AlGaAs laser	$P_l = 15$ dBm	
Degradation allowance	3 dB	
Source coupling loss	5 dB	
Fiber and coupler		$\tau_f = (\tau_{mod}^2 + \tau_{mat}^2 + \tau_c^2)^{1/2}$
GI (3 dB/km)	$a_f = 3 \times 0.5 \times 10 = 15$ dB	
Star coupler	$a_c = 10 \log 10 + 13 = 23$ dB	$= 1.6$ ns
Degradation allowance	3 dB	
Receiver		$\tau_r = 3$ ns
APD	$P_r = -55$ dBm	
Degradation allowance	5 dB	
Receiver coupling loss	10 dB	
System performance:	Power margin $= 7 - 41 + 40$	$\tau_s = (25 + 2.56 + 9)^{1/2}$
	$= 6$ dB	$= 6.05$ ns

[a] System requirements: $B = 6$ MHz, BER $= 10^{-9}$, 10 terminals at $L = 0.5$ km, $\tau_B \simeq 8$ ns.

tem rise time is about 14 ns, which is well within the time requirement (35 ns) specified by the system bandwidth. However, the power budget calculation indicates that the power margin is too small to allow for any excess loss. To increase the excess power margin, a lower-loss fiber with an attenuation coefficient of less than 5 dB/km must be considered. This example illustrates that a low-cost but reliable LED source is sufficient for use in a high-fidelity (BER $= 10^{-9}$) data transmission system operating at a bit rate of 20 Mb/s over a distance of 8 km without the need of a repeater. If a laser source is used, the length of the link can easily be increased by about a factor of two or more. A considerably longer data link can be made using components operating at 1.55 μm. An example of a longer-wavelength sys-

TABLE 14.4   Design Analysis for a Digital Data Transmission System[a]

	Power Budget		Rise Time
Source: LED	3 dBm		6 ns
Signal: NRZ	−3 dBm		
Fiber: G1: $\alpha = 5$ dB/km	−40 dB	Dispersion	10 ns
Splicing loss (3)	1.5 dB		
Detector: APD	59 dBm		3 ns
Source/fiber coupling	10 dB		
Fiber/detector coupling	1 dB		
Temperature degradation allowance	1 dB		
Other allowance	5 dB		
System requirement: $40 + 1.5 + 10 + 7 =$	58 dB	$0.7/20$ Mb/s $= 35$ ns	
System performance: $3 - 3 + 59 =$	59 dB	$1.11\sqrt{6^2 + 3^2 + 10^2} = 13.4$ ns	

[a] System requirements: BR $= 20$ Mb/s. BER $= 10^{-9}$; two terminals, terminal spacing $= 8$ km, data coding: PCM, NRZ.

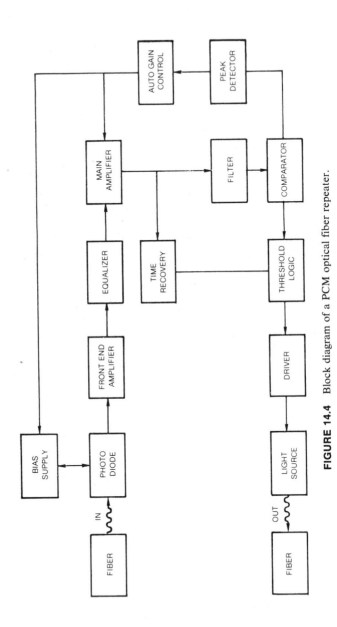

FIGURE 14.4 Block diagram of a PCM optical fiber repeater.

tem will be discussed in Section 14.7. The components used in this example are all commercially available. It is expected that the performance of these components operating in the range 0.82 to 0.85 μm will also be improved with time.

The last example given below describes an optical repeater. It basically consists of a photodiode, amplifier, equalizer, and signal regenerator. Figure 14.4 is a block diagram of a typical PCM, optical fiber repeater. The attenuated and distorted optical signal, after propagating along a long fiber, is collected by a photodiode, which converts the optical signal into an electrical signal. The amplifier, which is designed for a capacitive source typically of a few picofarads and having an input impedance equal to $[j\omega(C_{ampl} + C_{det}) + 1/R_{in}]^{-1}$, amplifies the received signal with as little added noise as possible. A low-noise amplifier front-end must be carefully designed. Typical values for the Z parameters vary from 1000 to 5000. The equalizer compensates for the effects of the detector and fiber dispersion by reducing the intersymbol interference at a penalty to the optical power. If the system is truly dispersion-limited, the repeater spacing can be increased with an equalizer. On the other hand, an equalizer should not be used if optical power is considered to be at a premium.

The signal regenerator is composed of a number of components, including a signal comparator, timing recovery circuit, waveform regenerating logic, driver, and source. The regenerator must be designed to reproduce the desired pulse-shape. This reproduction is usually accomplished using a feedback mechanism that responds to the output signal. The automatic gain control (AGC) circuit compensates for variations in the incoming signal levels, amplifier gain, and temperature effect in the APD device. To reproduce a desired pulse stream, it is ncessary to provide a periodic clock signal that is synchronized with the time slots of the received pulses. Such a clock signal can be generated from the receiver's output by using a phase-locked-loop timing recovery circuit, as shown in Figure 14.4. Variations in the pulse pattern could cause a phase jitter of the clock signal, which can generate an accumulative error in the output. A trade-off between the amount of narrow-bandwidth filtering required in the phase-locked-loop and the ability to lock onto the phase modulated component must be made in establishing a reliable phase-locked-loop.

14.4 TELECOMMUNICATION SYSTEMS

Analog signal processing, in particular, the frequency-division-multiplexing technique, was widely used in earlier telecommunication systems. In recent years, however, digital signal processing has been by far the fastest-growing technique, and will probably dominate the future telecommunication network. The reason for the rapid growth of digital transmission over the analog system has much to do with economics. The terminal cost for an FDM is considerably higher than that for a PCM, which requires a simpler circuitry, involving only an *A/D* converter and digitizer. For each voice channel, the analog speech signal is limited to a frequency range of 300 to 3400 Hz. Each voice signal is usually sampled at a rate of 8 kHz and each sample is converted into a digital "byte" involving 8 bits. Usually, the first bit gives the polarity of the signal and the remaining 7 bits provide the magnitude of the signal on a logarithmic scale. Within this framework, each voice channel requires a transmission rate of 64 Kb/s. For a time-domain multiplex system containing 24 voice channels, a bit stream of 1.54 Mb/s is required. Table 14.5 shows the standard transmission rate in the United States and Europe. The time-multiplexed channels are divided into consecutive time slots. Among these slots, at least one slot is used for the purpose of synchronization, redundancy, or transmitting the dealing number. Therefore, the total number of voice channels is slightly less than those allowed by the total bit rates of the system. These time-multiplexed blocks are classified by labeling them with T1, T2, T3, and so on, for U.S. telecommunication systems. For European networks, the standards are slightly different and correspond to the information rates shown in Table 14.5.

Transmission lines of present telecommunication systems utilize (1) twist wires, (2) coaxial cables, (3) terrestrial and satellite microwave, and (4) optical fibers. Each system has its advantages and disadvantages. This section presents a brief historical sketch of these systems.

The first transcontinental telecommunication occurred in 1915 and was transmitted on an open-wire pair. By 1930, open-wire multiplex systems with up to 12 voice channels were in widespread use. A major breakthrough was the development of the first coaxial cable transmission, which was put into service in 1940. This system was a 3-MHz system capable of transmitting 300 voice channels or a single TV channel. Since then, coaxial systems have evolved into a T4 (274 Mb/s) coaxial system, which was put into service in 1975.

Microwave links, unlike the radio waves that are reflected by the ionosphere, propagate in a line of sight. At microwave frequencies, much wider

TABLE 14.5   Standards for Digital Telecommunication Rates in the United States and Europe

Class	Bell Systems		European Systems	
	Channel	Rates (Mb/s)	Channel	Rates (Mb/s)
T1	24	1.544	32	2.048
T2	96	6.312	120	8.448
T3	672	46.304	480	34.368
T4	4032	274	1920	139.364

TABLE 14.6   Comparison between Coaxial, Microwave, and Fiber Optic
Communication Systems at a 45-Mb/s Bit Rate

System	Format	Gain (dB)	Loss (dB)	Length (km)
Coaxial (9.5-mm coax)	4B3T	90	85	7
Microwave (11 GHz)	16 QAM	109	100	50
Fiber optic (1.55 μm)	NRZ	51	45	50

transmission bandwidth becomes available. The first microwave system operating at 4 GHz, which linked New York and Boston, was put into service in 1948. It has been the workhorse of the T2 Bell systems serving coast to coast since 1951. The latest system introduced in 1973 has a capacity of carrying 12 two-way communication channels, each with approximately 1800 voice circuits.

After the breakthrough of low-loss glass fibers in 1970, the first major fiber optic T3 system was put into field trial at Atlanta in 1976 (Ref. 14.1). Since then many operating systems at rates ranging from T1 to T4 have been installed. Details on these systems are given in the following section. A comparison between coaxial, microwave, and fiber optics systems, can be made by arbitrarily selecting a 45-Kb/s digital communication system and listing the typical performance characteristics expected from each of the foregoing three technologies. Table 14.6 summarizes system performance characteristics for each of the three using state-of-the-art components. For comparison purposes the most important parameter is the maximum transmission length for each system without the need of a repeater. As shown in Table 14.6, the coaxial system is by far the shortest of the three. An optical transmission line can be made as long as a microwave link, provided that long-wavelength components ($\lambda = 1.55$ μm) are used. Looking toward the future, both the optical and microwave systems will likely continue to grow. However, it is anticipated that the growth in light-wave technology will be more significant because optical systems are considerably simpler and can be manufactured more economically in the long run. Microwave technology is now a mature field and most microwave systems are operating near their fundamental noise-limited capacity. Further improvements for these nearly ideal systems will be more difficult to achieve than for the newly emerging technology of fiber optics.

14.5 IN-SERVICE OPTICAL COMMUNICATION SYSTEMS

In the late seventies, many optical communication systems were in service both in the United States and overseas (Ref. 14.1). In the United States, American Telephone and Telegraph Corporation had installed two operating systems: one was an interoffice trunk that transmitted at a rate of 44.7

Mb/s, located within the Chicago metropolitan area; the other was an intercity transmission line service between New York City and Washington, D.C., which transmitted at a rate of 90 Mb/s. In Japan, one of the earlier projects was the Higashi-ikoma Optical Visual Information System (Hi-Ovis), which began operation in 1978, after completing a successful field trial in 1976 (Ref. 14.2). The Hi-Ovis project was sponsored by the Japanese government; its objective was to provide a two-way interactive CATV service in the model town of Higashi-ikoma, a suburb of Osaka, Japan.

Since 1977, many optical communication systems have been installed in the United States, Japan, and Europe, covering a wide range of bit rates, environmental conditions, and applications. In Europe, in-service systems are being developed at 8, 34, and 140 Mb/s. Among them, the 34-Mb/s system is especially attractive because it interfaces well with many existing networks and offers a high enough bit rate to make it cost-effective. Brief descriptions of the Chicago project, the Hi-Ovis project, and the British Post Office project are given below. These projects represent a major part of the applications that will probably occupy a large portion of the fiber optics market in the future.

The Chicago project, the first major optical fiber communication system put into service (in 1979), was designed to evaluate optical fiber technology for a wide range of Bell system service conditions. The system connected Illinois Bell customers in the Brunswick Building to the Franklin central office at a length of about 0.9 km, and also connected that office to the Wabash central office at a length of about 1.6 km. The system carried a variety of commercial traffic, including telephone channels, interoffice trunks at T3 rate, a picture phone, and a 4-MHz standard black-and-white video conference service. System parameters are listed in Table 14.7 and are very similar to those previously used in the field-trial experiment conducted in Atlanta, Georgia. The only exception is that the fibers used in the Chicago project had a lower attenuation coefficient (~4 dB/km) than those used in the Atlanta experiment. These cables were installed over a route having a total length of 2.5 km by pulling them through about 32 manholes with a total of five splices in the congested Chicago metropolitan area. Some of the manholes were partially filled with water, so this installation provided a real test for this new technology. A polyethylene duct was installed in the underground conduit for these fibers to maintain operation in a controlled environment. Cables were pulled through the duct by personnel with no special training, but the five splices were done by specially trained personnel. Results indicated that the system loss in the Chicago project was actually lower than that of the Atlanta experiment. This improvement was attributed to a decrease in both the fiber attenuation coefficient and microbending loss.

The Chicago project has been carrying commercial traffic since May 1977 without any outage. This performance has been outstanding, as indicated by a carefully monitored record. The most useful information emerging

TABLE 14.7   System Parameters Used in the Chicago Project

System	
Bit rate:	44.7 Mb/s
Total length:	2.5 km
Repeater spacing:	7 km
Channels:	144
Fiber	
Type:	Graded-index and ribbon structure
NA:	0.23
Dimension:	55-μm core, 110-μm cladding
Length:	0.66 km
Loss:	4 dB/km
Dispersion:	1.3 ns/$\sqrt{\text{km}}$
Splicing loss:	0.8 dB
Source	
Type:	AlGaAs DH laser
Wavelength:	0.82 μm
Power:	-3 dBm
Detector	
Type:	Si APD
Sensitivity:	-54 dBm

from this Chicago project is that, in dealing with the real world, repeater spacing cannot be derived simply from known parameters, such as transmitter power, fiber loss, and receiver sensitivity. The added unknown losses in cable pulling and splicing are dictated by route congestion, access restriction, and locations. These factors also play an important role in the design of a real system. Designers must allow sufficient margin to account for extremely hostile environments, especially in metropolitan areas, where underground ducts and conduits are often shared with other services, and the telephone transmission lines must withstand extremely rough treatment. In addition, water, soil, stream, hydrocarbon, and corrosive chemicals may be present in underground conduits and manholes.

Another important result that emerged from this Chicago project is the fact that the measured transmission loss of the installed system was very low. It is reassuring that very long repeater spacing is possible. The repeater is a very costly item and the number of repeaters, especially those located in manholes, must be reduced as much as possible. With low-loss fiber transmission, repeaters can either be eliminated or located in exchange centers instead of underground facilities. This trade-off is an important consideration in selecting new systems for future investment.

The Hi-Ovis project is a good representation of analog video CATV systems. This system is complex and provides computer-controlled video service to about 160 home subscribers and 8 local studio terminals located at public premises such as the city hall, schools, police and fire stations, and medical centers. The system uses approximately 400-km fibers. It consists of a 36-channel distribution network, each of which has a length of 6 km. In addition, there are 24 fibers 400 m in length, 18 fibers 500 m in length, 6 fibers 5.5 km in length, 4 fibers 1.5 km in length, and 2 fibers 31 km in length. The longest transmission distance for analog video and FM audio signals is about 4 km between a UHF receiving station located on top of Ikoma mountain and the control station in Higashi-ikoma. The system does not use repeaters.

The major feature of this system is the interactive programming between the users and the central station. At the station, a video switching network contains 32 inputs and 168 outputs. Two out of 32 inputs are used for video signal transmission from the home terminal video cameras to the station and the remaining 30 inputs are used for video signal transmission from the center to TV monitors of the subscribers. Eleven cables are distributed from the station, each of which can serve 16 subscribers with a two-way communication linkage involving two fibers. The components used in the Hi-Ovis project are: LED sources with an average output of 1 mW, plastic-clad fibers with an NA of 0.25 and a loss of 16 dB/km, pin photodiodes

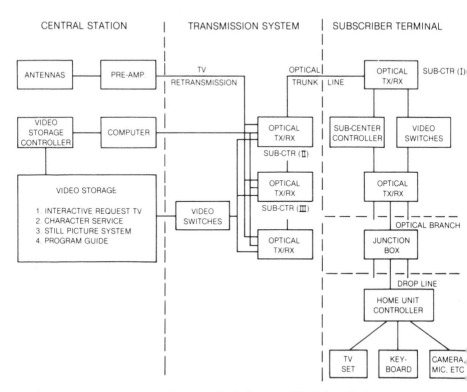

FIGURE 14.5   Block diagram of Hi-Ovis project.

with a capacitance of 2.5 pF, a quantum efficiency of 0.8, a time constant of 10 ns, and a S/N ratio of 56 dB for a video signal transmitted over a distance of 0.5 km.

Figure 14.5 is a block diagram of various components involved in the Hi-Ovis system. It contains basically three subsystems. The first is the front-end equipment, which includes a broadcast network and a computer controlled electronics to handle service requests of various subscribers. Services include alphanumeric data, movies, educational programs, local news, community events, and shopping information. The second is the information transmission subsystem, which consists of a video switching network designed to handle various services being offered. A shared frequency-division-multiplexed architecture is employed. To increase the system's bandwidth capability, a large number of fibers is used. The third subsystem is the subscriber equipment, which consists of a terminal controller, keyboard and TV receiver, video camera, and microphone. The drop cable to a subscriber's home consists of two optical fibers that connect directly to the distribution box. The optical transmitter and receiver were installed in a splicing box. The commercial market for interactive service at home is still in the distant future because the installation cost of the home terminal subsystem is very high. This situation could change when mass production of these items occurs.

TABLE 14.8   System Components and Parameters of British Post Office Field-Trial Experiments

	Bit Rate	
	8 Mb/s	140 Mb/s
System length	13 km	8 km
Telephone channel	120	1920
Source	LED	LD
Wavelength	0.82 μm	0.84 μm
Injection current	300 mA	30 mA
Output	65 μW	1 mW
Photodiode	APD	APD
Amplifier	Si J-FET	GaAs FET
Capacitance	10 pF	6 pF
Gain	40	80
AGC	25 dB	29 dB
Logic gate	Dual D-type bistable	D-type bistable
Receiver sensitivity	−59.7 dBm	−43 dBm
Fiber	Corning graded index	Corning graded index
Loss	4.5 dB/km	4.5 dB/km
Installation	Existing ducts	Existing ducts
Splicing	V-groove	V-groove

A field-trial experiment conducted by the British Post Office (Ref. 14.3) involved two 8-Mb/s optical fiber transmission lines over a 13-km route using LED and LD sources, and a 140-Mb/s optical fiber transmission line over a 8-km route using laser diodes. These routes were between the BPO Research Center at Martlesham Heath and the Ipswich Telephone exchange in Suffolk, England. The system parameters are listed in Table 14.8. The receiver sensitivity of the 8-Mb/s and the 140-Mb/s systems for an error rate of 10^{-9} is -60 dBm and -43 dBm, respectively. For the 8-Mb/s system where a LED is used, a repeater is placed in the middle of the 13-km path. No repeater is needed for both systems where an LD is used. At 140 Mb/s, results indicated that repeater spacing of at least 8 km was possible with the commercially available Corning graded-index fibers. The spacing could be extended by using lower-loss fibers and could be significantly extended by using components operating at 1.55 μm.

14.6 OPTICAL LOCAL AREA NETWORKS

The ever-increasing demand for vast data transmission and communication by automated offices, industrial facilities and medical centers, etc., can easily be met by optical Local Area Networks (LANs). Optical fiber LANs can provide a few hundred (200–300) megabits per second data over a span length up to a kilometer and are more than adequate to meet this demand. The most widely accepted topology for a LAN is the passive star configuration shown in Figure 14.6. Passive stars are more reliable than active stars and also offer greater flexibility without losing a significant amount of power. A fundamental limitation of a LAN is in the detection of a collision in the system where data streams are very congested and large differences exist in signal strength from various users. This variation in signal strength can cause problems. For example, a stronger signal can mask out a weaker signal. When this masking occurs, the receiver must be able to detect such a collision and redirect the data traffic. This problem is even more pronounced in a data bus topology because of the losses incurred in tapping off a bus.

Many methods have been suggested for collision detection in LANs. Three of the more relevant techniques are: (1) code rule violation collision detection (Ref. 14.4); (2) time-delay violation collison detection (Ref. 14.5); and, (3) sequence weight violation collision detection (Ref. 14.6). The first method employs a partial response code violation rule using the Manchester code, which offers an 8 dB dynamic range. The second technique requires each transmitter to have a priori knowledge of how long it takes for an optical signal to loop through a star coupler and return. The major drawback of this method is that the probability of collision detection usually is not 100%, even for strong colliding signals. The third method operates on a sequence weight

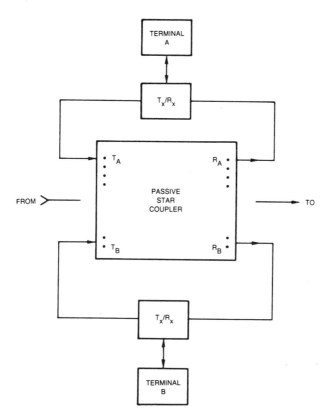

FIGURE 14.6   A passive star configured optical local area network (LAN).

violation rule using a cyclic error-correcting code. Each transmitter has its own unique sequence and all sequences have identical Hamming weight. With this method, a dynamic range of 17 dB is possible for a transmitter with an extinction ratio ϵ of 100 by using an APD receiver with a fixed threshold. Because of its detection accuracy, large dynamic range, and simplicity, the third method is the only one described in detail.

A block diagram of the optical LAN that uses a carrier sense multiple access with a sequence weight violation collision detector is shown in Figure 14.7. Every transmitter T_x is connected to a fiber via a LED and subsequently to the passive star coupler. The coupler output port is connected via a fiber to an APD at the receiving end. The choice of an APD affords a higher gain to increase the dynamic range of the receiver R_x. The APD output is digitized after the bit time synchronization is established and processed by a bit counter, which recognizes the occurrence of a collision. If a collision has occurred, the network is normally jammed and the transmitter is instructed to back off and retransmit at a later time. Otherwise, the system operates normally without interruption.

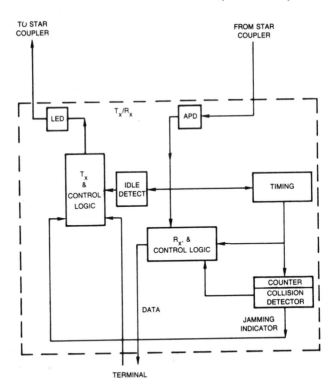

FIGURE 14.7   Collision detection circuit block diagram.

The Sequence Weight Violation rule assigns a short sequence, e.g., 100 bits to every packet of a few thousand bits length. The sequences are designed to have a given Hamming weight W_0 and a given minimum distance d_m between each other. Each station has a unique sequence placed in every packet transmitted from that station. A collision can occur only within the maximum end-to-end propagation delay T_m. Once the period T_m is over, normal transmission by a station commences. For this protocol to be effective, the total packet length that covers the T_m interval plus an extra length of one code work should be much longer than T_m. With this sequence, the Hamming weight, once the collision occurs, can be considerably larger than the normal weight of each sequence. Hence, by monitoring the packet header, a collision can easily be detected by counting the weight of the received sequence and comparing it to a collision weight threshold W_t. If the threshold is exceeded, a collision is declared.

The key feature of this method is that each transmitter is assigned with a unique sequence to each packet, which follows immediately the bit and word timing sequence in the packet header. The transmitting station also has full knowledge of the end-to-end delay through the star coupler. A colsion can only occur within a certain interval of time delay between two colliding packets. Within the time window of vulnerability, a collision involving a weak packet for a binary transmission system is identified when a minimum number of "ones" are visible in the received "zeros" of the strong packet. Otherwise, the weak packet is completely masked out and cannot be detected.

The proper way to construct a collision detection sequence is to use the theory of binary cyclic error-correcting code. It consists of words from a binary linear cyclic error-correcting code of length N bit intervals and a minimum Hamming distance d_m. By definition, all cyclic permutations of any code word in a cyclic code is also a code word in the same code. A collision is recognized if the code word is repeated at least once. The number of repeated code words and the code word length should be chosen so that the overlap within the collision detection window time equals at least the length of one code word. In this scheme, the collision detection sequences consist of at least two repeated identical code words.

To illustrate what happens in a collision, the bit synchronous collision at an otherwise arbitrary relative time delay is depicted in Figure 14.8. Note that the sequence is actually a part of the packet header, although it is shown separately from the header in Figure 14.8, for illustration purposes. Collision can, of course, happen at any relative time delay within an interval ΔT. For simplicity, this example deals with bit synchronous collision only. Thus, a collision between two packets occurring at any discrete bit time shift yields a full bit hit. Furthermore, assume that word C is within the time interval of words A and B. Let word A be a code word of Hamming weight W_0 from a binary cyclic error-correcting code with code word length N and a minimum Hamming distance d_{min}. Word B is identical to word A. Let word C belong to the same cyclic code as word A also with a Hamming weight W_0, but different from A and any cyclic permutation of A. Word D is identical to word C.

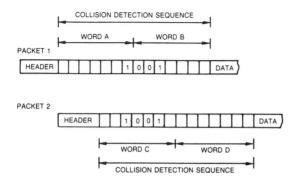

FIGURE 14.8 Collision with full bit overlap. (Ref. 14.6. Reprinted with permission of IEEE.)

By the rule, it is possible to guarantee a worst case separation between collision sequences at any relative time delay. Because A and C are words in the same cyclic code, the word C and any cyclic shift A are different in at least d_{min} positions for any delay difference in the specified time window. This situation follows from the fact that because A and B are identical words, the N bits covered by word C are always a cyclic permutation of word A.

If no collision occurs, the noiseless weight count over words A and B is $2W_0$. When a bit synchronous collision (full hit) occurs, the minimum number of "on" pulses contributed by the colliding packet sequence is

$$W_c = \frac{d_{min} + 1}{2} \tag{14.2}$$

which is the nearest integer smaller than or equal to $(d_{min} + 1)/2$. Thus, the weight count in the collision detector is at least $2W_0 + W_c$. The parameter W_0 is common to all sequences and known to all receivers. The positions in the two code words can be different in two ways, e.g., "1" in word C and "0" in word C_p (cyclic permutation) or "0" in word C and "1" in word C_p. Only the first type contributes to the weight count in the collision detector. Due to the symmetry in a constant weight code word case, equally as many of each type of difference exists. For an even minimum Hamming distance, this number is at least $d_{min}/2$ and for an odd minimum Hamming distance, this number is at least $(d_{min} + 1)/2$.

In the case of a fractional bit hit (non-bit synchronous collision), the time delay difference between the two packets is not an integer multiple of a bit time interval and a collision occurs in a fractional bit time interval. Two types of fractional bit hits contribute to the bit count in collision detection, namely, a "1" in word C overlapping ΔT of a "0" in word C_p and another "1" in word C overlapping the remaining part $T - \Delta T$, of a "0" in word C_p. At least $(d_{min} + 1)/2$ of each of these two types exist. Sometimes the two fractional types of hits appear in the same bit word, making it a full bit hit rather than two fractional bit hits.

Next, suitable sequences are selected by using the properties of binary cyclic error-correcting block codes (Ref. 14.7). The length of each sequence is given by the maximum relative propagation time difference in a collision and the system bit rate. At a given sequence length of N bits, two somewhat contradictory requirements are on the sequences. The largest possible minimum distance d_{min} between sequences should be chosen. However, a sufficiently large number of sequences N should be available in the network so that each user has his own unique, constant weight sequence. To obtain the largest number of sequences N at a given d_{min}, the best choice of sequence weight W_0 is $N/2$. In other words, only one of the cyclic permutation of a given code word of weight W_0 is a useful sequence. The weight distribution of the cyclic code is $A(j); j = 0, 1, \ldots, N$ where $A(j)$ is the number of code words of Hamming weight j. Thus, the number of useful sequences

M of weight W_0 is lower bounded by $A'(W_0)$ where

$$M \geq \frac{A(W_0)}{N} = A'(W_0) \tag{14.3}$$

Many different types of cyclic error-correcting block codes exist, e.g., Golay, Hamming, BCH, Repeat Golay, etc. For more information, consult References (14.6) and (14.7).

The performance of the sequence weight violation collision detection method can be analyzed by calculating the probability of failure P_F to detect a collision when a collided weak signal is masked out by a strong signal at a given receiver and the probability of a false alarm. For a binary transmission system with the bit streams in [0, 1] sequences and with a threshold setting γ, the average error probability is given by

$$P_{\text{error}} = \tfrac{1}{2}\,[p(0) + p(1)] \tag{14.4}$$

where $p(0)$ and $p(1)$ are the bit error probabilities for a transmitted "0" and a transmitted "1", respectively. For this case,

$$p(0) = \int_{-\infty}^{\infty} f_\zeta(z, 0)\text{Erfc}\left(\frac{\gamma - z}{\sigma_i}\right) dz \tag{14.5}$$

$$p(1) = \int_{-\infty}^{\infty} f_\zeta(z, 1)\text{Erfc}\left(\frac{z - \gamma)}{\sigma_i}\right) dz \tag{14.6}$$

where f_ζ is the conditional density function of normalized decision variables and Erfc(Q) is the error function as defined by Equation (13.35). To relate these functions to system parameters, denote the statistical average of ζ by $\langle\zeta\rangle = G\langle N_S + N_D\rangle$, where G is the average gain of the APD, N_S is the number of primary signal electrons satisfying the Poisson distribution, and N_D is the corresponding count in the absence of the signal or the dark current. The mean variance σ_i is given by

$$\sigma_i^2 = G^2\left[kG + \left(2 - \frac{1}{G}\right)(1 - k)\right]N_i + \sigma_n^2 \tag{14.7}$$

where $i = 1, 2, 3, 4,$ and 5, k is the ionization rate, and σ_n is the variance of an additive white Gaussian noise. Five different counts N_i are in the code sequences, which are described below.

The number N_S denotes the average number of primary electrons in a strong "1", and is the reference point for the system. Weaker signals in "1" are defined by means of fractions of N_S, e.g., αN_S ($\alpha \leq 1$), and a strong "0" is expressed by N_S/ϵ, where ϵ is the extinction ratio, which is the ratio of the number of N_S contributed by transmitting "1" over that for transmitting "0". The dynamic range in decibels is defined by

$$R_D = 10\log_{10}(1/\alpha)\text{ dB} \tag{14.8}$$

Figure 14.9 depicts two incidences of bits from colliding sequences over lapping a "0" of the reference sequence. The average number of primary electrons, N_1, contributed by the partially overlapped "1" bit (see Figure 14.9(a)) is

$$N_1 = \frac{\tau}{T} \alpha N_S + \frac{T - \tau}{T} \alpha \frac{N_S}{\epsilon} + \frac{N_S}{\epsilon} + N_D \qquad (14.9)$$

where T is the bit period and τ is a fraction of T during which the colliding "1" bit overlaps the reference "0" bit. The second type of collision incidence (see Figure 14.9(b)) involves the number of electrons, N_2, contributed by the "1" bit of the colliding sequence, which is

$$N_2 = \frac{\tau}{T} \frac{N_S}{\epsilon} + \frac{T - \tau}{T} \alpha N_S + \frac{N_S}{\epsilon} + N_D \qquad (14.10)$$

In the case that $\tau = T$ or $\tau = 0$, the collision incidence becomes a full hit. The number of primary electrons involved in this case is

$$N_3 = \alpha N_S + \frac{N_S}{\epsilon} + N_D \qquad (14.11)$$

which is the number for calculating the probability of error in detecting a colliding "1" bit. The number of primary electrons in the strong "0" bit of

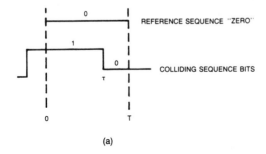

(a)

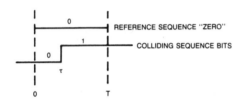

(b)

FIGURE 14.9 Two incidences of collision. (Ref. 14.6. Reprinted with permission of IEEE.)

the same sequence is

$$N_4 = \frac{N_S}{\epsilon} + N_D \tag{14.12}$$

which is the number for calculating the bit error probability for transmitting an uncollided strong "0". Finally, the number of primary electrons in a weak uncollided "1" is

$$N_5 = \alpha N_S + N_D \tag{14.13}$$

Figure 14.10 illustrates the density function associated with the decision variable ζ for a "1" and a "0" of the uncollided strong packet shown by the two solid curves. Also shown in Figure 14.10 are four dashed curves representing four density functions corresponding to a collision between a strong and weak packet. The separations δ_1 and δ_2 are directly related to the extinction ratio ϵ and the signal power level difference (or dynamic range) between a weak and strong packet.

The above definitions for various incidences of collision help to write the probability of failure to detect a collision. For the special case of full bit hits only, the probability of failure to detect a collision is

$$P_F = \sum_{i=W_c-W_T}^{W_c} \binom{W_c}{i} p_3^i (1 - p_3)^{W_c - i} \tag{14.14}$$

where W_T is the weight threshold and

$$p_3 = \text{Erfc} \left(\frac{N_3 G - \gamma}{\sigma_3} \right) \tag{14.15}$$

σ_i can be determined using Equation (14.7) and γ is an arbitrarily chosen

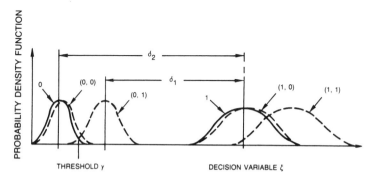

FIGURE 14.10 Probability density function of decision variable for an uncollided strong packet (solid curves) and for a collision with a weak packet (dashed curves). (0,1) denotes a strong "0" colliding with a weak "1", etc. (Ref. 14.6. Reprinted with permission of IEEE.)

threshold value for a desired bit error probability, e.g., $p(0) = 10^{-9}$. To a good approximation,

$$p(0) = \text{Erfc} \left(\frac{\gamma - N_4 G}{\sigma_4} \right)$$

For $p(0) = 10_g^{-9}$ $(\gamma - N_4 G)/\sigma_4 = 6$. Therefore,

$$\gamma = 6 \left[\frac{N_S}{\epsilon} \Gamma^2 + \sigma_n^2 \right]^{1/2} + \frac{N_S}{\epsilon} G \qquad (14.16)$$

where

$$\Gamma^2 = G^2 \left[kG + \left(2 - \frac{1}{G} \right)(1 - k) \right] \qquad (14.17)$$

Note that Equation (14.16) is obtained by assuming N_D is zero. The limit on α can be found by using γ from Equation (14.16) and assuming

$$p(1) = \text{Erfc} \left(\frac{N_5 G - \gamma}{\sigma_5} \right) = 10^{-9}$$

thus,

$$\alpha = \frac{1}{\epsilon} + A + \frac{18 N_S \Gamma^2}{N_S^3 G^2} + \left[\left(\frac{1}{\epsilon} + A + \frac{18 N_S \Gamma^2}{N_S^3 G^2} \right)^2 - \left(\frac{1}{\epsilon} + A \right)^2 + \frac{36 \sigma_n^2}{N_S^2 G^2} \right]$$

$$(14.18)$$

where

$$A = 6 \left[\frac{\frac{N_S}{\epsilon} \Gamma^2 + \sigma_n^2}{N_S^2 G^2} \right]^{1/2} \qquad (14.19)$$

Equation (14.18) shows that $\alpha > 1/\epsilon$. Thus, the value of the extinction ratio is a very important parameter for obtaining the desired dynamic range.

For example, system performance can be evaluated using LED with an extinction ratio $\epsilon = 100$ and an APD with $G = 40$ and $k = 0.1$. N_S is assumed to be 250,000. A Golay code is used and has the following parameters: $N = 23$, $W_0 = 12$, $A' = 56$, $W_c = 4$, and $W_t = 1$. The white Gaussian noise is assumed to be $\sigma_n^2 = 36 \times 10^6$, which corresponds to a rather noisy amplifier. The receiver decision threshold level γ is set so that the bit error probability $P(o)$ for received "0" with N_S/ϵ primary electrons is 10^{-9}. By using the Gaussian approximation, the probability of failure P_F to detect a collision can be calculated as a function of the dynamic range $1/\alpha$ in dB. Figure 14.11 shows, in addition to the case $\tau = 0$ and $\epsilon = 100$, the case $\tau = 0.5T$, $\epsilon = 100$; and, $\tau = 0.5T$, $\epsilon = 20$. The $\tau = 0.5T$ case corresponds

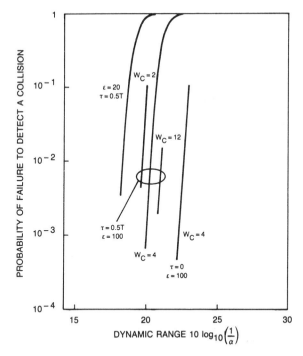

FIGURE 14.11   Probability of failure to detect a collision as a function of dynamic range for $G = 40$, $k = 0.1$, $W_T = 1$, $P_E = 10^{-9}$, $\sigma_n = 3.6 \times 10^{-6}$ and $N_S = 250,000$. (Ref. 14.6. Reprinted with permission of IEEE.)

to the worst case of collision detection. Figure 14.11 also shows the dependence of P_F on W_c, which determines d_{min}, the Hamming distance of the code generating the sequences. Variations of other parameters, e.g., W_T, G, N_S, σ_n^2, and γ result only in minor changes (within 1 dB) in the dynamic range value.

Figure 14.12 illustrates the variation of P_F with N_S in dB referred to 1 mW for $\tau = 0.5T$, $\alpha = 0.0009$, and $\tau = 0.5T$. $\alpha = 0.0006$ at a traffic rate of 50 M bit/s. In all cases, the key parameter for obtaining a large dynamic range is the optical power in terms of N_S. At a 10^{-9} probability of error level, the maximum dynamic range for $\epsilon = 100$ is 17 dB, and for $\epsilon = 20$, $R_D = 11.8$ dB using the system parameters $G = 40$, $k = 0.1$, $N_D = 5$, and $N_S = 250,000$. From Figure 14.11, it is clear that by using the sequence weight violation collision detection technique, system performance is limited by the uncollided weak packet rather than the collision detector. The system can tolerate far wider optical dynamic range variations than other available detection techniques. Collisions can be detected by any user's receiver throughout the network and can be subsequently broadcast to other users.

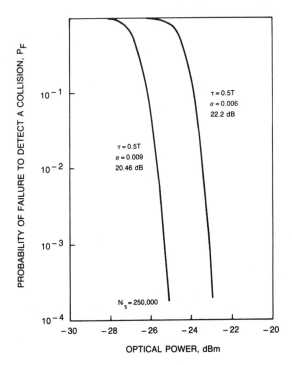

FIGURE 14.12   Probability of failure to detect a collision as a function of optical power in dBm for a data rate of 50 Mbit/s. (Ref. 14.6. Reprinted with permission of IEEE.)

Because this detection method is distributed and robust, the network can operate without signal jamming.

14.7 LONG-HAUL SYSTEMS

Design considerations for long-distance communication systems are different from those for interoffice and intracity trunks. Long-haul optical fiber communication systems require repeaters at regular spacings, which are determined primarily by transmitter power, receiver sensitivity, bandwidth, fidelity, and system losses. Their values are derived from parameters involving fiber attenuation and dispersion characteristics, system noise, signal distortion, and intersymbol interference. Figure 1.5 gives the ranges in which repeater spacing for various types of systems can be realized. The BER for a typical telecommunication system is 10^{-9}. For systems involving a large number of repeaters, the accumulated BER can be very high.

    For example, consider an optical repeater utilizing state-of-the-art components for a long-haul and high-data-rate (>1-Gb/s) communication link to be used in a transoceanic transmission system. Several systems are presently under development both in this country and overseas (see Table 14.9). Of most importance is system reliability for transoceanic operation. This requirement needs very careful selection of optical fiber components with proven performance and operating life. To achieve maximum repeater spacing, it is necessary to select a very low dispersive single mode fiber operating at 1.3 μm and having a loss figure at ≤0.5 dB/km. At this wavelength, fiber dispersion can be kept to less than a few ps/km-nm. An InGaAsP semiconductor laser, which is driven by a GaAs FET, must serve as the transmitter. The laser can be directly modulated in the PCM format by a pulse signal current superimposed on a dc pre-bias current. The modulated output is then coupled into the single mode fiber. Special cabling technology is required for transoceanic operation, where cable is under tremendous tension (≥10 tons) during the process of laying or raising it from a depth of about 6000 m. It is important to design the cable with a strength that can survive the undersea environment without deformation and breakage. The 1.3-μm detectors for the receiver must be either Ge APD or InGaAs APD, followed by an amplifier having a low-noise FET front-end. An InGaAs APD is expected to have a higher sensitivity than a Ge APD; however, InGaAs APDs are still in the research and development stage and not available commercially at the time when these projects were conducted.

Experimental results (Ref. 14.8) show that at a BER of 10^{-9}, Ge APD receiver sensitivity is -31.9 dBm, which is lower than the InGaAs APD sensitivity for a 2-Gb/s RZ system over a 44.3-km transmission length. The RZ signal waveform in general results in about 1 dB improvement in receiver sensitivity when compared to NRZ signals at the same data rate. To extend the repeater spacing of a 2-Gb/s communication system beyond 45 km, it is necessary to reduce fiber loss further, to below 0.3 dB/km, and also to decrease dispersion loss to less than 3 ps/km. To reach this level of performance, a zero-dispersive fiber with extremely low-loss (≤0.2 dB/km) must be used. Such a low-loss transmission line has been demonstrated

TABLE 14.9   System Design Parameters of Undersea Optical Fiber Systems

	United States (Bell System)	Japan (NTT)
Transmission length	8000 km	1000–10,000 km
Depth	7.5 m	8 km
Bit rate	274 Mb/s	260–400 Mb/s
Wavelength	1.3 μm	1.3 μm
Reliability	8 years	10 years
Repeater spacing	25–50 km	25–50 km
Power consumption	4 W	3–5 W
Fiber loss	<1 dB/km	<dB/km
Fiber strength	8 tons	7.5–10 tons

using an LD operating at 1.55 μm (Ref. 14.9). Fiber optic components operating at 1.55 μm are considered to be the most attractive for long-haul and high-data-rate systems; however, the technology at this wavelength must be developed, especially in the detector area, because the performance of a Ge APD degrades rapidly beyond 1.3 μm.

14.8 COHERENT LIGHT-WAVE COMMUNICATION SYSTEMS

With recent advances in optical fiber component technologies, coherent light-wave communication systems are becoming more attractive for long-haul transmission and wideband data distribution. Two major advantages that a coherent system offers are: (1) the receiver sensitivity of a coherent system can be improved significantly so that either the bit rates or repeater spacing can be greatly increased, and (2) the system is suitable for fine-grain wavelength multiplexing. However, coherent systems are relatively more complex than incoherent systems using direct detection. A typical coherent system, shown in Figure 14.13, consists of a laser transmitter with very high spectral purity and stability, a dispersion-free single mode fiber, and a coherent receiver that employs a local oscillator and a photomixer for heterodyne detection. An ideal coherent receiver needs only about 10 photons to recover 1 bit of information with a 10^{-9} BER. Such high receiver sensitivity may be compared to that of an incoherent receiver, which requires a minimum of 500 photons to recover 1 bit of information at the same BER.

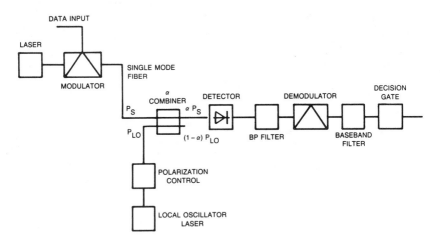

FIGURE 14.13 A coherent lightwave communication system whose receiver consists of a variable combiner of the transmitted and LO signals. (Ref. 14.10. Reprinted with permission of IEEE.)

As the bit rate increases, the incoherent system degrades at a much faster rate than the coherent system.

The improvement in receiver sensitivity of a coherent system is achieved by using a photomixing process between a very weak transmitted signal and a strong local oscillator signal in a square-law photodetector. This process, which is known as heterodyning, provides a conversion gain that allows the recovery of a very weak signal from the noise of the receiver. Let E_s and E_{LO} be the magnitudes of the electric fields of the signal and local oscillator incident on the surface of the photodiode. Each of the fields has its respective frequency. The number of electron-hole pairs generated at a depth x from the surface can be written as

$$N(x) \propto (E_s \cos \omega_s t + E_{LO} \cos \omega_{LO} t)^2 e^{-\alpha x} \tag{14.20}$$

where α is the absorption coefficient of the semiconductor. When the square term is expanded, the following results:

$$\tfrac{1}{2}E_s^2 + \tfrac{1}{2}E_{LO}^2 + E_s E_{LO} \cos (\omega_s - \omega_{LO})t$$

$$+ E_s E_{LO} \cos (\omega_S + \omega_{LO})t + \tfrac{1}{2}E_s \cos 2\omega_s t + \tfrac{1}{2}E_{LO} \cos \omega_{LO} t$$

Within the bandwidth B of the photodiode, only the first three terms are of interest. The first two terms are composed of the DC component

$$N_{dc} \propto (E_s^2 + E_{LO}^2)e^{-\alpha x} \tag{14.21}$$

and the third term is the mixing (IF) component

$$N_{\Delta \omega} \propto (2E_S E_{LO} \cos \Delta \omega t)e^{-\alpha x} \tag{14.22}$$

These results indicate that the ac (IF) component occurs at the difference frequency $\Delta \omega$ in the presence of the DC background. The peak AC current I_{ac} can be expressed by

$$I_{ac} = \frac{2E_S E_{LO}}{E_{LO}^2 + E_S^2} I_{dc} \tag{14.23}$$

if P_{LO} is sufficiently large so that the shot noise generated by the local oscillator dominates over all other noises. In this case,

$$I_{dc} = \eta e P_{LO}/h\nu \tag{14.24}$$

and the noise of the detector is given simply by Equation (12.35) as

$$N = 2e I_{dc} B \tag{14.25}$$

where B is the detector's bandwidth. In this case, the S/N ratio for shot-noise-limited operation is

$$\frac{S}{N} = 2 \frac{P_S}{P_{LO}} \frac{I_{dc}^2}{2e I_{dc} B} = \frac{\eta P_S}{h\nu B} \tag{14.26}$$

Equation (14.26) indicates that with sufficient P_{LO}, a heterodyne receiver is capable of detecting a single photon per unit bandwidth if $\eta = 1$. In practice, other limitations will degrade the conversion gain of a coherent receiver and will be discussed later. First the minimum P_{LO} that is required to overcome the thermal noise power generated at the 50 ohm input resistor of an amplifier is examined. Equation (12.44) yields

$$I_S I_{dc} R \geq 4kTB \qquad (14.27)$$

Substituting $I_s = 2eB$ and $I_{dc} = \eta e P_{LO}/hv$ into Equation (14.27) gives

$$P_{LO} \geq \frac{2kThv}{e^2 \eta R} = 0.43 \text{ mW}$$

for $T = 300°K$, $\eta = 1$, and $v = 3 \times 10^{14} \text{ sec}^{-1}$.

In practice, the anticipated heterodyne receiver gain as indicated by Equation (14.26) cannot be realized for many reasons. First, a much greater receiver bandwidth (by at least a factor of two) is required for heterodyne detection than that for direct detection. Consequently, the circuit noise, which is f^2-dependent in the heterodyne receiver, is much higher than that for the equivalent baseband direct receiver. In principle, this problem can be avoided if local oscillator power is increased to the required level to overcome the circuit noise. However, as local oscillator power increases, excess noise can be introduced into the system. To operate a coherent receiver within 1 dB of the quantum limit, it is necessary to raise the local oscillator power to a level so that the shot noise is equivalent to at least four times the circuit noise of the receiver. For a 1.5 micron receiver with a 50 Ω photodiode load and low-noise wideband amplifier, at least 10 mW of local oscillator power would be required. Second, the effect of nonlinear phase response in a coherent system can severely degrade system performance as a result of the finite laser linewidths. Clearly, the design of wide bandwidth coherent communication systems is influenced by several key factors:

1. the availability of local oscillator power
2. the excess noise generated by the local oscillator
3. the circuit noise
4. the laser linewidths
5. modulation and demodulation techniques

A range of system design options can be pursued either individually or in combination to increase the overall capability of a coherent transmission system. First, it is possible to use wavelength multiplexing (WDM) of several moderate (0.5 G-bit/s) data rate channels. This method would avoid the problem of a wide bandwidth receiver, high-speed modulators, and chromatic dispersion. However, with a multiple channel system it is necessary to consider both cross-talk from nonlinear transmission effects, such as Raman scattering and spurious interference signals produced by the coherent detection process. Three types of interference signals are recognized: direct detection terms at baseband, adjacent channel cross-products, and image band signals. If channel spacing is done properly, the penalties from these products can be avoided. For a simple heterodyne receiver, channel spacing should be three to five times the total channel bandwidth to avoid all unwanted components. Channel spacing may be reduced to only two times the channel bandwidth if a balanced heterodyne receiver that involves two detectors is employed.

Other means can be used to avoid the problem of receiver bandwidth. One is homodyne detection and the other is phase diversity reception. Homodyne detection requires normal direct detection receiver bandwidth and can provide an additional 3 dB improvement in receiver sensitivity over heterodyne detection. But a problem associated with the homodyne receiver is the difficulty involved with phase-locking a local oscillator laser at a remote location to a low-level modulated transmitting signal because both of these lasers are independent. Various techniques for phase-locking loops have been proposed. More details can be obtained in Ref. 14.10. Phase diversity reception is another technique that has been developed for receivers operating at microwave frequencies and can be applied to optical systems. It is a scheme by which a local oscillator laser is designed to operate at a frequency comparable to the incoming signal frequency, but is not phase-locked to it. One of the problems in realizing phase diversity receivers is in the imperfection of electrical demodulation. A slight imperfection in the square-law mixer can produce extraneous terms that appear at the baseband along with the recovered signal. The following discussion shall focus only on the design of a broadband heterodyne receiver.

Figure 14.13 shows a standard heterodyne receiver where a directional coupler of variable ratio α is used to combine an input signal and a local oscillator. The available local oscillator power on the detector is $P_{LO}(1 - \alpha)$ and the signal power is $P_s\alpha$. The system penalty relative to shot-noise-limited detection as the coupler ratio α is varied is shown in Figure 14.14. It demonstrates a compromise between local oscillator and signal path loss. The results of Figure 14.14 assume zero excess intensity noise introduced by the local oscillator laser power. Any excess intensity noise will become more pronounced as the local oscillator laser power incident on the detector increases beyond a level that is lower than the shot-noise-limited operation. Therefore, a modification in the choice of the optimum coupler ratio is necessary. Excess noise can be particularly important in multi-gigabit systems because it is likely to be a maximum at the lasers' relaxation oscillation frequencies, which fall within the IF passband.

A heterodyne receiver with low circuit noise at the IF requires very small amounts of local oscillator power (<1 mW). One way to reduce circuit

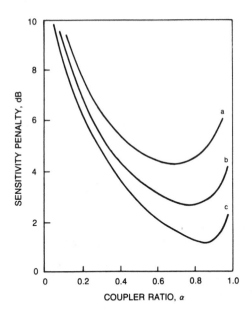

FIGURE 14.14 Choice of coupling ratio for a single detector receiver (10 mW local oscillator power, zero excess noise, 1.5 μm wavelength). (a) 50 Ω receiver with 10 mW LO power. (b) 10 pA/$\sqrt{\text{Hz}}$ circuit noise receiver with 10 mW LO power. (c) 5 pA/$\sqrt{\text{Hz}}$. (Ref. 14.10. Reprinted with permission of IEEE.)

noise is to use an inductor to resonate with the photodiode-amplifier input capacitance at the IF. This approach has been used (Ref. 14.11) to reduce circuit noise down to the 10 pA/$\sqrt{\text{Hz}}$ level while providing an 8 GHz bandwidth. At this level, the required local oscillator power is about three times less than the receiver using a 50 Ω input impedance corresponding to an equivalent circuit noise of 18 pA/$\sqrt{\text{Hz}}$. To make the best use of available local oscillator power without introducing unnecessary attenuation in the signal path, a technique has been introduced where a two detector or balanced receiver is used, as shown in Figure 14.15. In this arrangement, the local oscillator signal is combined with the input signal in a 3 dB directional coupler and the signal and local oscillator power is coupled to the two detectors. This coupling allows all of the local oscillator power to be used without introducing loss in the signal path. Furthermore, any excess noise from local oscillator is cancelled. At high frequency, it is necessary to achieve a good balance, not only for the gain of both arms but also the time delay. By this technique, a further reduction of the circuit noise from 10 pA/$\sqrt{\text{Hz}}$ level may be possible.

For example, a 2 Gbit/s optical frequency-shift key FSK heterodyne transmission system (Ref. 14.11) at λ = 1.52 microns is described here. This system has several distinct features including a directly modulated DFB laser transmitter that is frequency-stabilized to an etalon to reduce FM jitter, and a receiver that employs microwave matching techniques to achieve a bandwidth of 8 GHz. The system configuration is shown in Figure 14.16. The

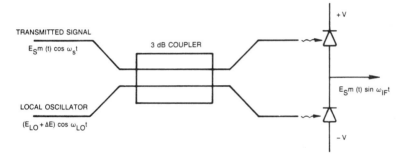

FIGURE 14.15   A schematic of a balanced receiver in which ΔE is excess noise. (Ref. 14.10. Reprinted with permission of IEEE.)

transmitter consists of a 1520 nm DFB laser diode with a linewidth of 12 MHz at 5 mW output power. A Continuous-Phase Frequency Shift Key (CPFSK) modulation with a frequency deviation of 1.4 GHz corresponding to a modulation index $m = 0.7$ is obtained by direct modulation of the laser injection current. The transmitter laser power of 1 dBm is coupled into a 101 km single mode fiber through two optical isolators with a total isolation of 60 dB. A fraction of the power is diverted through a 10:1 fiber coupler to a confocal etalon for the purpose of frequency locking the DFB laser. By this stabilization loop, a 20 dB reduction in FM noise is achieved for frequencies below 10 kHz.

The local oscillator is a tunable grating loaded external cavity laser. The received LO power $P_{LO} = 3.5$ dBm is obtained using a tapered lens fiber to couple light from the rear facet of the laser. An intracavity etalon maintains stable single mode operation. The outputs of these two lasers are combined using a variable directional coupler. An adjustable polarization controller following the LO is used to match the state of polarization of the two lasers.

The receiver consists of an InGaAs p-i-n photodiode connected to a three-stage wideband high impedance preamplifier. The receiver design employs interstage microwave matching circuitry to extend the receiver's bandwidth to 8 GHz. An equivalent input noise current of less than 12 pA/$\sqrt{\text{Hz}}$ has been achieved in the IF band (3.2 − 6.4 GHz) by reducing total input capacitance to 0.6 pF and by inductive peaking at the front-end. The average circuit noise is about 10 pA/$\sqrt{\text{Hz}}$, which accounts for the thermal noise from the input bias resistor $R_b = 5$ kΩ and the FET channel noise. An IF center frequency of 4.6 GHz is maintained by means of an AFC loop to a piezoelectric transducer controlling the grating angle of the LO. The IF signal is filtered using a microwave circulator with a bandpass of 3.3 GHz. A delay line discriminator that consists of a double-balance mixer with a differential delay between the two input ports of 270 ps is used to demodulate the CPFSK signal. The baseband filtering of the heterodyne signal is achieved by amplifying the down-converted frequency responses from the mixer with a baseband amplifier.

Because of a nonlinear phase of the FM response of the directly modulated semiconductor laser, an ideal square FM pulse-shape will be distorted at the system's output. This distortion results in a serious degradation in system performance in terms of receiver sensitivity and bit rate. Using state-of-the-art components, the system is restricted to 8 bit sequences modulated at a bit rate of 2 G bit/s with a corresponding receiver sensitivity of -36.7 dBm at 10^{-9} BER over a transmission length of 101 km. The following presents both the measured and calculated results for the system shown in Figure 14.16.

System performance can be analyzed by considering the CPFSK modulation signal consisting of a sequence (b_k) of $(2N + 1)$ bits, where b_k are statistically independent symbols taking on values ± 1 with equal probability. The transmitted signal can be expressed as

$$E_S(t) = E_S \cos\left[\omega_S t + m\pi \sum_{k=N}^{N} \int_{-\infty}^{\infty} b_k g(t' - kT)\, dt' + \phi_S(t) \right] \quad (14.28)$$

where T is the bit period, m is the modulation index, g is the pulse-shape function, and $\phi_s(t)$ is the laser phase noise. For a rectangular pulse,

$$g(t) = \begin{cases} 1/T & 0 < t < T \\ 0 & \text{at other times} \end{cases} \quad (14.29)$$

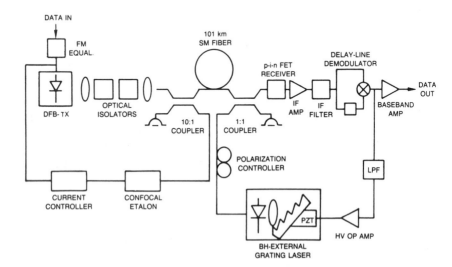

FIGURE 14.16 A 2 Gbit/s CPFSK coherent system block diagram. (Ref. 14.11. Reprinted with permission of IEEE.)

The post-detection voltage $V(t)$ after passing through the IF filter (see Figure 14.16) can be expressed as

$$V(t) = [V_S(t) + V_N(t)] * h_{IF}(t) \qquad (14.30)$$

where $h_{IF}(t)$ represents the impulse response of the IF filter and $*$ denotes the convolution. V_S is the signal component resulting from the photomixing process and can be expressed as

$$V_S(t) = A \cos[\Delta \omega t + \phi_m(t) + \Delta \phi] \qquad (14.31)$$

where

$$A = 2\sqrt{P_S P_{LO}} \left(\frac{e\eta}{hv} \right) Z \qquad (14.32)$$

and $\Delta \omega = \omega_S - \omega_{LO}$, $\Delta \phi = \phi_S - \phi_{LO}$, and $\phi_m(t)$ is the modulated phase component as given by the middle term in the right-hand side of Equation (14.28). In Equation (14.32), Z is the effective transimpedance of the receiver. $V_N(t)$ is the shot noise, which is assumed to be dominated by both the local oscillator noise and thermal noise of the receiver.

The signal $V(t)$ is fed to the demodulator, which consists of a balanced-mixer with a differential delay between its two paths. The output $V'(t)$ passes through a baseband filter and can be written as

$$V'(t) = [V(t)V(t - \tau)] * h_B(t) \qquad (14.33)$$

where h_B is the impulse response of the baseband filter. For an ideal square input pulse the output CPFSK signal is no longer a square pulse but has a sinusoidal rise and fall waveform with rise and fall times equal to τ. This distortion of the FM pulse-shape $g'(t)$ resulting from the nonlinear FM phase response on CPFSK signals can be calculated by performing an inverse Fourier transform of $G(\omega)F(\omega)$ where $G(\omega)$ is the transform of the ideal pulse shape $g(t)$. Various pseudo-random code sequences, e.g., 2^5, 2^8, and 2^{13} bit code sequences, yield a set of eye diagrams. Results (Ref. 14.11) indicate that the eye is nearly closed for a 2^{13} bit sequence. For 2^8 bit sequences, a penalty of only 2 dB results from intersymbol interference.

In the absence of phase noise, the optimum performance of the demodulator is achieved for an IF frequency chosen to be equal to the zero-crossing frequency of the demodulator with frequency deviation and delay selected to satisfy

$$\cos(\Delta \omega \tau + \pi m b_k \tau / T) = \pm 1 \qquad (14.34)$$

for the two symbols $b_k = \pm 1$. This condition is satisfied if, for a given modulation index m, $\Delta \omega$, and τ are chosen such that

$$\Delta \omega \tau = (n + \tfrac{1}{2})\pi \qquad (14.35)$$

where n is any integer and

$$\tau = T/2m \tag{14.36}$$

To determine the impact of noises on system performance, assume first that delay-line demodulation acts as an ideal linear FM discriminator of slope $K = 2T/m$ in converting Gaussian FM noise to Gaussian AM noise with a phase noise spectral density N_ϕ as given by

$$N_\phi = \frac{1}{2\pi} \Delta\nu (T/m)^2 A^4 \tag{14.37}$$

Next, consider the IF filter noise. For an IF filter that is symmetric about the zero-crossing frequency of the delay demodulation, $\langle V_n \rangle$ is equal to zero. The variance $\langle V_n^2 \rangle$, which consists of a phase noise plus other noise terms can be expressed as

$$\langle V_n^2 \rangle = [A^2 N_0 + N_\phi]2I_2/T + N_x^2 \tag{14.38}$$

where

$$N_0 = \left[P_{LO} \frac{e^2\eta}{h\nu} + \frac{1}{2} \langle (\delta I_c)^2 \rangle \right] Z^2 \tag{14.39}$$

and

$$N_x^2 = 4N_0^2 B_{IF}(I_2/T) \tag{14.40}$$

I_2 and B_{IF} in the above expressions are the baseband filter noise and IF bandwidth, respectively. $\langle (\delta I_c)^2 \rangle$ is nearly a constant, 10 pA/$\sqrt{\text{Hz}}$.

For convenience, the following calculation is carried out in the Gaussian approximation (see Equation 13.35). The BER is

$$\text{BER} = \text{Erfc}(Q)$$

where

$$Q = \frac{1}{2}\frac{S}{N} = \frac{\left(P_S \dfrac{e\eta}{h\nu} \right)^{1/2}}{\left[e(1 + \beta) + \dfrac{2\Delta\nu}{\pi}\left(\dfrac{T^2}{m}\right) P_S \left(\dfrac{e\eta}{h\nu}\right) + \dfrac{(1 + \beta)^2 e^2 B_{IF}}{2P_S \dfrac{e\eta}{h\nu}} \right]^{1/2} (2I_2/T)^{1/2}} \tag{14.41}$$

In Equation (14.41), β is the receiver thermal noise normalized to local oscillator shot noise as given by

$$\beta = \langle (\delta I_c)^2 \rangle / 2e P_{LO} \frac{e\eta}{h\nu} \tag{14.42}$$

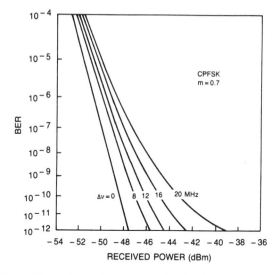

FIGURE 14.17   Theoretical performance of a 2 Gbit/s CPFSK heterodyne system as a function of IF linewidth $\Delta\nu$. (Ref. 14.11. Reprinted with permission of IEEE.)

For shot-noise-limited operation, consider the case of $m = 0.7$, $\eta = 1$, $\tau = T/2m$, $2I_2 = 1.33$, and $B_{IF} = 3.2$ GHz. The calculated BER is plotted in Figure 14.17 as a function of the received power P_R for linewidth $\Delta\nu$ varied from 0 to 20 MHz. Here, $\Delta\nu$ is taken to be the sum of the local oscillator and transmitter laser linewidths. The measured BER for a 2 Gbit/s FSK system is shown in Figure 14.18 as a function of the received power for very short (0.5 m) and very long (101 Km) single mode fibers. These results show

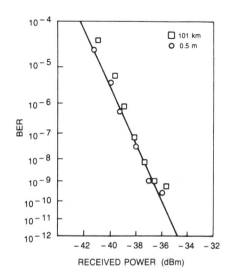

FIGURE 14.18   Theoretical and measured BER as a function of received power in dBm for a 2 Gbit/s CPFSK heterodyne system. (Ref. 14.11. Reprinted with permission of IEEE.)

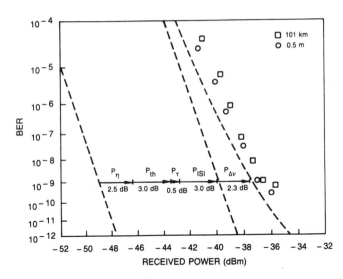

FIGURE 14.19 BER versus received power in dBm for a 2 Gbit/s FSK coherent system. □—101 km single mode fiber; ○—0.5 m single mode fiber. Left-most dashed curve is the theoretical result for $m = 0.7$. Various penalties associated with DFB phase noise $P_{\Delta v}$, intersymbol interference P_{ISI}, thermal noise P_{th}, and photodiode quantum efficiency P_η are indicated. (Ref. 14.11. Reprinted with permission of IEEE.)

that for $\Delta v = 0$, shot-noise-limited receiver sensitivity is calculated to be -49 dB at 10^{-9} BER. This level may be compared to the measured receiver sensitivity of -36.7 dB at the same BER. For $\Delta v = 12$ MHz, which corresponds to the measured linewidth, receiver sensitivity is degraded by 10 dB. This penalty is attributed to the nonlinear phase of IF and the baseband frequency responses that cause substantial intersymbol interference. System performance can be further degraded if local oscillator power is not sufficient to produce shot-noise-limited operation.

Figure 14.19 summarizes various penalities imposed on the system. The left-most dashed curve, representing shot-noise-limited performance, is calculated from Equation (14.41), assuming $\langle (\delta I_c)^2 \rangle = 0$, $\eta = 1$, $\tau = T/2m = 357$ ps, $2I_2 = 1.33$, and $B_{IF} = 3.2$ GHz. The measured receiver sensitivity is 12.3 dB away from the theoretical limit of -49 dB at 10^{-9} BER. Several sources contributing to the system penalty are indicated in this figure.

14.9 MULTITERMINAL CONTROL AND DATA DISTRIBUTION

Fiber optics can be used in many services other than telecommunication. Systems that transmit and distribute data within a building or a group of buildings can now make use of optical fiber technology to its best advantage.

Other applications include monitoring and controlling electrical power, managing energy, controlling and commanding data buses for aircraft and traffic, and so on. This section briefly describes some of these applications.

The intrafacility network within a building consists of an interface unit between a high-data-rate optical fiber transmission link and a subscriber's equipment, which is comprised of video terminal, telephone, computer, facsimile, and so on. Information and data can be distributed to a number of local optical fiber loops by means of a central control processor. Each loop contains a subscriber interface unit that supports several terminals. Time-division multiple access is a common technique used in this network, where time is divided into many slots that can be made available to various terminals. In this scheme, several protocals can be considered; for example, time slots within a recurring time frame can be dedicated to users, or time-slot access can be operated in contention among various users, who compete for the allocation of a time slot.

Optical fiber transmission lines have been utilized by the electric power utility industry in its power monitor and control systems. In a high-voltage (>200 kV) transmission network, many noise and electromagnetic interference problems are associated with the conventional communication circuits that cause difficulties in supervising, protecting, and controlling power delivery from the generating plant to a control center located in a metropolitan area. A system suitable for this purpose must be reliable and must have long-haul capability at bit rates as high as 30 Mb/s. From the control center, a multiterminal data distribution system is needed to communicate with a number of division offices and substations. A combination of technologies in microprocessors and fiber optics has provided the power utility industry with improved load management. One of the most important aspects is distribution automation, which provides all communication and load control functions, including time-of-day metering, remote meter reading, feeder switching, capacitor bank control, transformer temperature monitoring, fault location, and isolation.

In conclusion, fiber optic technology will continue to grow and will capture a major share of the market within this decade in the telecommunication, computer, and power utility industries, as well as other services, such as education, entertainment, home protection, banking, medical and health, business trade, and the military. It will also be used in instrumentation for automobiles, ships, and aircraft. As the technology matures, the costs of various components will be reduced in a way similar to that typified by the solid-state electronics industry. When this cost reduction occurs, optical fiber technology will become extremely competitive with already established radio-frequency and microwave technologies. An attractive application is to introduce services such as high-speed and interactive video in major population centers, which can be connected by long-haul fiber systems. Time-division and wavelength-division multiplexing can be used to relieve congestion as traffic increases. The potential for growth in this technology appears immense. Not only can optical fiber perform better and more economically than conventional transmission media in most applications, but it will also help to bring new services that are not possible or practical with the transmission media available today.

PROBLEMS

14.1. Calculate the system response and power budget for a digital link with the following specifications: BR = 50 Mb/s, BER = 10^{-9}, signal format is NRZ, terminal length is 10 km, and number of terminals is 2.

14.2. Design an optimum two-way telecommunication system, capable of transmitting a bit rate of 400 Mb/s, at an error rate of 10^{-9} over a distance of 10 km.

14.3. Design a distribution system that serves 100 parallel terminals each of which has a transmitter and a receiver and is located at a distance about 100 m from a star coupler. Calculate the total system loss.

14.4. For the system defined in Problem 14.3, what is the required transmitter power, assuming that 5-dB/km fibers with 0.2 NA are used?

14.5. Design a serial distribution system that serves 100 subscribers along a linear path of a total length 10 km. Calculate the total system loss.

14.6. Assuming that $W = 4$, calculate the probability of failure P_F to detect a full bit hit (e.g., $\tau = T$) for the limiting case when $p_1 = p_3 = \frac{1}{2}$. (Ans. $P_F = \frac{1}{16}$).

14.7. For a coherent communication system, various factors can degrade system performance. Discuss various sources that can affect the BER as indicated in Figure 14.19 and estimate the penalties associated with (1) the quantum efficiency of $\eta = 0.5$, and (2) the thermal noise $\langle (\delta I_c)^2 \rangle^{1/2} = 10 \text{ pA}/\sqrt{\text{Hz}}$.

REFERENCES

14.1. Third International Conference on Integrated Optics and Optical Fiber Communication, *Tech. Digest*, San Francisco, April 27–19, 1981.

14.2. First International Conference on Integrated Optics and Optical Fiber Communication, *Tech. Digest*, Tokyo, July 18–20, 1977.

14.3. R. W. Berry, D. J. Brace, and L. A. Ravenscroft, *IEEE Trans. Commun., COM-26*, 1020 (1978).

14.4. Y. Hakamoda and K. Oguchi, *IEEE J. Lightwave Tech., LT-3*, 511 (1985).

14.5. S. Monstakas and H. Witte, *Elect. Lett., 19*, 592 (1983).

14.6. M. Kavehrad and C. W. Sundberg, *IEEE J. Lightwave Tech., LT-5*, 1549 (1987).

14.7. G. C. Clark, Jr. and J. B. Cain, *Error-Correction Coding for Digital Communications*, New York Plenum Press, 1981.

14.8. J. Yamada and T. Kimura, *IEEE J. Quant. Elect.*, *QE-18*, 718 (1982).

14.9. T. Miya, Y. Terumuma, T. Hosaka, and T. Miyashita, *Elect. Lett.*, *15*, 106 (1979).

14.10. D. W. Smith, *IEEE J. Lightwave Tech.*, *LT-5*, 1466 (1987).

14.11. J. L. Gimlett, R. S. Vodhanel, M. M. Choy, A. F. Elrefaire, N. K. Cheung, and R. E. Wagner, *IEEE J. Lightwave Tech.*, *LT-5*, 1315 (1987).

Index

Abrupt heterojunction, 214–16
Absorption, 177–78, 181
Absorption coefficient, 183–84, 347
Alpha profile, 4, 64
Alternative mark inversion (AMI), 312, 314
AM noise, 326, 331
Ampere's law, 12
Amplifier noise, 383–87
Amplitude modulator, 319–20
Analog code, 311
Anderson model, 214
Astigmatic effect, 239, 249–50
Asymmetric guide, 32
Attennation coefficients, 143–45
Autocorrelation function, 98, 338
Automatic frequency control, 330
Avalanche gain, 352
Avalanche photodiode (APD), 7, 348–55

Bandgap energy, 160, 215
Bandtail states, 175
Band theory:
    Bloch Theorem, 162–63
    Bose-Einstein statistics, 178
    conduction band, 166
    density of state, 170
    effective mass, 166–67
    Fermi-Dirac statistics, 169

forbidden energy band, 161
Kronig-Penney model, 161
periodic potential, 161–62
Schrödinger equation, 160
valence band, 166
Bending loss, 95–99
Bessel functions, 43
Bias circuit, 309–10
Biconical tapered coupler, 156
Binary code, 312–14
Binary pulse code, 311
Bipolar amplifier, 384–86
Bipolar waveform, 315
Bit error rate (BER), 390–97
Bloch function, 160, 163–64
Boltzmann constant, 94, 169
Bose Einstein law, 178
Bragg condition, 162, 270
Bragg reflection, 162, 164
Bragg wavelength, 270
Brightness, 136
Buried heterostructure (BH) laser, 238, 259–67

Carrier confinement, 227
Catastrophic damage, 262
Caustic points, 67, 144
Central limit theorem, 335

Channel capacity theorem, 339
Channel waveguide, 37–38
Characteristic function, 337
Chemical vapor deposition (CVD), 127–32, 233–35
Cleaved Coupled-Cavity Laser (C^3):
    coupled-cavity mode, 276–77
    spectral spacing, 276
Coherent optical fiber communication, 9–10, 428–38
Collision detection, 416–26
Complemented mark inversion (CMI), 314–15
Composition fluctuation, 94
Confinement factor, 30–31
Confocal configuration, 322–23
Continuous-phase frequency shift key, 433–34
Convolution integral, 149, 337
Correction factor, 144, 147
Coupled mode theory, 97, 119
Coupled wave equation, 84, 270
Coupler, 156
Coupling efficiency:
    LED—multi-mode fiber, 136–38
    LED—single-mode fiber, 140–41
Current density vector, 12
Cutoff condition, 27
Cutoff frequency, 46
Cyclic code, 419–20
Cylindrical waveguide, 14, 17–18, 41–71

Dark current noise variance, 362
Dark noise, 359
deBroglie's relation, 167–70
Decay constant, 32
Decision rule, 391
Decision threshold, 391–92
Dehydration, 129, 132
Delay difference, 6
Delay time, 5
Density fluctuation, 94
Density of state, 170
Depletion region, 205, 209, 217
Dielectric constant, 11
Dielectric waveguide, 14
Differential phase shift keying, 330
Diffusion equation, 221
Diffusion length, 187, 221
Digital code, 311–14
Diode laser array:
    array mode, 278, 281
    complex propagation constant, 278, 281
    coupling coefficient, 279

eigenvalue equation, 279–80
    far field distribution, 283–85
Dirac delta function, 148
Direct bandgap, 176
Directional coupler, 325–26, 431
Dispersion:
    group delay, 75–76
    intermodel, 78–80
    intramodal, 78
    material, 79
    waveguide, 76
Dispersion-free fiber, 109
Dispersion relation:
    multimode fiber, 53–56, 70, 76
    planar waveguide, 32
Displacement current vector, 12
Distributed Bragg reflection laser. See Distributed feedback laser
Distributed coefficient, 187, 221
Distributed feedback laser (DFB), 268–74
    Bragg condition, 270
    Bragg wavelength, 270, 274
    complex refractive index, 269
    coupling coefficient, 271
    power gain coefficient, 269
    propagation constant, 269, 272–73
    resonance frequency, 273–74
    stop band, 272–74
Dopants, 173–74
Double crucible method, 133–35
Double heterostructure, 7, 238, 251–52, 257, 260–62, 265–68
Double cladded fibers, 105–6
Dynamic range, 424

Effective index method, 37–39
Effective mass, 166–67
EH modes, 46, 49
Eigen-polarization state, 120
Eigenvalue equation:
    double-cladded fiber, 107–8
    step-index fiber, 45, 59
    TE modes, 25–26
    TM modes, 25
Eikonal equation, 60
Electric field vector, 11
Electro luminescence, 159
Electron affinity, 168, 215
Electron density, 172–73
Electrooptic modulation, 316–26
Elliptical fiber, 115
Emission rate:
    spontaneous transition, 178, 181–83
    stimulated transition, 178

Energy gap, 224
Equation of indicatrix, 316–18
Equation of ray propagation, 62
Error function, 218
Error probability, 389, 393–95, 421–26
Evanescent wave, 23
Excess carrier concentration, 210
Excess noise factor, 352–53
Exponential pulse shape, 366
Extraordinary index, 318

Fabry-Perot cavity, 197
Fabry-Perot interferometer, 199
Faraday's law, 13
Feedback control, 311
Fermi-Dirac distribution, 169
FET amplifier, 383, 385–87
F-factor, 324–25
Fiber connectors, 156
Fiber splicing, 154
Field expressions for:
　cylindrical waveguide, 43
　elliptical fiber, 116
　planar waveguide, 27–29
Finesse, 200
F-number, 138
Forbidden energy band, 161
Forward-bias junction, 209–11
Fourier transform, 149, 357
Free particle approximation, 163–68
Frequency domain, 149
Frequency response:
　circuit parasites, 300–301
　photodiode, 367–69
　rate equation, 298
　relaxation frequency, 299
　semiconductor laser, 298–301
Frequency response of photodiode,
　367–69, 373, 377–78
Frequency-shift key (FSK), 432
Fresnel transmission coefficient, 139
Fundamental stretching vibration, 95

GaAs FET, 385–87
Gain coefficient, 188–89, 191–93
Gain compression, 298
Gain guided laser, 243–50
Gaussian distribution, 335
Gaussian pulse shape, 365
Gauss's law, 11
Goos Haenchen shift, 22
Graded heterojunction, 217
Graded index, 4, 64

Graded index fiber:
　index profile measurement, 142–47
　modal spacing, 70
　propagation constant, 70
　ray analysis, 60–66
　WKB method, 66–71
Group delay, 75–76
Group index, 110
Group velocity, 74

Half wave voltage, 320
Hamming distance, 419–20
Hamming weight, 418
HE_{11} mode:
　decay constant, 103
　eigenvalues, 102
　propagation constant, 103
　transverse field component, 102–4
Helmholtz equation, 43
Heterodyne detection, 429–30
Heterojunctions:
　double heterojunction, 225–30
　graded heterojunction, 217–21
　single heterojunction, 213–24
High power semiconductor lasers,
　BH laser, 259–60
　CDH laser, 259–62, 264
　CSP laser, 259–60, 264
　PCW laser, 259–60
　TJS laser, 259–61
　TRS laser, 264
Hole concentration, 171
Homodyne detection, 431
Hybrid modes, 18, 46, 48–49
Hyperbolic profile, 64

Impulse response, 77, 88–89, 148, 356
Impulse response function, 357
Impurity concentration, 175
Index ellipsoid, 316
Index guided laser:
　effective dielectric constant, 255
　high power laser, 259–64
　lateral field distribution, 256–57
　overlap parameter, 254
　propagation constant, 255–56
　single mode DH lasers, 260
Index profile:
　alpha profile, 4, 64
　double-cladded profile, 105–14
　graded index, 4, 64
　measurement techniques, 141–47
　parabolic profile, 144–46
　step index, 2, 105–6
　W-type, 106

Indirect bandgap, 176
Information carrying capacity, 339
Intensity modulation, 302–5
Interfacial trapping, 372
Intermodal dispersion, 78
Internal reflection, 19
Intramodal broadening, 78, 80–83
Intrinsic carrier concentration, 173
Inverse laplace transform, 87
Ionization energy, 174–75
Ionization rate, 349–51, 375

Johnson noise, 361

KDP, 150
Kell cell, 150
k-factor, 351–53
K-parameter, 253
Kronig-Penney model, 161

Lambertian, 136
Laplace transform, 85
Large optical confinement (LOC), 230
Laser diode (LD), 6, 212, 308
Laser diode array, 278–88
Lattice constant:
    binary compounds, 231–32
    quaternary compounds, 231–32
    ternary compounds, 231–32
Leaky waves, 96
Light emitting diode (LED), 6, 213, 308
Linear channel, 356
Line shape, 330–31
Liquid phase epitaxy (LPE), 233–34
Local area network (LAN), 416–26
Local numerical aperture, 142
Local oscillator, 429–30
Long wavelength laser:
    GaAsP/InP laser, 265
    InGaAsP/InP laser, 265–68
Long-haul system, 426–28
LP modes, 18, 51
Lump element, 324

Magnetic field vector, 12
Magnetic flux, 12
Magnetostatic law, 12
Magnification power, 138
Material dispersion, 79
Maxwell's equation, 11–14
Meridional rays, 63
M-fold degeneracy, 70

Microbending loss, 98
Mie scattering, 94
Minimum detectable power, 396
Minority carrier density, 190
Minority hole concentration, 210
Modal dispersion, 5, 26, 36
Modal power distribution, 56, 57
Modal spacing, 70, 83
Mode coupling, 83–89
Mode number, 67
Modulation bandwidth, 324
Molecular beam epitaxy (MBE), 233–35
Multimode, 5
Multiplication factor, 348–51

NP heterojunction, 226
Network former, 126–27
Network modifier, 126–27
NnP heterojunction, 226
Noise:
    background noise, 355, 359–60
    dark noise, 355, 359, 362
    Johnson noise, 361
    mean value, 355–56
    noise current spectral density, 357–59
    photodiode, 355–61
    quantum noise, 355, 360
    shot noise, 355, 359
    shunt noise, 361, 363
    thermal noise, 361
    transfer function, 359
    transmitter noise, 326–33
    variance, 355, 360
Nominal threshold current, 192
Non-absorbing mirror (NAM), 263
Normalized frequency:
    cylindrical waveguide, 43
    doubly cladded fibers, 106
    planar waveguide, 32
Normalized group delay, 117
Normalized index:
    cylindrical waveguide, 75
    double-cladded fibers, 110
    planar waveguide, 12, 34–35
Normalized length, 89–91
Normalized moment, 89
Normalized phase difference, 117
nP heterojunction, 224
npP heterojunction, 224
Numerical aperture, 6

Occupation factor, 171
Oersted, Hans C., 12
Optical fiber. *See* Cylindrical waveguide

Optical fiber system, 2, 411–40
Optical gain, 193–94
Optical local area network. *See* LAN
Optical receiver. *See* Receiver
Optical Source. *See* Laser diode, Light
    emitting diode and Stripe-geometry
    lasers
Ordinary index, 318

Parabolic profile, 4, 64
Passive star configuration, 217
Pauli exclusion principle, 169
Penetration depth, 209
Permeability, 12
Phase error variance, 329, 331
Phase lock loop, 329
Phase noise spectral density, 436
Phase retardation, 324
Phase shift, 321
Phase velocity, 14–15, 74
Photodiode:
    avalanche photodiode, 348–55
    conversion efficiency, 344
    frequency response, 367–69, 373, 377
    Ge photodiode, 347–48
    InGaAs photodiode, 354, 369–78
    InGaAsP photodiode, 354, 369–78
    photocurrent, 344
    pin, 8, 344, 346
    pn, 7, 205–9, 343–45
    quantum efficiency, 370, 376
    reverse-biased *pin* photodiode, 346
    reverse-biased *pn* junction, 345
    silicon photodiode, 345
Photon life time, 187, 201, 298
Photon spectral density, 178
pin photodiode, 8
Planar dielectric waveguide, 14, 16–17,
    19–39
Planck's relation, 167
Plasma augmented CVD, 130
pn junction, 7, 205–9
Poisson distribution, 335–37
Poisson equation, 218
Poisson's ratio, 119
Polarization-preserving fibers, 115, 119
Potential barrier, 209
Power coupling coefficient, 84
Power transmission coefficient, 121–24
Poynting vector, 30
Preform, 132
Principal crystal axes, 318
Principle of continuity, 21
Probability density formation, 334–38

Propagation constant:
    cylindrical waveguide, 43, 45, 70
    DFB laser, 272
    diode laser array, 278
    index guided laser, 255–56
    planar waveguide, 14, 17
    weakly guided fiber, 55
Pulse broadening, 82
Pulse code modulation, 311
Pulse distortion, 89–92
Pulse shape, 87–89
Pulse width, 77, 82

Quantum efficiency, 177, 370
Quantum noise, 360
Quantum well laser:
    band mixing, 290
    dipole moment, 289
    eigenvalue equation, 288
    k.p method, 289
    linewidth, 292–94
    optical gain, 291–92
    threshold current, 292–93
    wavefunction, 287–88
Quarter wave plate, 320
Quasi-fermi level, 171
Quaternary system, 7

Radial diffusion, 85
Radiative lifetime, 177, 186–87
Radiative recombination, 177
Raised cosine function, 364
Ratio test, 391
Ray analysis, 60–66
Rayleigh scattering, 93
Receiver:
    amplifier noise, 382–387
    equivalent circuit, 380–81
    front-end, 381, 384
    receiver design, 387–89
    receiver noise, 381–83
    sensitivity, 395–98
Recombination rate, 177, 186
Rectangular pulse shape, 365
Reflection coefficient, 21
Refractive index, 2
Relaxation frequency, 299
Repeater spacing, 9
Resonance–peak, 304
Responsivity, 376
Return to Zero (RZ), 312

Scalar field approximation, 42–43
Scaling rules, 32–37

Schrödinger equation, 160
Sequence weight violation, 417–18
Shot noise, 359
Shot noise limited operation, 429, 437
Shunt noise, 361
Shunt noise variance, 363
Sideband, 321
Signal to noise, 389, 429
Signal waveform effect, 361–67
Silicon tetrahedra, 125–26
Single heterojunction, 213–24
Single mode fiber:
　chromatic dispersion, 111
　cut-off frequency, 109–10
　dispersion-free fiber, 114
　dispersion parameters, 112–13
　eigenvalue equation, 107
　elliptical core fiber, 115
　field component, 102, 104
　material dispersion, 111
　modal power distribution, 105
　polarization preserving fiber, 119–21
　propagation constant, 103
Single mode laser, 258–64, 268–74
Skew rays, 63
Snell's law, 5
Soot process, 128
Spatial hole burning, 258–59
Spectral broadening, 258–59
Spectral width, 199–200
Spontaneous emission, 177
Star coupler, 157
Step index fiber:
　eigenvalue equation, 45
　exact solutions, 58–60
　HE modes, 48–50
　LP modes, 51–56
　normalized frequency, 43
　numerical aperture, 43
　power distribution, 56
　scaler field equation, 42
Step-index, 2
Stimulated emission, 177
Stirling approximation, 336
Stop band, 272
Strap fiber, 149
Stripe geometry laser:
　electron density profile, 241
　internal circulating power, 252
　K-Factor, 253
　output power, 252
　spectral characteristics, 252
　spontaneous emission profile, 242
Stripe width, 201

Susceptivity, 12
Symmetric waveguide, 33
System design, 403–9

Telecommunication system, 409–11
TE modes, 25, 48–49
TE waves, 16, 19–22
Thermal fluctuation, 94
Thermal noise, 359
Threshold current density, 196
Time domain, 149
Time Domain Reflectometry (TDR), 157
TM modes, 25, 48–49
TM waves, 16, 19, 22–23
Total internal reflection, 19–23
Transfer function:
　frequency modulation, 306
　intensity modulation, 302
Transit time, 324
Transition probability, 177
Transmission coefficient, 21
Transmitter noise:
　AM noise, 326–27, 331
　AM noise spectrum, 328
　carrier intensity fluctuation, 327
　feedback noise, 333
　FM noise, 326–28
　FM noise spectrum, 328
　intrinsic noise, 326, 332
　partition noise, 333
　phase error, 329
　phase noise spectrum, 328
　variance of phase error, 329
Traveling-wave modulator, 325
Tunneling coefficient, 145

Vapor Phase Axial Deposition (VAD), 131
Vapor Phase Epitaxy (VPE), 234

Wave equation, 16
Weakly-guided fiber, 50–56
Wentzel-Kramers-Brillouin (WKB)
　approximation, 66–68
Whispering gallery mode, 97
Work function, 216

Young's modulus, 119

Zero-crossing frequency, 435